Selected Titles in This Series

169 **S. K. Godunov,** Ordinary differential equations with constant coefficient, 1997
168 **Junjiro Noguchi,** Introduction to complex analysis, 1997
167 **Masaya Yamaguti, Masayoshi Hata, and Jun Kigami,** Mathematics of fractals, 1997
166 **Kenji Ueno,** An introduction to algebraic geometry, 1997
165 **V. V. Ishkhanov, B. B. Lur′e, and D. K. Faddeev,** The embedding problem in Galois theory, 1997
164 **E. I. Gordon,** Nonstandard methods in commutative harmonic analysis, 1997
163 **A. Ya. Dorogovtsev, D. S. Silvestrov, A. V. Skorokhod, and M. I. Yadrenko,** Probability theory: Collection of problems, 1997
162 **M. V. Boldin, G. I. Simonova, and Yu. N. Tyurin,** Sign-based methods in linear statistical models, 1997
161 **Michael Blank,** Discreteness and continuity in problems of chaotic dynamics, 1997
160 **V. G. Osmolovskiĭ,** Linear and nonlinear perturbations of the operator div, 1997
159 **S. Ya. Khavinson,** Best approximation by linear superpositions (approximate nomography), 1997
158 **Hideki Omori,** Infinite-dimensional Lie groups, 1997
157 **V. B. Kolmanovskiĭ and L. E. Shaĭkhet,** Control of systems with aftereffect, 1996
156 **V. N. Shevchenko,** Qualitative topics in integer linear programming, 1997
155 **Yu. Safarov and D. Vassiliev,** The asymptotic distribution of eigenvalues of partial differential operators, 1997
154 **V. V. Prasolov and A. B. Sossinsky,** Knots, links, braids and 3-manifolds. An introduction to the new invariants in low-dimensional topology, 1997
153 **S. Kh. Aranson, G. R. Belitsky, and E. V. Zhuzhoma,** Introduction to the qualitative theory of dynamical systems on surfaces, 1996
152 **R. S. Ismagilov,** Representations of infinite-dimensional groups, 1996
151 **S. Yu. Slavyanov,** Asymptotic solutions of the one-dimensional Schrödinger equation, 1996
150 **B. Ya. Levin,** Lectures on entire functions, 1996
149 **Takashi Sakai,** Riemannian geometry, 1996
148 **Vladimir I. Piterbarg,** Asymptotic methods in the theory of Gaussian processes and fields, 1996
147 **S. G. Gindikin and L. R. Volevich,** Mixed problem for partial differential equations with quasihomogeneous principal part, 1996
146 **L. Ya. Adrianova,** Introduction to linear systems of differential equations, 1995
145 **A. N. Andrianov and V. G. Zhuravlev,** Modular forms and Hecke operators, 1995
144 **O. V. Troshkin,** Nontraditional methods in mathematical hydrodynamics, 1995
143 **V. A. Malyshev and R. A. Minlos,** Linear infinite-particle operators, 1995
142 **N. V. Krylov,** Introduction to the theory of diffusion processes, 1995
141 **A. A. Davydov,** Qualitative theory of control systems, 1994
140 **Aizik I. Volpert, Vitaly A. Volpert, and Vladimir A. Volpert,** Traveling wave solutions of parabolic systems, 1994
139 **I. V. Skrypnik,** Methods for analysis of nonlinear elliptic boundary value problems, 1994
138 **Yu. P. Razmyslov,** Identities of algebras and their representations, 1994
137 **F. I. Karpelevich and A. Ya. Kreinin,** Heavy traffic limits for multiphase queues, 1994
136 **Masayoshi Miyanishi,** Algebraic geometry, 1994
135 **Masaru Takeuchi,** Modern spherical functions, 1994
134 **V. V. Prasolov,** Problems and theorems in linear algebra, 1994
133 **P. I. Naumkin and I. A. Shishmarev,** Nonlinear nonlocal equations in the theory of waves, 1994
132 **Hajime Urakawa,** Calculus of variations and harmonic maps, 1993
131 **V. V. Sharko,** Functions on manifolds: Algebraic and topological aspects, 1993

(Continued in the back of this publication)

Ordinary Differential Equations with Constant Coefficient

Translations of
MATHEMATICAL MONOGRAPHS

Volume 169

Ordinary Differential Equations with Constant Coefficient

S. K. Godunov

American Mathematical Society
Providence, Rhode Island

С. К. ГОДУНОВ

ОБЫКНОВЕННЫЕ ДИФФЕРЕНЦИАЛЬНЫЕ УРАВНЕНИЯ
С ПОСТОЯННЫМИ КОЭФФИЦИЕНТАМИ

ИЗ-ВО НОВОСИБИРСКОГО
УН-ТА, 1994

Translated from the Russian by Tamara Rozhkovskaya
with the participation of Scientific Books (RIMIBE NSU),
Novosibirsk, Russia

1991 *Mathematics Subject Classification.* Primary 34–01;
Secondary 34A30, 34Bxx, 34Dxx, 49J15, 49K15.

ABSTRACT. This book presents the theory of ordinary differential equations with constant coefficients. The exposition is based on the matrix calculus. Boundary-value problems, Green matrices, the Lopatinskii condition, and the Lyapunov stability are considered. Some qualitative aspects connected with the use of computers for the analysis and solution of boundary-value problems are discussed. Control problems reduced to systems of ordinary equations with constant coefficients are also discussed. The book can be used by researchers and students working in the theory of ordinary differential equations, in particular in its computational aspects.

Library of Congress Cataloging-in-Publication Data

Godunov, S. K. (Sergeĭ Konstantinovich)

[Obyknovennye different͡sial′nye uravnenii͡a s postoi͡annymi koeffit͡sientami. English]

Ordinary differential equations with constant coefficient / S. K. Godunov ; [translated from the Russian by Tamara Rozhkovskaya with the participation of Scientific Books (RIMIBE NSU), Novosibirsk, Russia].

p. cm. — (Translations of mathematical monographs, ISSN 0065-9282 ; v. 169)

"This book is a translation of the published first volume [Novosibirsk State University, 1994] and a chapter (Chapter 4 in this book) from the unpublished second volume in Russian"—Pref.

Includes bibliographical references.

ISBN 0-8218-0656-4 (hardcover : alk. paper)

1. Differential equations. I. Title. II. Series.

QA372.G56713 1997

515′.352—dc21 97-20182

CIP

∞ The paper used in this book is acid-free and falls within the guidelines
established to ensure permanence and durability.
Visit the AMS homepage at URL: http://www.ams.org/

10 9 8 7 6 5 4 3 2 1 02 01 00 99 98 97

Contents

Preface

This book presents the theory of ordinary differential equations with constant coefficients and follows lectures given by the author at Novosibirsk State University. The first version of the book was based on the ideas in the article [**14**] which develops the Gelfand–Shilov theorem [**6**] on the polynomial representation of a matrix exponential. The course was constantly improved by V. I. Kostin, V. M. Gordienko, G. V. Demidenko, S. I. Fadeev, E. V. Zolotareva, and G. A. Chumakov. Remarks by Yu. F. Borisov were also very useful.

Chapters 1 and 2 can be regarded as an extension of the books by the author [**8, 9**] published by Novosibirsk State University in 1982–1983. It turned out that the exposition of material suggested there is convenient for practical study of ordinary differential equations using computers. Therefore Chapter 3 is devoted to algorithmic and qualitative questions and contains computational procedures for solving boundary-value problems. In particular, the orthogonal sweep method is described in detail. Based on this method, some algorithms that allow us to clarify the dichotomy of the spectrum of the matrix of coefficients are also explained here. S. I. Fadeev provided substantial help in preparing this chapter.

Chapter 4 deals with stationary optimal control systems that are described by systems of differential equations with constant coefficients. In this chapter, the notions of controllability, observability, and stabilizability are analyzed. Some results of the analysis were published in [**3, 10, 11**], where some questions on the matrix Lur′e–Riccati equations were studied.

This book is a translation of the manuscript "Ordinary differential equations", submitted for publication to the Novosibirsk State University. The manuscript consists of two volumes. The publication of both volumes was approved by Novosibirsk State University, and they were scheduled to be published in 1989. However, Novosibirsk State University was able to publish only the first volume (Chapters 1–3), and only in 1994. The second volume remains unpublished.

This book is a translation of the published first volume and a chapter (Chapter 4 in this book) from the unpublished second volume in Russian. In the English version of the book some misprints and inaccuracies are corrected. Some style changes are also made in order to facilitate the reader's understanding.

S. K. Godunov
Novosibirsk, March 1997

CHAPTER 1

Matrix Exponentials, Green Matrices, and the Lopatinskii Condition

§1. Linear systems of equations with constant coefficients. The uniqueness and existence of a solution to the Cauchy problem for homogeneous equations

Simplest examples and definitions. Vector and matrix notation. Euclidean space. Vector and matrix norms. The infinite differentiability of solutions to equations with constant coefficients. The representation of a solution as the Taylor series. The Cauchy problem. The uniqueness and existence of a solution. The continuous dependence on parameters and initial conditions.

Linear equations with constant coefficients are the simplest differential equations. They relate a function $y(t)$ and its derivatives $y'(t), y''(t), \dots, y^{(n)}(t)$ by linear links. As examples we write the equations

$$y' = 2y, \quad y'' + 3y' + 2y = 0.$$

A system of several functions $y_1(t), y_2(t), \dots, y_k(t)$ can also be an unknown object in equations. The following example presents a system of two linear differential equations

$$y_1' = -y_2, \quad y_2' = y_1$$

with two unknown functions $y_1(t)$ and $y_2(t)$. Introducing the notation

$$y(t) = \begin{pmatrix} y_1(t) \\ y_2(t) \end{pmatrix},$$

for the vector-valued function $y(t)$ with the components $y_1(t)$ and $y_2(t)$, we can rewrite the above system in the vector form

$$y' = \begin{pmatrix} y_1 \\ y_2 \end{pmatrix}' = \begin{bmatrix} 0 & -1 \\ 1 & 0 \end{bmatrix} \begin{pmatrix} y_1 \\ y_2 \end{pmatrix}.$$

The formula becomes shorter if we denote the matrix of coefficients by a single letter A:

$$A = \begin{bmatrix} 0 & -1 \\ 1 & 0 \end{bmatrix}, \qquad y' = Ay.$$

Now we study systems of linear equations with constant coefficients with N unknown functions $y_1(t), \dots, y_N(t)$ which form the vector-valued function

$$y(t) = \begin{pmatrix} y_1 \\ y_2 \\ \vdots \\ y_N \end{pmatrix}.$$

The number of equations is assumed to coincide with the number of unknown functions.

The considered equations will be written as the system

$$\frac{d}{dt}y = Ay,$$

where A is a square matrix with elements a_{ij}:

$$A = \begin{bmatrix} a_{11} & a_{12} & \dots & a_{1N} \\ a_{21} & a_{22} & \dots & a_{2N} \\ \vdots & \vdots & \ddots & \vdots \\ a_{N1} & a_{N2} & \dots & a_{NN} \end{bmatrix}.$$

Componentwise, the system consists of the N equations

$$\frac{dy_i}{dt} = \sum_{j=1}^{N} a_{ij} y_j, \quad i = 1, 2, \dots, N.$$

We will also consider more complicated systems

$$\frac{dy_i}{dt} = \sum_{j=1}^{N} a_{ij} y_j + f_i(t), \quad i = 1, 2, \dots, N,$$

where given functions $f_j(t)$ appear on the right-hand sides of the equalities. Such systems are called nonhomogeneous to distinguish them from homogeneous systems with $f_j(t) = 0$.

Sometimes, we will consider matrix equations

$$\frac{d}{dt}Y = AY$$

with unknown functions combined in the matrix

$$Y = \begin{bmatrix} y_{11} & y_{12} & \dots & y_{1N} \\ y_{21} & y_{22} & \dots & y_{2N} \\ \vdots & \vdots & \ddots & \vdots \\ y_{N1} & y_{N2} & \dots & y_{NN} \end{bmatrix}.$$

Writing the matrix equation componentwise

$$\frac{dy_{ir}}{dt} = \sum_{j=1}^{N} a_{ij} y_{jr}, \quad i, r = 1, 2, \dots, N,$$

we see that these equalities can be regarded as a family of N independent but similar systems

$$\frac{d}{dt}y^{[k]}(t) = Ay^{[k]}$$

for unknown vector-valued functions

$$y^{[1]} = \begin{pmatrix} y_{11} \\ y_{21} \\ \vdots \\ y_{N1} \end{pmatrix}, \quad y^{[2]} = \begin{pmatrix} y_{12} \\ y_{22} \\ \vdots \\ y_{N2} \end{pmatrix}, \quad \dots, \quad y^{[N]} = \begin{pmatrix} y_{1N} \\ y_{2N} \\ \vdots \\ y_{NN} \end{pmatrix}.$$

Each of these N systems consists of N equations.

EXAMPLE 1. The matrix system

$$\frac{d}{dt}\begin{bmatrix} y_{11} & y_{12} \\ y_{21} & y_{22} \end{bmatrix} = \begin{bmatrix} 0 & -1 \\ 1 & 0 \end{bmatrix}\begin{bmatrix} y_{11} & y_{12} \\ y_{21} & y_{22} \end{bmatrix}$$

splits into two systems:

$$\frac{d}{dt}\begin{pmatrix} y_{11} \\ y_{21} \end{pmatrix} = \begin{bmatrix} 0 & -1 \\ 1 & 0 \end{bmatrix}\begin{pmatrix} y_{11} \\ y_{21} \end{pmatrix} \quad \left(\frac{d}{dt}y^{[1]} = \begin{bmatrix} 0 & -1 \\ 1 & 0 \end{bmatrix} y^{[1]}\right)$$
$$\frac{d}{dt}\begin{pmatrix} y_{12} \\ y_{22} \end{pmatrix} = \begin{bmatrix} 0 & -1 \\ 1 & 0 \end{bmatrix}\begin{pmatrix} y_{12} \\ y_{22} \end{pmatrix} \quad \left(\frac{d}{dt}y^{[2]} = \begin{bmatrix} 0 & -1 \\ 1 & 0 \end{bmatrix} y^{[2]}\right)$$

which can be combined in the following vector system for four dimensional composite vectors:

$$\frac{d}{dt}\begin{pmatrix} y^{[1]} \\ y^{[2]} \end{pmatrix} \equiv \frac{d}{dt}\begin{pmatrix} y_{11} \\ y_{21} \\ y_{12} \\ y_{22} \end{pmatrix} = \begin{bmatrix} 0 & -1 & 0 & 0 \\ 1 & 0 & 0 & 0 \\ 0 & 0 & 0 & -1 \\ 0 & 0 & 1 & 0 \end{bmatrix}\begin{pmatrix} y_{11} \\ y_{21} \\ y_{12} \\ y_{22} \end{pmatrix}.$$

EXAMPLE 2. Similarly, an arbitrary matrix system of order N can be written as a vector system of order N^2

$$\frac{d}{dt}\begin{pmatrix} y^{[1]} \\ y^{[2]} \\ \vdots \\ y^{[N]} \end{pmatrix} = \begin{bmatrix} A & & & \mathbf{0} \\ & A & & \\ & & \ddots & \\ \mathbf{0} & & & A \end{bmatrix}\begin{pmatrix} y^{[1]} \\ y^{[2]} \\ \vdots \\ y^{[N]} \end{pmatrix}.$$

For any t the values of the vectors

$$y(t) = \begin{pmatrix} y_1(t) \\ y_2(t) \\ \vdots \\ y_N(t) \end{pmatrix}$$

belong to a linear N-dimensional space which is equipped with the Euclidean norm

$$\|y\| = \sqrt{(y, y)} = \sqrt{y_1^2 + y_2^2 + \cdots + y_N^2}.$$

By the norm of a matrix (operator) we mean the number[1]

$$\|A\| = \sup_{\|y\|=1} \|Ay\| = \sqrt{\sup_{\|y\|=1} (Ay, Ay)} = \sqrt{\sup_{\|y\|=1} (A^*Ay, y)} = \sqrt{\lambda_{\max}(A^*A)}.$$

The norm $\|A\|$ is usually called the spectral norm of the matrix A. From the definition the following important properties of the norm follow:

$$\|Ax\| \leqslant \|A\|\,\|x\|,$$
$$\|ABx\| = \|A[Bx]\| \leqslant \|A\|\,\|Bx\| \leqslant \|A\|\,\|B\|\,\|x\|,$$
$$\|AB\| \leqslant \|A\|\,\|B\|, \text{ in particular, } \|A^n\| \leqslant \|A\|^n.$$

[1]Cf. [**4**, Chapter 10, §7].

Together with the spectral norm, we will use the norm

$$\|A\|_E = \sqrt{\sum_{i,j=1}^{N} a_{ij}^2},$$

which is called the Euclidean (Frobenius) norm of the matrix A. It is easy to verify that $\|A\| \leqslant \|A\|_E$. Indeed,

$$\|Ay\| = \sqrt{\sum_{i=1}^{N}\left(\sum_{j=1}^{N} a_{ij}y_j\right)^2}.$$

By the Cauchy–Schwarz–Bunyakovskii inequality, we have

$$\left(\sum_{j=1}^{N} a_{ij}y_j\right)^2 \leqslant \left(\sum_{j=1}^{N} a_{ij}^2\right)\left(\sum_{j=1}^{N} y_j^2\right),$$

$$\sum_{i=1}^{N}\left(\sum_{j=1}^{N} a_{ij}y_j\right)^2 \leqslant \sum_{i=1}^{N}\left(\sum_{j=1}^{N} a_{ij}^2\right)\left(\sum_{j=1}^{N} y_j^2\right) = \left(\sum_{i,j=1}^{N} a_{ij}^2\right)\|y\|^2.$$

Therefore,

$$\|Ay\| \leqslant \sqrt{\sum_{i,j=1}^{N} a_{ij}^2}\,\|y\|, \qquad \frac{\|Ay\|}{\|y\|} \leqslant \sqrt{\sum_{i,j=1}^{N} a_{ij}^2}, \qquad \|A\| \leqslant \sqrt{\sum_{i,j=1}^{N} a_{ij}^2} = \|A\|_E^2.$$

Similarly, because of the obvious chain of relations

$$\|A\| = \sqrt{\lambda_{\max}(A^*A)} \geqslant \sqrt{\frac{1}{N}\sum_{i=1}^{N}\lambda_i(A^*A)} = \sqrt{\frac{\operatorname{tr}(A^*A)}{N}}$$

and the fact that the trace $\operatorname{tr}(A^*A)$, which is equal to the sum of the diagonal elements b_{ii} of the matrix $B = A^*A$, can be found by the formulas

$$b_{ii} = \sum_{j=1}^{N} a_{ji}a_{ji} = \sum_{j=1}^{N} a_{ji}^2, \quad \operatorname{tr}(A^*A) = \sum_{i=1}^{N} b_{ii} = \sum_{i,j=1}^{N} a_{ij}^2 = \|A\|_E^2,$$

we conclude that

$$\frac{\|A\|_E}{\sqrt{N}} \leqslant \|A\| \leqslant \|A\|_E.$$

After the above preliminary remarks, we finally proceed to properties of solutions to differential equations.

DEFINITION 1. We say that a vector-valued function $y(t)$ satisfies an equation $y' = Ay$ on the interval $t_1 < t < t_2$, if, in addition to the above equation, each component $y_j(t)$ of the vector-valued function $y(t)$ is differentiable (therefore continuous) for any t in the interval.

If t_1', t_2' are the interior points of the interval $t_1 < t_1' < t_2' < t_2$, then each of the continuous functions $y_j(t)$ is bounded on $t_1' \leqslant t \leqslant t_2'$. Since the number of these functions is finite, there is a constant M such that

$$|y_j(t)| \leqslant M, \qquad \|y(t)\| = \sqrt{\sum_{j=1}^{N} |y_j|^2} \leqslant \sqrt{N} M$$

for $j = 1, 2, \ldots, N$ and $t_1' \leqslant t \leqslant t_2'$. Since $y' = Ay$ ($y_i' = \sum\limits_{j=1}^{N} a_{ij} y_j$, $i = 1, 2, \ldots, N$), the functions y_j', being linear combinations of differentiable functions, are differentiable, i.e, the derivative $y'(t)$ of $y(t)$ is differentiable. Therefore, the differentiation of the equality $y'(t) = Ay(t)$ is legitimate and we can deduce that $y''(t) = Ay'(t) = A^2 y(t)$. By the continuity and differentiability of $y(t)$, this equality allows us to deduce that $y''(t)$ is continuous and differentiable and to obtain the new equality $y'''(t) = A^2 y'(t) = A^3 y(t)$, which, in turn, implies the continuity and differentiability of $y'''(t)$.

Continuing these arguments, we easily establish that the solution $y(t)$ is infinitely differentiable and its derivatives can be found from $y(t)$ by the equalities

$$y^{(k)}(t) = A^k y(t).$$

These equalities imply the following inequalities:

$$\|y^{(k)}(t)\| \leqslant \|A^k\| \, \|y(t)\| \leqslant \|A\|^k \|y(t)\|.$$

In particular, for $t_1' \leqslant t \leqslant t_2'$ we have

$$\|y^{(k)}\| \leqslant \|A\|^k \sqrt{N} M, \quad k = 1, 2, 3, \ldots,$$

$$|y^{(k)}| \leqslant \sqrt{\sum_{j=1}^{N} \left[y_j^{(k)}\right]^2} = \|y^{(k)}(t)\| \leqslant \|A\|^k \sqrt{N} M.$$

Thus, we have proved the following proposition.

PROPOSITION 1. *A solution $y(t)$, $t_1 < t < t_2$, to the system $y' = Ay$ has infinitely differentiable components $y_j(t)$ at all interior points of the interval $t_1 < t < t_2$. Furthermore, for any segment $[t_1', t_2']$ inside the interval (t_1, t_2) there exists a constant $M = M(t_1', t_2')$ such that the derivatives $y_j^{(k)}(t)$ at t from the segment $t_1' \leqslant t \leqslant t_1'$ are estimated as follows:*

$$|y_j^{(k)}| \leqslant \|A\|^k \sqrt{N} M(t_1', t_2').$$

We choose a point t_0 from the interval $[t_1', t_2']$ and expand $y_j(t)$ by the Taylor formula:

$$\begin{aligned} y_j(t) &= y_j(t_0) + \frac{(t-t_0)}{1!} y_j'(t_0) + \frac{(t-t_0)^2}{2!} y_j''(t_0) + \ldots \\ &\quad + \frac{(t-t_0)^k}{k!} y_j^{(k)}(t_0) + \frac{(t-t_0)^{k+1}}{(k+1)!} y_j^{(k+1)}(t_0 + \theta(t - t_0)) \\ &= S_{kj}(t) + \frac{(t-t_0)^{k+1}}{(k+1)!} y_j^{(k+1)}(t_0 + \theta(t-t_0)), \qquad 0 \leqslant \theta \leqslant 1, \end{aligned}$$

where S_{kj} denotes the sum of the $k+1$ first terms of the series. It is obvious that for $t_1' \leqslant t \leqslant t_2'$ the difference between $y_j(t)$ and $S_{kj}(t)$ satisfies the following estimate:

$$\begin{aligned}|y_j(t) - S_{kj}(t)| &\leqslant \frac{|t-t_0|^{(k+1)}}{(k+1)!}\left|y_j^{(k+1)}(t_0+\theta(t-t_0))\right| \\ &\leqslant \frac{\left[|t-t_0|\,\|A\|\right]^{k+1}}{(k+1)!}\sqrt{N}M(t_1',t_2') \leqslant \frac{\left[|t_2-t_1|\,\|A\|\right]^{k+1}\sqrt{N}M}{(k+1)!}.\end{aligned}$$

It is easy to verify that the right-hand side of this inequality tends to zero as $k \to \infty$. Indeed, we first choose some integer k so as $k > k^* = 2|t_2 - t_1|\|A\|$ and consider only the values of k greater than k^*. Then

$$\begin{aligned}&\frac{\left[(t_2-t_1)\|A\|\right]^{k+1}}{(k+1)!}\sqrt{N}M(t_1',t_2') \\ &\qquad < \frac{\left[(t_2-t_1)\|A\|\right]^{k^*}}{k^*!}\frac{\left[(t_2-t_1)\|A\|\right]^{k+1-k^*}}{[k^*]^{k+1-k^*}}\sqrt{N}M(t_1',t_2') \\ &\qquad < \frac{\left[(t_2-t_1)\|A\|\right]^{k^*}}{k^*!}\left(\frac{1}{2}\right)^{k+1-k^*}\sqrt{N}M(t_1',t_2').\end{aligned}$$

It is clear that the right-hand side of the last inequality tends to zero as $k \to \infty$. Therefore the convergence $S_{kj} \to y_j$ is uniform on the segment $[t_1', t_2']$. Thus, we have proved the following proposition.

PROPOSITION 2. *Each component $y_j(t)$ of a solution $y(t)$, $t_1 < t < t_2$, to the system $y'Ay$ can be represented as the sum of the infinite series (the Taylor series)*

$$y_j(t) = y_j(t_0) + \frac{(t-t_0)}{1!}y_j'(t_0) + \frac{(t-t_0)^2}{2!}y_j''(t_0) + \frac{(t-t_0)^3}{3!}y_j'''(t_0) + \dots$$

which converges uniformly on any segment $[t_1', t_2']$ inside the interval (t_1, t_2) on which the solution $y(t)$ is defined.

It is convenient to combine the series representing $y_1(t), y_2(t), \dots, y_N(t)$ in the vector notation

$$y(t) = y(t_0) + \frac{(t-t_0)}{1!}y'(t_0) + \frac{(t-t_0)^2}{2!}y''(t_0) + \dots.$$

Noting that $y^{(k)}(t_0) = A^k y(t_0)$, we can rewrite the series as follows:

$$y(t) = y(t_0) + \frac{(t-t_0)}{1!}Ay(t_0) + \frac{(t-t_0)^2}{2!}A^2y(t_0) + \dots.$$

This shows that the values $y(t)$ at t are uniquely determined from the value $y(t_0)$ at an arbitrary point t_0 on the considered interval. Thus, we have proved the following assertion.

PROPOSITION 3. *If $y(t)$, $t_1 < t < t_2$, is a solution to the system $y'(t) = Ay(t)$ and $y(t_0) = 0$ at some point t_0, $t_1 < t_0 < t_2$, then $y(t)$ is identically zero on the whole existence interval (t_1, t_2).*

Proposition 3 is referred to as the uniqueness theorem.

It turns out that for any given value of the vector a and any point t_0 ($-\infty < t_0 < \infty$) it is always possible to construct a vector-valued function $y(t)$ that is defined for all t, $-\infty < t < \infty$, is differentiable, and satisfies the condition $y(t_0) = a$ and the equation $y'(t) = Ay(t)$. This means that such a vector-valued function (if it exists) is unique. The question about its existence is still open. But we now give arguments which prove, in fact, that a solution exists. The proof we suggest uses the explicit representation of a solution in the form of the Taylor series which appeared in the proof of uniqueness. The expression can be formally written even if we know nothing about the existence of a solution. After that we must verify that the function defined by this expression indeed exists and is a solution to the problem.

We construct the vector-valued function $z(t)$ as the sum of the following series:

$$z(t) = a + \frac{(t-t_0)}{1!}Aa + \frac{(t-t_0)^2}{2!}A^2a + \frac{(t-t_0)^3}{3!}A^3a + \dots .$$

The vector-valued terms of the series satisfy the estimates

$$\left\| \frac{(t-t_0)^k}{k!}A^k a \right\| \leqslant \frac{|t-t_0|^k}{k!}\|A\|^k\|a\|.$$

For any T the series converges uniformly on the segment $|t-t_0| \leqslant T$. The proof essentially coincides with the proof of the convergence of the Taylor series for $y(t)$. By the uniform convergence, the vector-valued function $z(t)$ is continuous on the segment $|t-t_0| \leqslant T$. Since T is arbitrary, $z(t)$ is a continuous function of t on the whole real axis $-\infty < t < \infty$. To show that $z(t)$ is differentiable, we formally differentiate the series

$$a + \frac{(t-t_0)}{1!}Aa + \frac{(t-t_0)^2}{2!}A^2a + \frac{(t-t_0)^3}{3!}A^3a + \dots .$$

The result

$$0 + Aa + \frac{(t-t_0)}{1!}A^2a + \frac{(t-t_0)^2}{2!}A^3a + \dots$$

is also the series with vector-valued terms satisfying the estimates

$$\left\| \frac{(t-t_0)^{k-1}}{(k-1)!}A^k a \right\| \leqslant \frac{(t-t_0)^{k-1}}{(k-1)!}\|A\|^k\|a\|.$$

Again, from these estimates it follows that the formally differentiated series converges on any segment $|t-t_0| \leqslant T$. By the well-known theorem in analysis, $z(t)$ is differentiable and its derivative is the sum of the series

$$\begin{aligned} z' &= Aa + \frac{(t-t_0)}{1!}A^2a + \frac{(t-t_0)^2}{2!}A^3a + \dots \\ &= \lim_{k\to\infty}\left[\sum_{j=0}^{k}\frac{(t-t_0)^j}{j!}A^{j+1}a\right] = \lim_{k\to\infty} A\left[\sum_{j=0}^{k}\frac{(t-t_0)^j}{j!}A^j a\right] = Az(t). \end{aligned}$$

We have established that $z' = Az$ for $|t-t_0| \leqslant T$. Since T is arbitrary, $z(t)$ is differentiable for all t and the equality $z'(t) = Az(t)$ holds everywhere.

To calculate the value $z(t_0)$, we again use the above series and verify that $z(t_0) = a$. By setting $y(t) = z(t)$, we arrive at a solution to our problem, i.e., we have constructed a solution to the equation $y' = Ay$ which takes the value $y(t_0) = a$ at $t = t_0$. Having constructed such a solution, we have proved its existence.

Let us present a more precise version of the assertion just proved. We note that the partial sums

$$S_k(t) = \left[I + \frac{(t-t_0)}{1!}A + \frac{(t-t_0)^2}{2!}A^2 + \dots \frac{(t-t_0)^k}{k!}A^k\right] y(t_0)$$

of the Taylor series are continuous vector-valued functions of t, t_0, the components $y_j(t_0)$ of the initial vector $y(t_0)$, and the elements a_{ij} of the matrix A. By I we denote the identity matrix. The remainder of the Taylor series satisfies the estimate[2]

$$\|S_k(t) - y(t)\| \leqslant \frac{|t-t_0|^{k+1}}{(k+1)!}\|A\|^{k+1}\|y(t_0)\|\, e^{|t-t_0|\,\|A\|}.$$

For any bounded domain, e.g.,

$$|t| < T, \quad |t_0| < T, \quad \|A\| < M, \quad \|y(t_0)\| < M,$$

we can use the following less precise estimate:

$$\|S_k(t) - y(t)\| < \frac{(M\cdot 2T)^{k+1}}{(k+1)!} M e^{2TM},$$

which allows us to conclude that the convergence of $S_k(t)$ to $y(t)$ is uniform in this domain. The uniform convergence implies that $y(t) \equiv y(t, t_0, y(t_0), A)$ is a continuous vector-valued function not only of the variable t, but of t_0, $y(t_0)$, and A in this domain. Thus, we have shown that a solution $y(t)$ to the homogeneous linear system $y' = Ay$ that takes the value $y(t_0)$ at a point t_0 depends continuously on this value $y(t_0)$, the point t_0, and the coefficients a_{ij} of the matrix of the system, i.e., we have proved the following theorem.

THEOREM 1. *The system of differential equations*

$$\frac{d}{dt}y = Ay, \quad -\infty < t < \infty,$$

has a unique solution $y = y(t) = y(t, t_0, a, A)$ such that $y(t_0) = a$. The solution y is a continuous function and has continuous derivatives of all orders with respect to t. Moreover, y is a continuous function of the arguments t, t_0, a, $a_{11}, \dots, a_{NN}$, i.e., the solution y depends continuously on the initial conditions t_0, a and the elements a_{ij} of the matrix A.

---

[2]Here is another proof of this estimate ($0 \leqslant \theta \leqslant 1$):

$$\begin{aligned}\|S_k(t) - y(t)\| &= \left\| \sum_{j=k+1}^{\infty} \frac{(t-t_0)^j}{j!} A^j y(t_0)\right\| \leqslant \sum_{j=k+1}^{\infty} \frac{|t-t_0|^j}{j!}\|A\|^j\|y(t_0)\| \\ &= \left[e^{|t-t_0|\,\|A\|} - \left(1 + \sum_{j=1}^{k}\frac{|t-t_0|^j}{j!}\|A\|^j\right)\right]\|y(t_0)\| \\ &= \frac{|t-t_0|^{k+1}}{(k+1)!}\|A\|^{k+1}e^{\theta|t-t_0|\,\|A\|}\|y(t_0)\| \leqslant \frac{|t-t_0|^{k+1}}{(k+1)!}\|A\|^{k+1}e^{|t-t_0|\,\|A\|}\|y(t_0)\|.\end{aligned}$$

DEFINITION 2. The problem of finding a solution to the vector equation $\frac{dy}{dt} = Ay$ satisfying the initial condition $y(t_0) = a$ is called the Cauchy problem.

Theorem 1 is, in fact, the existence and uniqueness theorem for the Cauchy problem. Moreover, it means that the solution depends continuously on the coefficients of the system (the elements of the matrix A) and the initial conditions.

REMARK 1. The above arguments show that the just-proved theorem on the existence, uniqueness, and continuous dependence on parameters can be expanded to the matrix differential equation

$$\frac{dY}{dt} = AY, \quad Y(t_0) = Y_0.$$

In this case, elements of the matrices A, Y_0, and the initial point t_0 can be regarded as parameters.

EXERCISE 1. Construct a vector-valued series representing a solution to the system $y' = Ay$, where A is one of the following matrices:

$$\begin{bmatrix} 0 & -1 \\ 1 & 0 \end{bmatrix}, \qquad \begin{bmatrix} 2 & 2 & 0 \\ 0 & 2 & 3 \\ 0 & 0 & 2 \end{bmatrix}.$$

§2. Fundamental matrices and matrix exponentials

The Liouville formula for the determinant of a matrix solution. Fundamental matrices and their properties. Matrix exponentials and their properties.

We consider a square matrix $Y(t)$ satisfying the differential equation

$$\frac{d}{dt}Y(t) = AY(t) \tag{1}$$

and study how the determinant $\Delta(t) = \det Y(t)$ changes when t changes. We analyze the expression for the derivative:

$$\Delta'(t) = \frac{d}{dt}\det\begin{bmatrix} y_{11} & y_{12} & \cdots & y_{1N} \\ y_{21} & y_{22} & \cdots & y_{2N} \\ \cdots & \cdots & \cdots & \cdots \\ y_{N1} & y_{N2} & \cdots & y_{NN} \end{bmatrix} = \det\begin{bmatrix} y'_{11} & y'_{12} & \cdots & y'_{1N} \\ y_{21} & y_{22} & \cdots & y_{2N} \\ \cdots & \cdots & \cdots & \cdots \\ y_{N1} & y_{N2} & \cdots & y_{NN} \end{bmatrix}$$
$$+ \det\begin{bmatrix} y_{11} & y_{12} & \cdots & y_{1N} \\ y'_{21} & y'_{22} & \cdots & y'_{2N} \\ \cdots & \cdots & \cdots & \cdots \\ y_{N1} & y_{N2} & \cdots & y_{NN} \end{bmatrix} + \cdots + \det\begin{bmatrix} y_{11} & y_{12} & \cdots & y_{1N} \\ y_{21} & y_{22} & \cdots & y_{2N} \\ \cdots & \cdots & \cdots & \cdots \\ y'_{N1} & y'_{N2} & \cdots & y'_{NN} \end{bmatrix}.$$

The derivative of the determinant of a matrix is equal to the sum of the determinants of matrices that are obtained from the initial matrix by differentiating one of the rows. The number of terms of such a sum is equal to the number of rows N.

This rule can be easily derived from the formula for the derivative of the product and the representation of the determinant as the sum of products of elements of the matrix.

We now consider a term of the expression for the derivative of the determinant, say, the first one:

$$\det \begin{bmatrix} y'_{11} & y'_{12} & y'_{13} & \cdots & y'_{1N} \\ y_{21} & y_{22} & y_{23} & \cdots & y_{2N} \\ \cdots & \cdots & \cdots & \cdots & \cdots \\ y_{N1} & y_{N2} & y_{N3} & \cdots & y_{NN} \end{bmatrix}.$$

The matrix differential equation (1) implies that

$$y'_{1j} = a_{11}y_{1j} + a_{12}y_{2j} + \cdots + a_{1N}y_{Nj}.$$

Therefore,

$$\begin{aligned}(y'_{11}, y'_{12}, \ldots, y'_{1N}) &= a_{11}(y_{11}, y_{12}, \ldots, y_{1N}) \\ &\quad + a_{12}(y_{21}, y_{22}, \ldots, y_{2N}) + \cdots + a_{1N}(y_{N1}, y_{N2}, \ldots, y_{NN}).\end{aligned}$$

From these expressions and the properties of determinant we obtain the following chain of equalities:

$$\begin{aligned}\det \begin{bmatrix} y'_{11} & y'_{12} & \cdots & y'_{1N} \\ y_{21} & y_{22} & \cdots & y_{2N} \\ \cdots & \cdots & \cdots & \cdots \\ y_{N1} & y_{N2} & \cdots & y_{NN} \end{bmatrix} &= \det \begin{bmatrix} a_{11}y_{11} & a_{11}y_{12} & \cdots & a_{11}y_{1N} \\ y_{21} & y_{22} & \cdots & y_{2N} \\ \cdots & \cdots & \cdots & \cdots \\ y_{N1} & y_{N2} & \cdots & y_{NN} \end{bmatrix} \\ &= a_{11} \det \begin{bmatrix} y_{11} & y_{12} & \cdots & y_{1N} \\ y_{21} & y_{22} & \cdots & y_{2N} \\ \cdots & \cdots & \cdots & \cdots \\ y_{N1} & y_{N2} & \cdots & y_{NN} \end{bmatrix} = a_{11} \det Y(t).\end{aligned}$$

The expressions for the remaining terms can be handled in the same way. So, it is obvious that

$$\begin{aligned}\Delta'(t) &= \frac{d}{dt}[\det Y(t)] = (a_{11} + a_{22} + \cdots + a_{NN}) \det Y(t) \\ &= (a_{11} + a_{22} + \cdots + a_{NN})\Delta(t).\end{aligned}$$

We have established that $\Delta(t)$ satisfies the differential equation

$$\Delta' = (a_{11} + a_{22} + \cdots + a_{NN})\Delta.$$

This equation has the following partial solution:

$$\Delta(t) = \Delta(t_0) \exp((a_{11} + a_{22} + \cdots + a_{NN})(t - t_0)).$$

By the uniqueness theorem (cf. Proposition 3, §1), there are no other solutions $\Delta(t)$ taking the value $\Delta(t_0)$ at $t = t_0$. Thus, we have proved the following assertion.

PROPOSITION 1. *A solution $Y(t)$ to the matrix equation* (1) *satisfies the relation* (*the Liouville formula*)

$$\det Y(t) = e^{(t-t_0)\operatorname{tr} A} \det Y(t_0). \tag{2}$$

The Liouville formula (2) means that if the determinant of the matrix $Y(t)$ does not vanish at some point $t = t_0$, then it does not vanish at all points t.

DEFINITION 1. A matrix $Y(t)$ satisfying the differential equation (1) and the condition $\det Y(t_0) \neq 0$ at least at one point t_0 is called a fundamental matrix of solutions to the system $y' = Ay$.

The columns of the fundamental matrix

$$Y(t) = \begin{bmatrix} y_{11} & y_{12} & \cdots & y_{1N} \\ y_{21} & y_{22} & \cdots & y_{2N} \\ y_{31} & y_{32} & \cdots & y_{3N} \\ \cdots & \cdots & \cdots & \cdots \\ y_{N1} & y_{N2} & \cdots & y_{NN} \end{bmatrix}$$

are linearly independent solutions to the system $y' = Ay$.

The following proposition explains why $Y(t)$ is referred to as a fundamental matrix of solutions.

PROPOSITION 2. *Every solution $y(t)$ to the system $y' = Ay$ can be represented as a linear combination of the columns of the fundamental matrix $Y(t)$, i.e.,*

$$y(t) = Y(t)c \tag{3}$$

or, in the matrix notation,

$$y(t) = \begin{pmatrix} y_1(t) \\ y_2(t) \\ y_3(t) \\ \vdots \\ y_N(t) \end{pmatrix} = c_1 \begin{pmatrix} y_{11} \\ y_{21} \\ y_{31} \\ \vdots \\ y_{N1} \end{pmatrix} + c_2 \begin{pmatrix} y_{12} \\ y_{22} \\ y_{32} \\ \vdots \\ y_{N2} \end{pmatrix} + \cdots + c_N \begin{pmatrix} y_{1N} \\ y_{2N} \\ y_{3N} \\ \vdots \\ y_{NN} \end{pmatrix}$$
$$= \begin{bmatrix} y_{11} & y_{12} & \cdots & y_{1N} \\ y_{21} & y_{22} & \cdots & y_{2N} \\ \cdots & \cdots & \cdots & \cdots \\ y_{N1} & y_{N2} & \cdots & y_{NN} \end{bmatrix} \begin{pmatrix} c_1 \\ c_2 \\ \vdots \\ c_N \end{pmatrix} = Y(t) \begin{pmatrix} c_1 \\ c_2 \\ \vdots \\ c_N \end{pmatrix} = Y(t)c.$$

PROOF. We begin with the remark that for any constant vector c (independent of t) the vector $z(t) = Y(t)c$ satisfies the vector equation $z' = Az$. This can be checked by the direct differentiation as follows:

$$\frac{d}{dt} z(t) = \frac{d}{dt} Y(t)c = AY(t)c = Az(t).$$

By the uniqueness theorem (cf. Proposition 3, §1), two solutions $z(t)$ and $y(t)$ coincide if $y(t_0)$ and $z(t_0)$ coincide. The latter is true if the vector c satisfies the equation $y(t_0) = Y(t_0)c$. Since $\det Y(t_0) \neq 0$, the vector c can be uniquely found from this equation. □

It is obvious that the representation

$$y(t) = Y(t)c, \qquad \det Y(t_0) \neq 0$$

implies that solutions $y(t)$ of the vector equation $y' = Ay$ form a linear space of dimension N.

Let $Y_1(t)$ and $Y_2(t)$ be two different fundamental matrices of solutions to the system $y' = Ay$, i.e.,

$$\frac{d}{dt}Y_1(t) = AY_1(t), \quad \det Y_1(t) \neq 0,$$
$$\frac{d}{dt}Y_2(t) = AY_2(t), \quad \det Y_2(t) \neq 0.$$

In addition,

$$Y_2(t_0) = Y_1(t_0)[Y_1^{-1}(t_0)Y_2(t_0)] = Y_1(t_0)B,$$

where $B = Y_1^{-1}(t_0)Y_2(t_0)$ is a constant matrix such that

$$\det B = \frac{\det Y_2(t_0)}{\det Y_1(t_0)} \neq 0.$$

We consider the matrix

$$Z(t) = Y_2(t) - Y_1(t)B.$$

It is obvious that $Z(t_0) = Y_2(t_0) - Y_1(t_0)B = 0$ and

$$\begin{aligned}\frac{d}{dt}Z(t) &= \frac{d}{dt}Y_2(t) - \left[\frac{d}{dt}Y_1(t)\right]B \\ &= AY_2(t) - AY_1(t)B = A[Y_2(t) - Y_1(t)B] = AZ(t).\end{aligned}$$

By the uniqueness theorem (cf. Proposition 3, §1), we conclude that $Z(t) = 0$ or that any fundamental matrices $Y_2(t)$ and $Y_1(t)$ of solutions to the system (1) are connected by the equality $Y_2(t) = Y_1(t)B$, where B is a nonsingular constant matrix. On the other hand, if $Y(t)$ is a fundamental matrix ($Y' = AY$, $\det Y(t) \neq 0$) and B is a nonsingular constant matrix, then

$$\det[Y(t)B] = \det Y(t) \det B \neq 0,$$
$$\frac{d}{dt}[Y(t)B] = [AY(t)]B = A[Y(t)B],$$

i.e., $Y(t)B$ is also a fundamental matrix. Thereby, we have described all fundamental matrices in terms of one of them. Thus, every solution to the system $y' = Ay$ can be represented in terms of the fundamental matrix $Y(t)$ of solutions by the formula (3).

We recall that to solve the Cauchy problem is to find a solution $y(t)$ to the system $y' = Ay$ that takes the given value g at $t = t_0$, i.e., $y(t_0) = g$. Hence to solve the Cauchy problem

$$\frac{d}{dt}y(t) = Ay(t), \quad y(t_0) = g, \tag{4}$$

we must determine a constant vector c from the system $Y(t_0)c = g$. It is clear that $c = Y^{-1}(t_0)g$. The inverse matrix $Y^{-1}(t_0)$ exists since $\det Y(t_0) \neq 0$ in view of the assumption that $Y(t)$ is fundamental. A solution $y(t)$ to the Cauchy problem (4) can be written in the form

$$y(t) = Y(t)Y^{-1}(t_0)g.$$

Among all fundamental matrices of the considered system, we distinguish fundamental matrices $Y(t)$ such that $Y(0) = I$, where I is the identity matrix. Such matrices $Y(t)$ are defined by the conditions

$$\frac{d}{dt}Y(t) = AY(t), \qquad Y(0) = I.$$

As we know, a solution to this problem can be obtained using the matrix Taylor series

(5) $$Y(t) = \left[I + \frac{t}{1!}A + \frac{t^2}{2!}A^2 + \dots\right] Y(0) = \lim_{k\to\infty} [S_k(t)],$$

where

$$S_k(t) = I + \frac{t}{1!}A + \frac{t^2}{2!}A^2 + \dots \frac{t^k}{k!}A^k.$$

This series looks like the Taylor series for the exponential function

$$e^{ta} = 1 + \frac{t}{1!}a + \frac{t^2}{2!}a^2 + \frac{t^3}{3!}a^3 + \dots.$$

This fact explains the following definition.

DEFINITION 2. By a matrix exponential we mean the matrix $Y(t)$ of the form (5); we denote

(6) $$e^{tA} \equiv I + \frac{t}{1!}A + \frac{t^2}{2!}A^2 + \dots.$$

Since $S_k(t)$ is a matrix polynomial in A, we have $AS_k(t) = S_k(t)A$. Passing to the limit as $k \to \infty$, we see that

$$A \lim_{k\to\infty} S_k(t) = Ae^{tA} = e^{tA}A = [\lim_{k\to\infty} S_k(t)]A.$$

Many properties of matrix exponentials are similar to those of the scalar exponential function. However, it is necessary to be very careful. For example, a matrix exponential does not satisfy, in general, the equality $e^{t(B+C)} = e^{tB}e^{tC}$.

EXERCISE 1. Verify the equalities

$$\exp\left(t\begin{bmatrix}1 & 0\\ 0 & 2\end{bmatrix}\right) = \begin{bmatrix}e^t & 0\\ 0 & e^{2t}\end{bmatrix},$$

$$\exp\left(t\begin{bmatrix}0 & 1\\ 0 & 0\end{bmatrix}\right) = \begin{bmatrix}1 & t\\ 0 & 1\end{bmatrix},$$

$$\exp\left(t\begin{bmatrix}1 & 1\\ 0 & 2\end{bmatrix}\right) = \begin{bmatrix}e^t & e^{2t} - e^t\\ 0 & e^{2t}\end{bmatrix}$$

and show that $e^{t(B+C)} \neq e^{tB}e^{tC}$.

We prove that $e^{tA}e^{-tA} = I$ or, which is equivalent, $(e^{tA})^{-1} = e^{-tA}$. Indeed, let $Y(t) = e^{tA}$. Then we have

$$\frac{d}{dt}Y(t) = AY(t), \qquad Y(0) = I,$$

$$\frac{d}{dt}Y(-t) = -AY(-t), \qquad Y(0) = I,$$

$$\begin{aligned}\frac{d}{dt}[Y(t)Y(-t)] &= \left[\frac{d}{dt}Y(t)\right]Y(-t) + Y(t)\frac{d}{dt}[Y(-t)]\\ &= AY(t)Y(-t) + Y(t)[-AY(-t)] = Ae^{tA}e^{-tA} - e^{tA}Ae^{-tA}\\ &= Ae^{tA}e^{-tA} - Ae^{tA}e^{-tA} = 0.\end{aligned}$$

The equality $e^{tA}A = Ae^{tA}$ was used here. Thus, $Y(t)Y(-t)$ is a constant matrix:

$$Y(t)Y(-t) = Y(0)Y(0) = I^2 = I, \qquad e^{tA}e^{-tA} = I.$$

EXERCISE 2. Prove that $e^{(t+s)A} = e^{tA}e^{sA}$.

We indicate one more interesting and useful property of the matrix exponential. By definition, $e^{tA} = Y(t)$ if $Y(t)$ satisfies the initial condition $Y(0) = I$ and the matrix equation (1). We know that $\det Y(t) = e^{t(\operatorname{tr} A)} \det Y(0)$ for every solution $Y(t)$ to the matrix equation (1). In the considered case, $Y(0) = I$ and $\det I = 1$. Therefore,

$$\det[e^{tA}] = e^{t(\operatorname{tr} A)}. \tag{7}$$

Thus, we have established a formula for the determinant of the matrix exponential.

Together with the fundamental matrix $Y(t) = e^{tA}$ satisfying the conditions

$$\frac{d}{dt}Y(t) = AY(t), \qquad Y(0) = I,$$

one often considers the fundamental matrix $Z(t) = Y(t - t_0) = e^{(t-t_0)A}$. We have

$$\frac{d}{dt}Z(t) = AZ(t), \qquad Z(t_0) = I.$$

It is convenient to express a solution to the Cauchy problem

$$\frac{d}{dt}y(t) = Ay(t), \qquad y(t_0) = g$$

in terms of $Z(t)$:

$$y(t) = Z(t)y(t_0) = Y(t - t_0)g = e^{(t-t_0)A}g.$$

§3. Estimates for the matrix exponential. Polynomial representation of the matrix exponential

The simplest estimate for the matrix exponential. The closeness estimate for two matrix exponentials. The matrix exponential as a polynomial in A with coefficients depending on an independent variable t. The Gelfand–Shilov estimate for the matrix exponential and some simplifications of the estimate.

For further considerations it is interesting and important to obtain various estimates for the norm of the matrix exponential. Using the inequalities (cf. §1)

$$\|A^k\| \leqslant \|A\|^k, \qquad k = 1, 2, \dots,$$

it is easy to derive the relation

$$\begin{aligned}\left\|I + \frac{t}{1!}A + \frac{t^2}{2!}A^2 + \dots\right\| &\leqslant \|I\| + \left\|\frac{t}{1!}A\right\| + \left\|\frac{t^2}{2!}A^2\right\| + \dots \\ &\leqslant 1 + \frac{|t|}{1!}\|A\| + \frac{|t^2|}{2!}\|A\|^2 + \dots = e^{|t|\,\|A\|},\end{aligned}$$

i.e., $\|e^{tA}\| \leqslant e^{|t|\,\|A\|}$. By the above equality,

$$1 = \|I\| = \|e^{tA}e^{-tA}\| \leqslant \|e^{tA}\|\|e^{-tA}\|,$$

i.e., $\|e^{tA}\| \geqslant 1/\|e^{-tA}\|$. Replacing A by $-A$, we obtain

$$\|e^{-tA}\| \leqslant e^{|t|\,\|A\|}, \qquad \|e^{tA}\| \geqslant \frac{1}{\|e^{-tA}\|} \geqslant e^{-|t|\|A\|}.$$

Thus, we have proved the following two-sided estimate for the norm of the matrix exponential:

$$e^{-|t|\,\|A\|} \leqslant \|e^{tA}\| \leqslant e^{|t|\,\|A\|}. \tag{1}$$

We give one more simple and convenient inequality for the difference of matrix exponentials (cf. Proposition 1 below), which will be often used. To prove this inequality, we first establish two lemmas.

LEMMA 1. *For every integer $k \geqslant 1$ the following inequality holds:*

$$\|(A+B)^k - A^k\| \leqslant (\|A\| + \|B\|)^k - \|A\|^k. \tag{2}$$

PROOF. For $k = 1$ the required inequality follows from the equality

$$\|(A+B) - A\| = \|B\| = (\|A\| + \|B\|) - \|A\|.$$

For $k > 1$ we proceed by induction. Assume that we have proved the inequality

$$\|(A+B)^{k-1} - A^{k-1}\| \leqslant (\|A\| + \|B\|)^{k-1} - \|A\|^{k-1},$$

where $k \geqslant 2$. Using the equality

$$(A+B)^k - A^k = (A+B)\big[(A+B)^{k-1} - A^{k-1}\big] + BA^{k-1},$$

we see that

$$\begin{aligned}\|(A+B)^k - A^k\| &\leqslant (\|A\| + \|B\|)\big\|(A+B)^{k-1} - A^{k-1}\big\| + \|B\|\,\|A\|^{k-1} \\ &\leqslant (\|A\| + \|B\|)\big[(\|A\| + \|B\|)^{k-1} - \|A\|^{k-1}\big] + \|B\|\,\|A\|^{k-1} \\ &= (\|A\| + \|B\|)^k - \|A\|^k.\end{aligned}$$

From the last relation we consequently derive the required inequality (2) for $k = 2, 3, 4, \dots$, i.e., for all positive integers k. □

LEMMA 2. *The following inequality holds:*

$$\|e^{A+B} - e^A\| \leqslant e^{\|A\|+\|B\|} - e^{\|A\|} = (e^{\|B\|} - 1)e^{\|A\|}. \tag{3}$$

PROOF. The representations

$$e^{A+B} = I + \sum_{k=1}^{\infty} \frac{1}{k!}(A+B)^k, \quad e^A = I + \sum_{k=1}^{\infty} \frac{1}{k!} A^k$$

imply the representation

$$e^{A+B} - e^A = \sum_{k=1}^{\infty} \frac{1}{k!}\left[(A+B)^k - A^k\right]$$

and the inequality

$$\|e^{A+B} - e^A\| \leqslant \sum_{k=1}^{\infty} \frac{1}{k!}\left\|(A+B)^k - A^k\right\|.$$

(To verify the above assertion we change the position of terms in the infinite sums. We recommend the reader to justify the validity of such permutations.) Using Lemma 1, we find

$$\begin{aligned}\|e^{A+B} - e^A\| &\leqslant \sum_{k=1}^{\infty} \frac{1}{k!}\left[(\|A\| + \|B\|)^k - \|A\|^k\right] = \sum_{k=1}^{\infty} \frac{1}{k!}(\|A\| + \|B\|)^k - \sum_{k=1}^{\infty} \frac{1}{k!}\|A\|^k \\ &= e^{\|A\|+\|B\|} - e^{\|A\|} = (e^{\|B\|} - 1)e^{\|A\|}. \quad \square\end{aligned}$$

PROPOSITION 1. *The following estimate holds:*

$$\|e^{A+B} - e^A\| \leqslant \|B\| e^{\|A\|+\|B\|}. \tag{4}$$

PROOF. The inequality (4) follows from the relation

$$e^{\|B\|} - 1 = \|B\|\left(1 + \sum_{k=1}^{\infty} \frac{\|B\|^k}{(k+1)k!}\right) \leqslant \|B\|\left(1 + \sum_{k=1}^{\infty} \frac{\|B\|^k}{k!}\right) = \|B\| e^{\|B\|}. \quad \square$$

Studying matrix exponentials, we used the representation of the matrix exponential as a matrix series. As is shown in linear algebra, rather general functions of matrices, in particular, matrix exponentials, can be represented as matrix polynomials. We will not use results of a general theory, but give an independent derivation of the polynomial representation.

We consider some special scalar functions $\psi_1(t), \psi_2(t), \ldots, \psi_N(t)$ that form a solution to the following Cauchy problem:

$$\begin{aligned} \frac{d\psi_1(t)}{dt} &= \tau_1 \psi_1(t), \quad \psi_1(0) = 1, \\ \frac{d\psi_2(t)}{dt} &= \tau_2 \psi_2(t) + \psi_1(t), \quad \psi_2(0) = 0, \\ \frac{d\psi_3(t)}{dt} &= \tau_3 \psi_3(t) + \psi_2(t), \quad \psi_3(0) = 0, \\ &\cdots\cdots\cdots\cdots\cdots\cdots \\ \frac{d\psi_N(t)}{dt} &= \tau_N \psi_N(t) + \psi_{N-1}(t), \quad \psi_N(0) = 0. \end{aligned} \tag{5}$$

Introducing the vector

$$\psi = \begin{pmatrix} \psi_N \\ \psi_{N-1} \\ \vdots \\ \psi_1 \end{pmatrix},$$

we can rewrite the Cauchy problem (5) in the matrix form as follows:

$$\frac{d}{dt}\psi(t) = \begin{bmatrix} \tau_N & 1 & & & & \mathbf{0} \\ & \tau_{N-1} & 1 & & & \\ & & \tau_{N-2} & 1 & & \\ & & & \ddots & \ddots & \\ \mathbf{0} & & & & \tau_2 & 1 \\ & & & & & \tau_1 \end{bmatrix} \psi(t), \quad \psi(0) = \begin{pmatrix} 0 \\ 0 \\ \vdots \\ 0 \\ 0 \\ 1 \end{pmatrix}. \tag{6}$$

The existence and uniqueness of such functions $\psi_1(t), \psi_2(t), \dots, \psi_N(t)$ and continuous dependence of $\psi_1(t), \psi_2(t), \dots, \psi_N(t)$ on the parameters $\tau_1, \tau_2, \dots, \tau_N$ follow from §1.

Assuming that $\tau_1, \tau_2, \dots, \tau_N$ coincide with the eigenvalues of the matrix A, i.e., the roots of the characteristic equation

$$\det[A - \tau I] = 0, \tag{7}$$

and using $\psi_j(t)$, it is not hard to construct the matrix polynomial in A which satisfies the same differential equation and initial conditions as the matrix exponential e^{tA}. Namely, the polynomial

$$\begin{aligned} \mathcal{P}(t, A) = \psi_1(t)I + \psi_2(t)[A - \tau_1 I] + \psi_3(t)[A - \tau_1 I][A - \tau_2 I] + \dots \\ + \psi_N(t)[A - \tau_1 I][A - \tau_2 I] \cdots [A - \tau_{N-1} I] \end{aligned}$$

possesses the required properties. To verify this assertion it is convenient to introduce auxiliary matrix polynomials

$$\begin{aligned} &\mathcal{P}_1(A) = [A - \tau_1 I], \\ &\mathcal{P}_2(A) = [A - \tau_1 I][A - \tau_2 I], \\ &\dots\dots\dots\dots\dots\dots \\ &\mathcal{P}_{N-1}(A) = [A - \tau_1 I][A - \tau_2 I] \cdots [A - \tau_{N-1} I], \\ &\mathcal{P}_N(A) = [A - \tau_1 I][A - \tau_2 I] \cdots [A - \tau_N I] \end{aligned}$$

and check the relations

$$\begin{aligned} &\mathcal{P}_1(A) + \tau_1 I = AI, \\ &\mathcal{P}_2(A) + \tau_2 \mathcal{P}_1(A) = A\mathcal{P}_1(A), \\ &\dots\dots\dots\dots\dots\dots \\ &\mathcal{P}_{N-1}(A) + \tau_{N-1}\mathcal{P}_{N-2}(A) = A\mathcal{P}_{N-2}(A), \\ &\tau_N \mathcal{P}_{N-1}(A) \equiv \mathcal{P}_N(A) + \tau_N \mathcal{P}_{N-1}(A) = A\mathcal{P}_{N-1}(A). \end{aligned}$$

The last equality is obtained by collecting similar terms and using the equality $\mathcal{P}_N(A) = 0$ which is valid since any square matrix satisfies its characteristic equation in view of the Cayley–Hamilton theorem. We find the derivative of $\mathcal{P}(t, A)$ with respect to t:

$$\begin{aligned}
\frac{d}{dt}\mathcal{P}(t,A) &= \frac{d\psi_1}{dt}I + \frac{d\psi_2}{dt}\mathcal{P}_1(A) + \frac{d\psi_3}{dt}\mathcal{P}_2(A) + \cdots + \frac{d\psi_N}{dt}\mathcal{P}_{N-1}(A) \\
&= \tau_1\psi_1 I + (\tau_2\psi_2 + \psi_1)\mathcal{P}_1(A) + (\tau_3\psi_3 + \psi_2)\mathcal{P}_2(A) + \dots \\
&\quad + (\tau_N\psi_N + \psi_{N-1})\mathcal{P}_{N-1}(A) \\
&= \psi_1\big(\mathcal{P}_1(A) + \tau_1 I\big) + \psi_2\big(\mathcal{P}_2(A) + \tau_2\mathcal{P}_1(A)\big) + \dots \\
&\quad + \psi_{N-1}\big(\mathcal{P}_{N-1}(A) + \tau_{N-1}\mathcal{P}_{N-2}(A)\big) + \psi_N\tau_N\mathcal{P}_{N-1}(A) \\
&= \psi_1 A + \psi_2 A\mathcal{P}_1(A) + \cdots + \psi_{N-1}A\mathcal{P}_{N-2}(A) + \psi_N A\mathcal{P}_{N-1} = A\mathcal{P}(t,A).
\end{aligned}$$

Moreover,

$$\begin{aligned}
\mathcal{P}(0,A) &= \psi_1(0)I + \psi_2(0)\mathcal{P}_1(A) + \cdots + \psi_N(0)\mathcal{P}_{N-1}(A) \\
&= 1\cdot I + 0\cdot\mathcal{P}_1(A) + \cdots + 0\cdot\mathcal{P}_{N-1}(A) = I.
\end{aligned}$$

By the uniqueness theorem, $\mathcal{P}(t, A) = e^{tA}$.

Thus, the matrix exponential e^{tA} can be represented in the form of an infinite series (cf. Definition 2 in §2)

$$e^{tA} = I + tA + \frac{t^2}{2!}A^2 + \dots \tag{8}$$

and as a polynomial, i.e., in the form of a finite sum

$$e^{tA} = \psi_1(t)I + \psi_2(t)[A - \tau_1 I] + \cdots + \psi_N(t)[A - \tau_1 I]\cdots[A - \tau_{N-1}I]. \tag{9}$$

The equivalence of these two representations (8) and (9) can be proved by direct but cumbersome computation which is based on the possibility of expressing $A^N, A^{N+1}, A^{N+2}, \dots$ as linear combinations of $I, A, A^2, \dots, A^{N-1}$ since, by the Cayley–Hamilton theorem,

$$\mathcal{P}_N(A) = (A - \tau_1 I)(A - \tau_2 I)\cdots(A - \tau_N I) = 0.$$

We gave an indirect proof of the polynomial representation (9) of the matrix exponential e^{tA}. Since $|\tau_j| \leqslant \|A\|$, from (9) we obtain the inequality

$$\|e^{tA}\| \leqslant |\psi_1(t)| + 2\|A\|\,|\psi_2(t)| + (2\|A\|)^2|\psi_3(t)| + \cdots + (2\|A\|)^{N-1}|\psi_N(t)|.$$

We assume that

$$\operatorname{Re}\tau_1 \leqslant \alpha, \quad \operatorname{Re}\tau_2 \leqslant \alpha, \quad \dots, \quad \operatorname{Re}\tau_N \leqslant \alpha.$$

We will show below that under these assumptions, the following inequalities hold ($t \geqslant 0$):

$$|\psi_1(t)| \leqslant e^{\alpha t}, \ |\psi_2(t)| \leqslant te^{\alpha t}, \ |\psi_3(t)| \leqslant \frac{t^2}{2!}e^{\alpha t}, \dots, \ |\psi_j(t)| \leqslant \frac{t^{j-1}}{(j-1)!}e^{\alpha t}.$$

For $t \geqslant 0$ these inequalities lead to one more estimate for the matrix exponential which will be often used:

$$\|e^{tA}\| \leqslant \left[1 + 2\|A\|t + (2\|A\|)^2\frac{t^2}{2!} + \cdots + (2\|A\|)^{N-1}\frac{t^{N-1}}{(N-1)!}\right]e^{\alpha t}.$$

We now justify the estimate

$$|\psi_k(t)| \leqslant \frac{t^{k-1}}{(k-1)!}e^{\alpha t}, \quad t \geqslant 0,$$

for $\operatorname{Re}\tau_j \leqslant \alpha$, $j = 1, 2, \dots, k$. Note that

$$\psi_1(t) = e^{\tau_1 t}, \quad |\psi_1(t)| \leqslant e^{\alpha t} \text{ (for } t > 0).$$

We proceed by induction. We consider the integral (for $k \geqslant 2$)

$$\mathcal{J}(t) = \int_0^t e^{\tau_k(t-s)}\psi_{k-1}(s)\,ds$$

and verify that $\mathcal{J}(0) = 0$ and

$$\frac{d\mathcal{J}(t)}{dt} = \psi_{k-1}(t) + \tau_k \int_0^t e^{\tau_k(t-s)}\psi_{k-1}(t)\,ds = \tau_k\mathcal{J}(t) + \psi_{k-1}(t),$$

i.e., $\mathcal{J}(t)$ satisfies the problem

$$\frac{\mathcal{J}(t)}{dt} = \tau_k\mathcal{J}(t) + \psi_{k-1}(t), \quad \mathcal{J}(0) = 0.$$

The function $\psi_k(t)$ satisfies the same equation and the same initial condition:

$$\frac{d\psi_k(t)}{dt} = \tau_k\psi_k(t) + \psi_{k-1}(t), \quad \psi_k(0) = 0.$$

The difference $z(t) = \psi_k(t) - \mathcal{J}(t)$ satisfies the homogeneous equation and the zero initial condition:

$$\frac{dz(t)}{dt} = \tau_k z(t), \quad z(0) = 0.$$

By the uniqueness theorem (cf. Proposition 3, §1), $z(t) = 0$. Hence $\psi_k(t) = \mathcal{J}(t)$. Thereby, for $\psi_k(t)$ we have established the integral representation

$$\psi_k(t) = \int_0^t e^{\tau_k(t-s)}\psi_{k-1}(s)\,ds$$

which implies the inequality ($t > 0$, $\operatorname{Re}\tau_k < \alpha$)

$$|\psi_k(t)| \leqslant \int_0^t e^{\alpha(t-s)}|\psi_{k-1}(s)|\,ds.$$

Assuming that $k \geqslant 2$ and

$$|\psi_{k-1}(s)| \leqslant \frac{s^{k-2}}{(k-2)!}e^{\alpha s}$$

for $s > 0$, we easily obtain

$$|\psi_k(t)| \leqslant \int_0^t e^{\alpha(t-s)}\frac{s^{k-2}}{(k-2)!}ds = \frac{t^{k-1}}{(k-1)!}e^{\alpha t}.$$

Since we have proved the inequality

$$|\psi_{2-1}(t)| = |\psi_1(t)| \leqslant e^{\alpha t} = \frac{t^{2-2}}{(2-2)!}e^{\alpha t},$$

we can conclude by induction that the estimate

$$|\psi_k(t)| \leqslant \frac{t^{k-1}}{(k-1)!}e^{\alpha t}$$

is valid for $t > 0$, $k = 1, 2, 3, \ldots$.

Thus, we have proved the following assertion.

PROPOSITION 2. *If all eigenvalues τ_j of a matrix A lie in the half-plane* ($\operatorname{Re}\tau_j \leqslant \alpha$), *then for $t \geqslant 0$*

$$\|e^{tA}\| \leqslant \left[1 + 2\|A\|t + (2\|A\|)^2\frac{t^2}{2!} + \cdots + (2\|A\|)^{N-1}\frac{t^{N-1}}{(N-1)!}\right] e^{\alpha t}. \tag{10}$$

The estimate (10) was originally obtained by Gelfand and Shilov [**6**].

In the sequel, we often use the following consequence of the estimate (10) for matrices whose eigenvalues lie in the left half-plane.

PROPOSITION 3. *If the eigenvalues τ_j of a matrix A lie in the left half-plane* ($\operatorname{Re}\tau_j \leqslant -\delta$, $\delta > 0$), *then for $t \geqslant 0$ we have*

$$\|e^{tA}\| \leqslant \gamma(N)\left(\frac{\|A\|}{\delta}\right)^{N-1} e^{-t\delta/2}. \tag{11}$$

PROOF. In Proposition 2 we take $\alpha = -\delta$ ($\delta > 0$), i.e., $\operatorname{Re}\tau_j \leqslant -\delta$ ($\delta > 0$). Taking into account that $t^j e^{-t\delta/2}$ is bounded for $t > 0$, we estimate the maximum of this expression. If the maximal value is attained at a point t_0, then

$$\frac{d}{dt}\left(t^j e^{-t\delta/2}\right) = \left(j - \frac{\delta}{2}t\right) t^{j-1}e^{-t\delta/2} = 0.$$

Hence $t_0 = 2j/\delta$. Consequently,

$$t^j e^{-t\delta/2} \leqslant \left(\frac{2j}{\delta}\right)^j e^{-j},$$

$$\begin{aligned}
&e^{-t\delta/2} + 2\|A\|te^{-t\delta/2} + (2\|A\|)^2\frac{t^2}{2!}e^{-t\delta/2} + \cdots + (2\|A\|)^{N-1}\frac{t^{N-1}}{(N-1)!}e^{-t\delta/2} \\
&\leqslant 1 + \left(\frac{4\|A\|}{\delta}\right)\frac{1}{1!}e^{-1} + \left(\frac{4\|A\|}{\delta}\right)^2\frac{(2e^{-1})^2}{2!} + \cdots + \left(\frac{4\|A\|}{\delta}\right)^{N-1}\frac{[(N-1)e^{-1}]^{N-1}}{(N-1)!} \\
&\leqslant \gamma(N)\left(\frac{\|A\|}{\delta}\right)^{N-1},
\end{aligned}$$

where $\gamma(N)$ is a constant depending only on N. We have used the inequality $\|A\|/\delta \geqslant 1$, which is valid because

$$0 < \delta \leqslant |\operatorname{Re}\tau_j| \leqslant |\tau_j| \leqslant \|A\|.$$

By the proved inequality, we rewrite the estimate for the matrix exponential as follows ($t \geqslant 0$, $\operatorname{Re}\tau_j \leqslant -\delta$, $\delta > 0$):

$$\begin{aligned}\|e^{tA}\| &\leqslant \left[1 + 2\|A\|t + \frac{(2\|A\|t)^2}{2!} + \dots + \frac{(2\|A\|t)^{N-1}}{(N-1)!}\right] e^{-\delta t} \\ &= \left[1 + 2\|A\|t + \frac{(2\|A\|t)^2}{2!} + \dots + \frac{(2\|A\|t)^{N-1}}{(N-1)!}\right] e^{-t\delta/2}e^{-t\delta/2} \\ &\leqslant \gamma(N)\left(\frac{\|A\|}{\delta}\right)^{N-1} e^{-t\delta/2}.\end{aligned} \qquad \square$$

§4. Fundamental systems of solutions to a linear equation of higher order

The reduction of a single Nth order equation to a system of equations. The Cauchy problem for a single equation of an arbitrary order. The existence and uniqueness of a solution. A fundamental system, the Wronski determinant, and the Liouville formula. Three special fundamental systems.

To this point we have studied a system of equations that relate the values of unknown functions and the first order derivatives of these functions. Such a system was written as a single vector equation $y' = Ay$. It is often necessary to consider equations containing derivatives of the second, third, and higher orders.

As a typical and important example, we indicate the Nth order equation

$$x^{(N)} + p_1 x^{(N-1)} + p_2 x^{(N-2)} + \dots + p_{N-1}x' + p_N x = 0. \tag{1}$$

We introduce the notation

$$\begin{aligned} &y_1 = x, \quad y_2 = x' = y_1', \quad y_3 = x'' = y_2', \quad \dots, \\ &y_{N-1} = x^{(N-2)} = y_{N-2}', \quad y_N = x^{(N-1)} = y_{N-1}'. \end{aligned}$$

Now we write the above equations in another order and add the original equation written in the new notation. We obtain the system

$$\begin{aligned} y_1' &= y_2, \\ y_2' &= y_3, \\ &\cdots\cdots \\ y_{N-1}' &= y_N, \\ y_N' &= -p_1 y_N - p_2 y_{N-1} - \dots - p_{N-1} y_2 - p_N y_1. \end{aligned}$$

It is obvious that for every solution $x = x(t)$ to equation (1), where $x(t)$ is a scalar function, we can construct the N-dimensional vector $y = y(t)$ satisfying the above system. Conversely, if we know a solution $(y_1(t), y_2(t), \dots, y_N(t))^T$ to the system, then, denoting $y_1(t)$ by $x(t)$, from the first $N-1$ equations we subsequently find

$$y_2(t) = x'(t), \quad y_3(t) = x''(t), \quad \dots, \quad y_N(t) = x^{(N-1)}(t).$$

The last equation of the system can be rewritten as the equality

$$x^{(N)}(t) = -p_1 x^{(N-1)}(t) - p_2 x^{(N-2)}(t) - \dots - p_{N-1}x'(t) - p_N x(t),$$

which is equivalent to the considered equation. Thus, we have proved that there is a one-to-one correspondence

$$(x, x', \dots, x^{(N-1)}) = (y_1, y_2, \dots, y_N)$$

between solutions to equation (1) and solutions to the system

$$\text{(2)} \quad \frac{d}{dt}\begin{pmatrix} y_1 \\ y_2 \\ y_3 \\ \vdots \\ y_{N-1} \\ y_N \end{pmatrix} = \begin{bmatrix} 0 & 1 & 0 & 0 & \dots & 0 & 0 \\ 0 & 0 & 1 & 0 & \dots & 0 & 0 \\ 0 & 0 & 0 & 1 & \dots & 0 & 0 \\ \dots & \dots & \dots & \dots & \dots & \dots & \dots \\ 0 & 0 & 0 & 0 & \dots & 0 & 1 \\ -p_N & -p_{N-1} & -p_{N-2} & -p_{N-3} & \cdots & -p_2 & -p_1 \end{bmatrix} \begin{pmatrix} y_1 \\ y_2 \\ y_3 \\ \vdots \\ y_{N-1} \\ y_N \end{pmatrix}$$

constructed from equation (1).

Solutions to the system (2) form an N-dimensional linear space. Let us show that solutions to equation (1) also form a linear space of dimension N. Let solutions $x_1(t), x_2(t), \dots, x_N(t)$ to equation (1) be linearly dependent. By definition, there exist numbers $\alpha_1, \alpha_2, \dots, \alpha_N$, not all of them vanishing, and such that

$$\alpha_1 x_1(t) + \alpha_2 x_2(t) + \dots + \alpha_N x_N(t) = 0$$

for all t. Differentiating this equality $N-1$ times, for the corresponding solutions to the system (2) we obtain the equality

$$\alpha_1 y^{[1]}(t) + \alpha_2 y^{[2]}(t) + \dots + \alpha_N y^{[N]}(t) = 0,$$

i.e., the solutions to the system (2) corresponding to linearly dependent solutions to equation (1) are also linearly dependent. Therefore, for N linearly independent solutions to the system (2) (such solutions exist), their first components will be linearly independent solutions to equation (1).

DEFINITION 1. If we deal with a system of equations, by a "solution" we mean a vector (N functions) and components of N linearly independent solutions form a fundamental matrix. In the case of a single equation, by a "solution" we mean a single function and the set of N linearly independent solutions $x_1(t), x_2(t), \dots, x_N(t)$ is called a fundamental system of solutions.

The existence and uniqueness theorem for a system of equations provides the validity of the same theorem in the case of a single equation. Let us give an accurate statement of this assertion.

DEFINITION 2. By the Cauchy problem for equation (1) we mean the problem of finding a function $x(t)$ satisfying equation (1) and the initial condition

$$\text{(3)} \quad \begin{aligned} x(t_0) &= b_1, \\ x'(t_0) &= b_2, \\ &\dots\dots \\ x^{(N-1)}(t_0) &= b_N. \end{aligned}$$

THEOREM 1. *For any t_0, $b_1, b_2, \dots, b_N$ a solution $x(t)$ to the Cauchy problem (1), (3) exists and is unique. The solution $x(t)$ is defined for all real t, is continuous, and has continuous derivatives of all orders for any t. In addition, $x(t)$ continuously depends on the parameters, i.e., on the initial data t_0, $b_1, b_2, \dots, b_N$ in (3) and the coefficients $p_1, p_2, \dots, p_N$ in (1).*

We consider the notion of a fundamental matrix of solutions to equation (1). Introducing the vector y with components $x, x', \dots, x^{(N-1)}$, we can rewrite the equation as the equivalent system $y' = Ay$. It is natural to regard a fundamental matrix for this system as a fundamental matrix for equation (1). The first row of this matrix

$$\Phi(t) = \begin{bmatrix} x_1 & x_2 & \dots & x_N \\ x_1' & x_2' & \dots & x_N' \\ \dots & \dots & \dots & \dots \\ x_1^{(N-1)} & x_2^{(N-1)} & \dots & x_N^{(N-1)} \end{bmatrix}$$

is a fundamental system of solutions, i.e., N linearly independent solutions $x_i(t)$, the second row consists of the first order derivatives $x_j'(t)$, and so on, so that the $(k+1)$th row contains the kth order derivatives $x_j^{(k)}(t)$ of these solutions. By the linear independence, $\det \Phi(t) \neq 0$. It suffices to verify the last relation only at a single point because the values of $\det \Phi(t)$ at different points are connected by the relation

$$\det \Phi(t) = e^{(t-t_0)\operatorname{tr} A} \det \Phi(t_0).$$

It is interesting to note that almost all elements of the principal diagonal of the matrix A, which is associated with equation (1), vanish. Only the diagonal element located in the right lower corner of the matrix can be nonzero. This element is $-p_1$. Therefore, the trace of the matrix A, which is the sum of the diagonal element of A, is equal to this element:

$$\operatorname{tr} A = -p_1.$$

The formula

$$\det \Phi(t) = e^{-p_1(t-t_0)} \det \Phi(t_0), \tag{4}$$

which expresses the determinant of a fundamental matrix of solutions, is called the Liouville formula (cf. (2) in §2).

DEFINITION 3. The determinant $\det \Phi(t)$ is usually referred to as the Wronski determinant or Wronskian and is denoted by $W(t)$, i.e.,

$$W(t) = \det \begin{bmatrix} x_1 & x_2 & \dots & x_N \\ x_1' & x_2' & \dots & x_N' \\ \dots & \dots & \dots & \dots \\ x_1^{(N-1)} & x_2^{(N-1)} & \dots & x_N^{(N-1)} \end{bmatrix}, \tag{5}$$

$$W(t) = e^{-p_1(t-t_0)} W(t_0).$$

We emphasize one more important property of the system $x_1(t), x_2(t), \dots, x_N(t)$. By definition, if some linear combination of $x_j(t)$ with constant coefficients vanishes at all points of the segment $t' \leqslant t \leqslant t''$ $(t' < t'')$, i.e.,

$$\alpha_1 x_1(t) + \alpha_2 x_2(t) + \dots + \alpha_N x_N(t) = 0,$$

then $\alpha_1 = \alpha_2 = \cdots = \alpha_N = 0$. Therefore, if

$$\int_{t'}^{t''} \left[\alpha_1 x_1(t) + \alpha_2 x_2(t) + \cdots + \alpha_N x_N(t)\right]^2 dt = 0,$$

then $\alpha_1 = \alpha_2 = \cdots = \alpha_N = 0$. We introduce the notation

$$x_{ij} = \int_{t'}^{t''} x_i(t) x_j(t)\, dt. \tag{6}$$

Since

$$\int_{t'}^{t''} \left[\alpha_1 x_1(t) + \alpha_2 x_2(t) + \cdots + \alpha_N x_N(t)\right]^2 dt = \sum_{i,j=1}^{N} x_{ij}\alpha_i\alpha_j \geqslant 0,$$

we can assert that the matrix

$$X = \begin{bmatrix} x_{11} & x_{12} & \dots & x_{1N} \\ x_{21} & x_{22} & \dots & x_{2N} \\ \dots & \dots & \dots & \dots \\ x_{N1} & x_{N2} & \dots & x_{NN} \end{bmatrix} \tag{7}$$

is strictly positive definite. This is true because $\sum x_{ij}\alpha_i\alpha_j = 0$ implies $\alpha_j = 0$ for all j. The matrix X of the form (7) with elements (6) is called the Gram matrix of the functions $x_1(t), x_2(t), \dots, x_N(t)$ on the segment $[t', t'']$. Thus, we have proved the following theorem.

THEOREM 2. *The Gram matrix of any fundamental system of solutions to equation* (1) *on an arbitrary segment is strictly positive definite.*

We introduce the following polynomial of degree k in a variable τ:

$$\mathcal{P}_k(\tau) = \tau^k + p_1\tau^{k-1} + p_2\tau^{k-2} + \cdots + p_{k-1}\tau + p_k.$$

It can be factorized as follows:

$$\mathcal{P}_k(\tau) = (\tau - \tau_1)(\tau - \tau_2)\cdots(\tau - \tau_k).$$

The quantities $\tau_1, \tau_2, \dots, \tau_k$ are roots of the polynomial $\mathcal{P}_k(\tau)$ and the coefficients $p_1, p_2, \dots, p_k$ are symmetric functions of these roots:

$$\begin{aligned} p_1 &= -(\tau_1 + \tau_2 + \cdots + \tau_k), \\ p_2 &= \tau_1\tau_2 + \tau_1\tau_3 + \cdots + \tau_{k-1}\tau_k, \\ p_3 &= -(\tau_1\tau_2\tau_3 + \cdots + \tau_{k-2}\tau_{k-1}\tau_k), \\ &\dots\dots\dots\dots \\ p_k &= (-1)^k\tau_1\tau_2\cdots\tau_k. \end{aligned}$$

With the equation

$$\frac{d^k}{dt^k}x + p_1\frac{d^{k-1}}{dt^{k-1}}x + p_2\frac{d^{k-2}}{dt^{k-2}}x + \cdots + p_k x = 0 \tag{8}$$

we associate the symbolic notation

$$\mathcal{P}_k\left(\frac{d}{dt}\right)x = 0. \tag{9}$$

The polynomial $\mathcal{P}_k(\tau)$ is usually called the symbol of the operator $\mathcal{P}_k\left(\frac{d}{dt}\right)$. The possibility of constructing the symbol is based on the commutativity of products of differential operators with constant coefficients. Indeed, if

$$\mathcal{P}_{k-1}(\tau) = (\tau - \tau_1)(\tau - \tau_2)\cdots(\tau - \tau_{k-1}) = \tau^{k-1} + b_1\tau^{k-2} + \cdots + b_{k-2}\tau + b_{k-1},$$

then the symbols satisfy the equality

$$\mathcal{P}_k(\tau) = \mathcal{P}_{k-1}(\tau)(\tau - \tau_k) = (\tau - \tau_k)\mathcal{P}_{k-1}(\tau),$$

which implies the equality

$$\mathcal{P}_k\left(\frac{d}{dt}\right)x = \mathcal{P}_{k-1}\left(\frac{d}{dt}\right)\left[\frac{dx}{dt} - \tau_k x\right] = \left(\frac{d}{dt} - \tau_k\right)\left[\mathcal{P}_{k-1}\left(\frac{d}{dt}\right)x\right].$$

We describe an approach to the construction of a fundamental system of solutions.

The fundamental system of quasimonomials. Consider the equation

$$x^{(k)} + q_1 x^{(k-1)} + \cdots + q_{k-1}x' + q_k x = 0.$$

The characteristic polynomial has the form

$$\tau^k + q_1\tau^{k-1} + \cdots + q_{k-1}\tau + q_k \equiv (\tau - \tau_0)^k.$$

The equation can be rewritten as follows:

$$\left(\frac{d}{dt} - \tau_0\right)^k x(t) = 0. \tag{10}$$

Introduce the functions

$$x_1(t) = e^{\tau_0 t}, \quad \ldots, \quad x_j(t) = \frac{t^{j-1}}{(j-1)!}e^{\tau_0 t}, \quad \ldots, \quad x_k(t) = \frac{t^{k-1}}{(k-1)!}e^{\tau_0 t}$$

and study how they are transformed under the action of the differential operator $\frac{d}{dt} - \tau_0$:

$$\left(\frac{d}{dt} - \tau_0\right)x_1(t) = \left(\frac{d}{dt} - \tau_0\right)e^{\tau_0 t} = 0,$$

$$\left(\frac{d}{dt} - \tau_0\right)x_j(t) = \left(\frac{d}{dt} - \tau_0\right)\left[\frac{t^{j-1}}{(j-1)!}e^{\tau_0 t}\right] = \frac{t^{j-2}}{(j-2)!}e^{\tau_0 t} = x_{j-1}(t), \quad j \geqslant 2.$$

It is easy to derive the following relations ($j = 1, 2, \ldots, k$):

$$\left(\frac{d}{dt} - \tau_0\right)^j x_j(t) = 0, \quad \left(\frac{d}{dt} - \tau_0\right)^{j-1} x_j(t) = x_1(t),$$

$$\left(\frac{d}{dt} - \tau_0\right)^r x_j(t) = x_{j-r}(t),$$

$$\left(\frac{d}{dt} - \tau_0\right)^k x_j(t) = \left(\frac{d}{dt} - \tau_0\right)^{k-j}\left(\frac{d}{dt} - \tau_0\right)^j x_j(t) = 0.$$

We have proved that the functions

$$e^{\tau_0 t},\ \frac{t}{1!}e^{\tau_0 t},\ \frac{t^2}{2!}e^{\tau_0 t},\ \dots,\ \frac{t^{k-1}}{(k-1)!}e^{\tau_0 t}$$

are solutions of the equation under consideration. It is obvious that

$$x_1(0) = 1,\ x_2(0) = 0,\ x_3(0) = 0,\ \dots,\ x_k(0) = 0.$$

For $j - 1 \geqslant r \geqslant 1$, using the equality

$$\left(\frac{d}{dt} - \tau_0\right)^r x_j(t) = x_{j-r}(t),$$

we can prove that

$$\left(\frac{d}{dt} - \tau_0\right)^r x_j(t)\bigg|_{t=0} = \begin{cases} 0 & \text{if } r \leqslant j-2, \\ 1 & \text{if } r = j-1 \end{cases}$$

and, by induction,

$$x_j^{(j-1)}(0) = 1, \quad x_j^{(r)}(0) = 0 \quad \text{if } r \leqslant j-2.$$

EXERCISE 1. Give a detailed justification of the above assertion.

It is now clear that the determinant of the matrix

$$\Phi(0) = \begin{bmatrix} x_1(0) & x_2(0) & \dots & x_k(0) \\ x_1'(0) & x_2'(0) & \dots & x_k'(0) \\ \dots & \dots & \dots & \dots \\ x_1^{(k-1)}(0) & x_2^{(k-1)}(0) & \dots & x_k^{(k-1)}(0) \end{bmatrix} = \begin{bmatrix} 1 & 0 & \dots & 0 \\ x_1'(0) & 1 & \dots & 0 \\ \dots & \dots & \dots & \dots \\ x_1^{(k-1)}(0) & x_2^{(k-1)}(0) & \dots & 1 \end{bmatrix}$$

is equal to 1 ($\det \Phi(0) = 1$). Hence the constructed functions

$$x_1(t) = e^{\tau_0 t},\ x_2(t) = \frac{t}{1!}e^{\tau_0 t},\ \dots,\ x_k(t) = \frac{t^{k-1}}{(k-1)!}e^{\tau_0 t}$$

form a fundamental system of solutions to equation (10).

The characteristic polynomial of an arbitrary equation

$$\mathcal{P}\left(\frac{d}{dt}\right)x(t) = x^{(N)}(t) + p_1 x^{(N-1)}(t) + \dots + p_{N-1}x'(t) + p_N x(t)$$

can be factorized:

$$\mathcal{P}(t) = \tau^N + p_1\tau^{N-1} + \dots + p_{N-1}\tau + p_N = (\tau - \tau_1)^{k_1}(\tau - \tau_2)^{k_2}\cdots(\tau - \tau_s)^{k_s}$$
$$(k_1 + k_2 + \dots + k_s = N, \quad \tau_l \neq \tau_m \text{ for } l \neq m)$$

and the equation

$$\mathcal{P}\left(\frac{d}{dt}\right)x(t) = 0 \tag{11}$$

can be written in the form

$$\mathcal{P}\left(\frac{d}{dt}\right)x(t) = \left(\frac{d}{dt} - \tau_1\right)^{k_1}\left(\frac{d}{dt} - \tau_2\right)^{k_2}\cdots\left(\frac{d}{dt} - \tau_s\right)^{k_s}x(t) = 0.$$

If $z(t)$ satisfies one of the equations

$$\left(\frac{d}{dt} - \tau_l\right)^{k_l} z(t) = 0,$$

then it also satisfies the equation $\mathcal{P}\left(\frac{d}{dt}\right)z(t) = 0$. Therefore, the functions

$$x_{11} = e^{\tau_1 t},\ x_{12}(t) = \frac{1}{1!}e^{\tau_1 t},\ \ldots,\ x_{1k_1}(t) = \frac{t^{(k_1-1)}}{(k_1-1)!}e^{\tau_1 t},$$
$$x_{21} = e^{\tau_2 t},\ x_{22}(t) = \frac{1}{1!}e^{\tau_2 t},\ \ldots,\ x_{2k_2}(t) = \frac{t^{(k_2-1)}}{(k_2-1)!}e^{\tau_2 t},$$
$$\cdots\cdots\cdots\cdots\cdots\cdots\cdots\cdots$$
$$x_{s1} = e^{\tau_s t},\ x_{s2}(t) = \frac{1}{1!}e^{\tau_s t},\ \ldots,\ x_{sk_s}(t) = \frac{t^{(k_s-s)}}{(k_s-1)!}e^{\tau_s t}$$

and all linear combinations of these functions are solutions to equation (11). The number $k_1 + k_2 + \cdots + k_s$ of the indicated partial solutions coincides with the dimension N of the space of all solutions to the considered equation. The natural hypothesis is that these partial solutions form a fundamental system of solutions. We see below that the hypothesis is true. To prove this assertion we need to show that any solution $x(t)$ to equation (11) can be represented in the form

$$x(t) = \sum_{l=1}^{s}\sum_{r=1}^{k_l} c_{lr}x_{lr}(t).$$

To prove this fact we will use the following theorem about polynomials.

THEOREM 3. *If polynomials $\mathcal{R}_1(\tau), \ldots, \mathcal{R}_s(\tau)$ are relatively prime, i.e., their greatest common divisor is* 1, *then there exist polynomials $\mathcal{Q}_1(\tau), \ldots, \mathcal{Q}_s(\tau)$ such that*

$$\mathcal{R}_1(\tau)\mathcal{Q}_1(\tau) + \mathcal{R}_2(\tau)\mathcal{Q}_2(\tau) + \cdots + \mathcal{R}_s(\tau)\mathcal{Q}_s(\tau) = 1. \tag{12}$$

For the proof in the case of two polynomials $\mathcal{R}_1(\tau)$ and $\mathcal{R}_2(\tau)$ see [**16**].

If the characteristic polynomial has the form

$$\mathcal{P}(\tau) = \prod_{l=1}^{s}(\tau - \tau_l)^{k_l},$$

we introduce the polynomials

$$\mathcal{R}_m(\tau) = \frac{\mathcal{P}(\tau)}{(\tau - \tau_m)^{k_m}}, \quad m = 1, 2, \ldots, s.$$

It is obvious that $\mathcal{R}_1(\tau), \mathcal{R}_2(\tau), \ldots, \mathcal{R}_s(\tau)$ are relatively prime.

Let $\mathcal{Q}_1(\tau), \mathcal{Q}_2(\tau), \ldots, \mathcal{Q}_s(\tau)$ be such that (12) holds. Substituting in (12) for τ the differentiation operator $\frac{d}{dt}$, we obtain

$$\mathcal{R}_1\left(\frac{d}{dt}\right)\mathcal{Q}_1\left(\frac{d}{dt}\right) + \mathcal{R}_2\left(\frac{d}{dt}\right)\mathcal{Q}_2\left(\frac{d}{dt}\right) + \cdots + \mathcal{R}_s\left(\frac{d}{dt}\right)\mathcal{Q}_s\left(\frac{d}{dt}\right) = 1.$$

We apply the result to an arbitrary solution $x(t)$ to equation (11), using the relation

$$x(t) = \mathcal{R}_1\left(\frac{d}{dt}\right)\mathcal{Q}_1\left(\frac{d}{dt}\right)x(t) + \mathcal{R}_2\left(\frac{d}{dt}\right)\mathcal{Q}_2\left(\frac{d}{dt}\right)x(t) + \cdots + \mathcal{R}_s\left(\frac{d}{dt}\right)\mathcal{Q}_s\left(\frac{d}{dt}\right)x(t).$$

Then $x(t)$ is represented as a sum of s terms:

$$x(t) = x_1(t) + x_2(t) + \cdots + x_s(t),$$
$$x_l(t) = \mathcal{R}_l\left(\frac{d}{dt}\right)\mathcal{Q}_l\left(\frac{d}{dt}\right)x(t).$$

All derivatives $x'(t), x''(t), \ldots$, as well as any linear combination of these derivatives, satisfy the same differential equation with constant coefficients as $x(t)$. Similarly, $\mathcal{Q}_k\left(\frac{d}{dt}\right)x(t)$ is a solution to the equation

$$\mathcal{P}\left(\frac{d}{dt}\right)\mathcal{Q}_k\left(\frac{d}{dt}\right)x(t) = 0.$$

Consequently,

$$\left(\frac{d}{dt} - \tau_l\right)^{k_l} x_l(t) = \left(\frac{d}{dt} - \tau_l\right)^{k_l} \mathcal{R}_l\left(\frac{d}{dt}\right)\mathcal{Q}_l\left(\frac{d}{dt}\right)x(t) = \mathcal{P}\left(\frac{d}{dt}\right)\mathcal{Q}_l\left(\frac{d}{dt}\right)x(t) = 0.$$

As was shown, from the equation

$$\left(\frac{d}{dt} - \tau_l\right)^{k_l} x_l(t) = 0$$

it follows that $x_l(t)$ can be represented as the linear combination

$$x_l(t) = \sum_{r=1}^{k_l} c_{lr}\frac{t^{r-1}}{(r-1)!}e^{\tau_l t}.$$

Now, it is clear that any solution $x(t)$ admits the representation

$$x(t) = \sum_{l=1}^{s}\sum_{r=1}^{k_l} c_{lr}\frac{t^{r-1}}{(r-1)!}e^{\tau_l t}.$$

Hence the constructed systems of partial solutions to equation (11) is fundamental. This system consists of $N = k_1 + k_2 + \cdots + k_s$ functions

$$e^{\tau_l t}, \frac{t}{1!}e^{\tau_l t}, \ldots, \frac{t^{(k_l-1)}}{(k_l-1)!}e^{\tau_l t}, \quad l = 1, 2, \ldots, s.$$

The fundamental system of solutions $\psi_1, \psi_2, \ldots, \psi_N$. We show that the functions $\psi_1(t), \psi_2(t), \ldots, \psi_N(t)$ introduced in §3 are solutions to the Cauchy problems

$$\mathcal{P}_k\left(\frac{d}{dt}\right)x(t) = 0, \quad x(0) = 0,\ x'(0) = 0,\ \ldots,\ x^{(k-2)}(0) = 0,\ x^{(k-1)}(0) = 1.$$

Indeed, the equalities

$$\begin{aligned}
&\left(\frac{d}{dt}-\tau_1\right)\psi_1 \equiv \frac{d\psi_1}{dt}-\tau_1\psi_1=0,\\
&\left(\frac{d}{dt}-\tau_2\right)\psi_2=\psi_1,\\
&\dots\dots\dots\dots\\
&\left(\frac{d}{dt}-\tau_k\right)\psi_k=\psi_{k-1}
\end{aligned}$$

imply

$$\begin{aligned}
&\left(\frac{d}{dt}-\tau_1\right)\psi_1=0,\\
&\left(\frac{d}{dt}-\tau_1\right)\left(\frac{d}{dt}-\tau_2\right)\psi_2=\left(\frac{d}{dt}-\tau_1\right)\psi_1=0,\\
&\dots\dots\dots\dots\dots\dots\dots\dots\dots\dots\dots\dots\\
&\left(\frac{d}{dt}-\tau_1\right)\dots\left(\frac{d}{dt}-\tau_{k-1}\right)\left(\frac{d}{dt}-\tau_k\right)\psi_k=\left(\frac{d}{dt}-\tau_1\right)\dots\left(\frac{d}{dt}-\tau_{k-1}\right)\psi_{k-1}=0,
\end{aligned}$$

i.e.,

$$\mathcal{P}_1\left(\frac{d}{dt}\right)\psi_1=0,\quad \mathcal{P}_2\left(\frac{d}{dt}\right)\psi_2=0,\quad \dots,\mathcal{P}_k\left(\frac{d}{dt}\right)\psi_k=0.$$

Furthermore, taking into account the conditions

$$\psi_1(0)=1,\ \psi_2(0)=0,\ \dots\ ,\ \psi_k(0)=0,$$

from the same equalities for $t=0$ we obtain

$$\psi_2'(0)=1,\quad \psi_3'(0)=0,\dots,\psi_k'(0)=0.$$

Differentiating, we find

$$\psi_3''(0)=1,\ \psi_4''(0)=0,\dots,\psi_k''(0)=0$$

and so on, by induction.

Let $j\leqslant N$. We show that $\psi_j(t)$ satisfies the equation $\mathcal{P}_N\left(\frac{d}{dt}\right)x(t)=0$. We use the factorization

$$\begin{aligned}
\mathcal{P}_N(\tau)=\big[(\tau-\tau_N)&(\tau-\tau_{N-1})\cdots(\tau-\tau_{j+1})\big]\\
&\times\big[(\tau-\tau_j)(\tau-\tau_{j-1})\cdots(\tau-\tau_1)\big]=\mathcal{Q}_{N-j}(\tau)\mathcal{P}_j(\tau),
\end{aligned}$$

where

$$\begin{aligned}
\mathcal{Q}_{N-j}(\tau)&=(\tau-\tau_N)(\tau-\tau_{N-1})\cdots(\tau-\tau_{j+1})\\
\mathcal{P}_j(\tau)&=(\tau-\tau_j)(\tau-\tau_{j-1})\cdots(\tau-\tau_1).
\end{aligned}$$

Since $\mathcal{P}_j\left(\frac{d}{dt}\right)\psi_j(t)=0$, we have

$$\mathcal{P}_N\left(\frac{d}{dt}\right)\psi_j(t)=\mathcal{Q}_{N-j}\left(\frac{d}{dt}\right)\left[\mathcal{P}_j\left(\frac{d}{dt}\right)\psi_j(t)\right]=0.$$

The proof is complete.

Now we show that the solutions $\psi_1(t), \psi_2(t), \dots, \psi_N(t)$ form a fundamental system of solutions to the equation $\mathcal{P}_N\left(\frac{d}{dt}\right)x = 0$. It suffices to form the fundamental matrix

$$Y(t) = \begin{bmatrix} \psi_1(t) & \psi_2(t) & \dots & \psi_N(t) \\ \psi_1'(t) & \psi_2'(t) & \dots & \psi_N'(t) \\ \dots & \dots & \dots & \dots \\ \psi_1^{(N-1)}(t) & \psi_2^{(N-1)}(t) & \dots & \psi_N^{(N-1)}(t) \end{bmatrix}$$

and prove that the determinant $W(t) = \det Y(t)$ differs from zero at least at one point. We compute the determinant for $t = 0$. Using the known initial conditions for $\psi_j(t)$, we easily conclude that the matrix $Y(0)$ has the following triangular form:

$$Y(0) = \begin{bmatrix} 1 & & & & \\ \times & 1 & & \mathbf{0} & \\ \times & \times & 1 & & \\ \vdots & \vdots & \vdots & \ddots & \\ \times & \times & \times & \times & 1 \end{bmatrix}.$$

Crosses denote some, generally speaking nonzero, elements whose values are not essential for our purposes. Now, it is obvious that $W(0) = \det Y(0) = 1$, where $W(t)$ denotes the Wronski determinant (cf. (5)). By the Liouville formula (cf. (4)) about the Wronski determinant (i.e., the determinant of a fundamental matrix), we can see that

$$W(t) = \det Y(t) = \exp\{t(\tau_1 + \tau_2 + \dots + \tau_N)\}W(0) = \exp\{t(\tau_1 + \tau_2 + \dots + \tau_N)\}.$$

The fundamental system of solutions $\omega_1, \omega_2, \dots, \omega_N$. We introduce one more fundamental system of solutions $\omega_1(t), \omega_2(t), \dots, \omega_N(t)$ as solutions to the following Cauchy problems:

$$\begin{gathered} \omega_i^{(N)} + p_1\omega_i^{(N-1)} + \dots + p_N\omega_i = 0, \\ \omega_i(0) = 0, \ \dots, \ \omega_i^{(i-2)}(0) = 0, \ \omega_i^{(i-1)}(0) = 1, \\ \omega_i^{(i)}(0) = 0, \ \dots, \ \omega_i^{(N-1)}(0) = 0. \end{gathered}$$

The functions $\omega_1(t), \omega_2(t), \dots, \omega_N(t)$ are linearly independent because the fundamental matrix of solutions

$$\Omega(t) = \begin{bmatrix} \omega_1(t) & \omega_2(t) & \dots & \omega_N(t) \\ \omega_1'(t) & \omega_2'(t) & \dots & \omega_N'(t) \\ \dots & \dots & \dots & \dots \\ \omega_1^{(N-1)}(t) & \omega_2^{(N-1)}(t) & \dots & \omega_N^{(N-1)}(t) \end{bmatrix}$$

is nonsingular for $t = 0$ ($\Omega(0) = I$).

We already emphasized that a fundamental matrix (denoted here by $\Omega(t)$) of solutions to equation (1) is simultaneously a fundamental matrix of solutions to the system (2). Denote the matrix in (2) by P. Since

$$\frac{d}{dt}\Omega(t) = P\Omega(t), \quad \Omega(0) = I,$$

we have $\Omega(t) = e^{tP}$. For the further considerations we need to obtain one more system of equalities which defines $\omega_j(t)$. Since P and e^{tP} commute (cf. §2):

$$P\Omega(t) = Pe^{tP} = e^{tP}P = \Omega(t)P,$$

we see that

$$\frac{d}{dt}\Omega^*(t) = [P\Omega(t)]^* = [\Omega(t)P]^* = P^*\Omega^*(t), \quad \Omega^*(0) = I. \tag{13}$$

The first column of $\Omega^*(t)$ consists of the components $\omega_1(t), \omega_2(t), \dots, \omega_N(t)$:

$$\Omega^*(t) = \begin{bmatrix} \omega_1(t) & \times & \dots & \times \\ \omega_2(t) & \times & \dots & \times \\ \dots & \dots & \dots & \dots \\ \omega_N(t) & \times & \dots & \times \end{bmatrix}$$

and the matrix P^* has the form

$$P^* = \begin{bmatrix} 0 & 0 & 0 & \dots & 0 & -p_N \\ 1 & 0 & 0 & \dots & 0 & -p_{N-1} \\ 0 & 1 & 0 & \dots & 0 & -p_{N-2} \\ \dots & \dots & \dots & \dots & \dots & \dots \\ 0 & 0 & 0 & \dots & 1 & -p_1 \end{bmatrix}.$$

From (13) we immediately conclude that

$$\begin{aligned} \frac{d\omega_1}{dt} &= -p_N\omega_N, \quad \omega_1(0) = 1, \\ \frac{d\omega_2}{dt} &= \omega_1 - p_{N-1}\omega_N, \quad \omega_2(0) = 0, \\ &\dots\dots\dots\dots\dots\dots \\ \frac{d\omega_N}{dt} &= \omega_{N-1} - p_1\omega_N, \quad \omega_N(0) = 0. \end{aligned}$$

We recall (cf. (9) in §3) the representation of the matrix exponential e^{tA} in the form of the matrix polynomial:

$$e^{tA} = \psi_1(t)I + \psi_2(t)[A - \tau_1 I] + \dots + \psi_N(t)[A - \tau_1 I][A - \tau_2 I]\cdots[A - \tau_{N-1}I].$$

Sometimes it is convenient to perform all the multiplications in this representation and, collecting the terms at the powers of the matrix A, to write out the polynomial in the standard form. We show that it is possible to arrive at the same form by indirect way which is less tedious than the above method.

Let the characteristic polynomial of the matrix A have the form

$$\det[A - \tau I] = (-1)^N[\tau^N + p_1\tau^{N-1} + p_2\tau^{N-2} + \dots + p_{N-1}\tau + p_N].$$

We construct a fundamental system of solutions $\omega_1(t), \omega_2(t), \dots, \omega_N(t)$ to equation (1). Using the functions $\omega_1(t), \omega_2(t), \dots, \omega_N(t)$ we construct the matrix polynomial

$$\mathcal{Q}(t, A) = \omega_1(t)I + \omega_2(t)A + \dots + \omega_N(t)A^{N-1}$$

and compute its derivative in t:

$$\begin{aligned}\frac{d}{dt}\mathcal{Q}(t,A) &= \omega_1'(t)I + \omega_2'(t)A + \omega_3'(t)A^2 + \dots + \omega_N'(t)A^{N-1} \\ &= -p_N\omega_N I + (\omega_1 - p_{N-1}\omega_N)A + (\omega_2 - p_{N-2}\omega_N)A^2 + \dots \\ &\quad + (\omega_{N-1} - p_1\omega_N)A^{N-1} \\ &= A[\omega_1 I + \omega_2 A + \dots + \omega_{N-1}A^{N-2}] - \omega_N[p_N I + p_{N-1}A + \dots \\ &\quad + p_1 A^{N-1}] + \omega_N A A^{N-1} - \omega_N A^N \\ &= A[\omega_1 I + \omega_2 A + \dots + \omega_{N-1}A^{N-2} + \omega_N A^{N-1}] \\ &\quad - \omega_N[p_N I + p_{N-1}A + \dots + p_1 A^{N-1} + A^N] \\ &= A\mathcal{Q}(t,A) - \omega_N[p_N I + p_{N-1}A + \dots + p_1 A^{N-1} + A^N].\end{aligned}$$

By the Cayley–Hamilton theorem, any square matrix satisfies its characteristic equation. Therefore,

$$p_N I + p_{N-1}A + \dots + p_1 A^{N-1} + A^N = 0.$$

Consequently, $\frac{d}{dt}\mathcal{Q}(t,A) = A\mathcal{Q}(t,A)$. Moreover,

$$\begin{aligned}\mathcal{Q}(0,A) &= \omega_1(0)I + \omega_2(0)A + \omega_3(0)A^2 + \dots + \omega_N(0)A^{N-1} \\ &= 1\cdot I + 0\cdot A + 0\cdot A^2 + \dots + 0\cdot A^{N-1} = I.\end{aligned}$$

It is now clear that $\mathcal{Q}(t,A) = e^{tA}$ and

(14) $$e^{tA} = \omega_1(t)I + \omega_2(t)A + \omega_3(t)A^2 + \dots + \omega_N(t)A^{N-1}.$$

We recall that the functions $\omega_j(t)$ are linearly independent. Hence the Gram matrix H consisting of the elements

$$h_{ij} = \int_{t'}^{t''} \omega_i(t)\omega_j(t)\,dt$$

is nonsingular and is strictly positive definite for any segment $[t', t'']$ $(t' < t'')$.

§5. The continuation of the study of fundamental systems of solutions to a linear equation of higher order

The special functions $\psi_j(t)$ and $\omega_j(t)$ appearing in the polynomial representation of the matrix exponential and in fundamental systems from the previous section. Exponential decrease of solutions to equations and components of solutions to systems which converge to zero as $t \to \infty$.

In §4, we described the construction of three fundamental systems of solutions to the Nth order equation

(1) $$x^{(N)} + p_1 x^{(N-1)} + p_2 x^{(N-2)} + \dots + p_{N-1}x' + p_N x = 0.$$

They are:

1) the fundamental system of quasimonomials;
2) the fundamental system of solutions $\psi_1, \psi_2, \dots, \psi_N$;
3) the fundamental system of solutions $\omega_1, \omega_2, \dots, \omega_N$.

The fundamental system of quasimonomials is rather simple and can be expressed by explicit formulas. However it has the essential deficiency: it does not depend continuously on the roots of the characteristic polynomial

$$\tau^N + p_1\tau^{N-1} + \cdots + p_{N-1}\tau + p_N = 0.$$

Indeed, let $N = 2$ and let $\tau_1 = \tau_0 - \delta$, $\tau_2 = \tau_0 + \delta$. Then

$$x_1(t) = e^{(\tau_0-\delta)t}, \quad x_2(t) = e^{(\tau_0+\delta)t}, \quad \delta \neq 0,$$

$$x_1(t) = e^{\tau_0 t}, \quad x_2(t) = te^{\tau_0 t}, \quad \delta = 0.$$

The fundamental systems $\psi_j(t)$ and $\omega_j(t)$, being solutions to the Cauchy problem, depend continuously on $\tau_1, \tau_2, \dots, \tau_N$, but their construction is not effective because we did not derive explicit formulas for their representation. We have only reduced the construction of these functions to solving some specific problems for equations and proved that these problems are uniquely solvable. Therefore, we need to continue the study of the special solutions $\psi_j(t)$ and $\omega_j(t)$. In some cases we will obtain formulas for the solutions, while in other cases we will only derive some recurrence relations between the considered functions.

We consider the Cauchy problem for the equation

$$(2) \qquad \mathcal{P}_k\left(\frac{d}{dt}\right)x(t) \equiv x^{(k)}(t) + p_1x^{(k-1)}(t) + \cdots + p_{k-1}x'(t) + p_kx(t) = 0$$

with the initial data

$$(3) \qquad x(0) = 0,\ x'(0) = 0,\ \dots,\ x^{(k-2)}(0) = 0,\ x^{(k-1)}(0) = 1.$$

As we know, a solution $x(t)$ to the problem (2), (3) exists, is unique, and is a continuous function of t and the coefficients p_j. Consequently, the solution $x(t)$ can be regarded as a continuous function of t, $\tau_1, \tau_2, \dots, \tau_k$.

We introduce the notation[3]

$$x(t) = \psi_k(t) \equiv \psi(t, \tau_1, \tau_2, \dots, \tau_k).$$

We note that $\psi(t, \tau_1, \tau_2, \dots, \tau_k)$ is invariant under permutations of $\tau_1, \tau_2, \dots, \tau_k$, i.e., ψ is a symmetric function of $\tau_1, \tau_2, \dots, \tau_k$.

In some special cases we will indicate explicit formulas for $\psi(t, \tau_1, \tau_2, \dots, \tau_k)$.

If $k = 1$, then $\mathcal{P}_1(\tau) = \tau - \tau_j$. Consequently, $\psi(t, \tau_j)$ must be determined from the conditions

$$\frac{d}{dt}\psi(t, \tau_j) - \tau_j\psi(t, \tau_j) = 0, \quad \psi(0, \tau_j) = 1.$$

It is easy to see that $\psi(t, \tau_j) = e^{\tau_j t}$.

Let $k > 1$ and let $\tau_1, \tau_2, \dots, \tau_k$ be pairwise distinct. Then $x_j(t)$ takes the form

$$x_1(t) = e^{\tau_1 t},\ x_2(t) = e^{\tau_2 t},\ \dots,\ x_k(t) = e^{\tau_k t}.$$

---

[3]This notation for the functions $\psi_k(t)$ will be more convenient to use hereinafter.

With $x_j(t)$ we associate the fundamental matrix of solutions

$$X(t) = \begin{bmatrix} x_1(t) & x_2(t) & \dots & x_k(t) \\ x_1'(t) & x_2'(t) & \dots & x_k'(t) \\ \dots & \dots & \dots & \dots \\ x_1^{(k-1)}(t) & x_2^{(k-1)}(t) & \dots & x_k^{(k-1)}(t) \end{bmatrix} = \begin{bmatrix} e^{\tau_1 t} & e^{\tau_2 t} & \dots & e^{\tau_k t} \\ \tau_1 e^{\tau_1 t} & \tau_2 e^{\tau_2 t} & \dots & \tau_k e^{\tau_k t} \\ \dots & \dots & \dots & \dots \\ \tau_1^{(k-1)} e^{\tau_1 t} & \tau_2^{(k-1)} e^{\tau_2 t} & \dots & \tau_k^{(k-1)} e^{\tau_k t} \end{bmatrix}.$$

The Wronskian

$$W(t) = \det X(t) = \exp\{t(\tau_1 + \tau_2 + \dots + \tau_k)\} \begin{bmatrix} 1 & 1 & \dots & 1 \\ \tau_1 & \tau_2 & \dots & \tau_k \\ \tau_1^2 & \tau_2^2 & \dots & \tau_k^2 \\ \dots & \dots & \dots & \dots \\ \tau_1^{k-1} & \tau_2^{k-1} & \dots & \tau_k^{k-1} \end{bmatrix}$$
$$= \exp\{t(\tau_1 + \tau_2 + \dots + \tau_k)\} \prod_{k \geqslant i > j \geqslant 1} (\tau_i - \tau_j)$$

does not vanish only if all τ_j are distinct. In this case, any solution $x(t)$ to equation (2) admits the representation $x(t) = c_1 e^{\tau_1 t} + c_2 e^{\tau_2 t} + \dots + c_k e^{\tau_k t}$. In particular, we can assert that

$$\psi(t, \tau_1, \tau_2, \dots, \tau_k) = c_1 e^{\tau_1 t} + c_2 e^{\tau_2 t} + \dots + c_k e^{\tau_k t}.$$

The constants $c_1, c_2, \dots, c_k$ are found from the following conditions:

$$\psi(0, \tau_1, \tau_2, \dots, \tau_k) = 0, \qquad \psi'(0, \tau_1, \tau_2, \dots, \tau_k) = 0, \ \dots,$$
$$\psi^{(k-2)}(0, \tau_1, \tau_2, \dots, \tau_k) = 0, \quad \psi^{(k-1)}(0, \tau_1, \tau_2, \dots, \tau_k) = 1.$$

It is easy to check that all these conditions are satisfied if

$$\psi(t, \tau_1, \tau_2, \dots, \tau_k) = \frac{\begin{vmatrix} 1 & 1 & \dots & 1 \\ \tau_1 & \tau_2 & \dots & \tau_k \\ \tau_1^2 & \tau_2^2 & \dots & \tau_k^2 \\ \dots & \dots & \dots & \dots \\ \tau_1^{k-2} & \tau_2^{k-2} & \dots & \tau_k^{k-2} \\ e^{\tau_1 t} & e^{\tau_2 t} & \dots & e^{\tau_k t} \end{vmatrix}}{\begin{vmatrix} 1 & 1 & \dots & 1 \\ \tau_1 & \tau_2 & \dots & \tau_k \\ \tau_1^2 & \tau_2^2 & \dots & \tau_k^2 \\ \dots & \dots & \dots & \dots \\ \tau_1^{k-2} & \tau_2^{k-2} & \dots & \tau_k^{k-2} \\ \tau_1^{k-1} & \tau_2^{k-1} & \dots & \tau_k^{k-1} \end{vmatrix}}.$$

Now we indicate explicit formulas for $\psi(t, \tau_1, \tau_2, \dots, \tau_k)$ if all τ_j coincide. More exactly, we show that

$$\psi_k(t) \equiv \psi(t, \underbrace{\tau, \tau, \dots, \tau}_{k \text{ times}}) = \frac{t^{k-1}}{(k-1)!} e^{\tau t}.$$

We recall that in §3, the functions

$$\psi_1(t) = \psi(t,\tau_1),\ \psi_2(t) = \psi(t,\tau_1,\tau_2),\ \ldots,\ \psi_k(t) = \psi(t,\tau_1,\tau_2,\ldots,\tau_k)$$

were defined as a solution to the Cauchy problem

$$\frac{d\psi(t,\tau_1)}{dt} = \tau_1\psi(t,\tau_1), \quad \psi(0,\tau_1) = 1,$$
$$\frac{d\psi(t,\tau_1,\tau_2)}{dt} = \tau_2\psi(t,\tau_1,\tau_2) + \psi(t,\tau_1), \quad \psi(0,\tau_1,\tau_2) = 0,$$
$$\cdots\cdots\cdots\cdots\cdots\cdots$$
$$\frac{d\psi(t,\tau_1,\tau_2,\ldots,\tau_k)}{dt} = \tau_k\psi(t,\tau_1,\tau_2,\ldots,\tau_k) + \psi(t,\tau_1,\ldots,\tau_{k-1}),$$
$$\psi(0,\tau_1,\ldots,\tau_k) = 0.$$

We introduce the notation

$$\chi_m(t) = \frac{t^{m-1}}{(m-1)!}e^{\tau t}.$$

By direct calculations, we can check that

$$\frac{d}{dt}\chi_m(t) = \frac{d}{dt}\left[\frac{t^{m-1}}{(m-1)!}e^{\tau t}\right] = \frac{t^{m-2}}{(m-2)!}e^{\tau t} + \tau\frac{t^{m-1}}{(m-1)!}e^{\tau t} = \tau\chi_m(t) + \chi_{m-1}(t),$$
$$\frac{d}{dt}\chi_1(t) = \frac{d}{dt}[e^{\tau t}] = \tau\psi_1(t)$$

and

$$\chi_1(0) = 1, \chi_2(0) = \chi_3(0) = \cdots = \chi_m(0) = 0.$$

Hence $\chi_m(t) = \psi_m(t)$. Thus, we have shown that if $\tau_j \equiv \tau$ for all j, then

$$\psi_1(t) = \psi(t,\tau) = x_1(t) = e^{\tau t},$$
$$\cdots\cdots\cdots$$
$$\psi_k(t) = \psi(t,\underbrace{\tau,\tau,\ldots,\tau}_{k \text{ times}}) = \frac{t^{k-1}}{(k-1)!}e^{\tau t}.$$

We will not write out explicit formulas for $\psi(t,\tau_1,\tau_2,\ldots\tau_k)$ in the case when only some of parameters τ_j coincide, because such formulas are too cumbersome. However, we recall once more that $\psi(t,\tau_1,\tau_2,\ldots\tau_k)$ continuously depend on t as well as on $\tau_1,\tau_2,\ldots,\tau_k$.

We derive one more series of recurrence relations. First, we prove that

$$\psi^{(k)}(0,\tau_1,\tau_2,\ldots,\tau_k) = \tau_1 + \tau_2 + \cdots + \tau_k.$$

Indeed, $x(t) = \psi(t,\tau_1,\tau_2,\ldots\tau_k)$ is an infinitely differentiable function of t satisfying the equation

$$x^{(k)}(t) - (\tau_1+\tau_2+\cdots+\tau_k)x^{(k-1)}(t) + p_2x^{(k-2)}(t) + \cdots + p_kx(t) = 0.$$

Since

$$x(0) = x'(0) = \cdots = x^{(k-2)}(0) = 0,\ x^{(k-1)}(0) = 1,$$

we arrive at the required relation

$$\psi^{(k)}(0,\tau_1,\tau_2,\ldots,\tau_k) = x^{(k)}(0) = \tau_1 + \tau_2 + \cdots + \tau_k.$$

It was shown in §4 that the functions $\psi(t,\tau_1,\tau_2,\dots\tau_j)$ for $j \leqslant k$ are solutions to the equation

$$\mathcal{P}_k\left(\frac{d}{dt}\right)y = 0.$$

Since $\psi(t,\tau_1,\tau_2,\dots\tau_j)$ is a symmetric function of τ_j, we obtain the equalities

$$\mathcal{P}_k\left(\frac{d}{dt}\right)\psi(t,\tau_1,\tau_2,\dots,\tau_{k-1}) = 0,$$
$$\mathcal{P}_k\left(\frac{d}{dt}\right)\psi(t,\tau_2,\tau_3,\dots,\tau_k) = 0.$$

Assume that $\tau_1 \neq \tau_k$. Defining the function

$$\gamma(t) = \frac{\psi(t,\tau_2,\tau_3,\dots,\tau_k) - \psi(t,\tau_1,\tau_2,\dots,\tau_{k-1})}{\tau_k - \tau_1},$$

we easily conclude that

$$\mathcal{P}_k\left(\frac{d}{dt}\right)\gamma(t) = 0.$$

We derive the initial conditions for $\gamma(t)$:

$$\begin{aligned}
\gamma(0) &= \big(\psi(t,\tau_2,\tau_3,\dots,\tau_k) - \psi(t,\tau_1,\tau_2,\dots,\tau_{k-1})\big)/(\tau_k-\tau_1) = 0/(\tau_k-\tau_1) = 0,\\
\gamma'(0) &= \big(\psi'(t,\tau_2,\tau_3,\dots,\tau_k) - \psi'(t,\tau_1,\tau_2,\dots,\tau_{k-1})\big)/(\tau_k-\tau_1) = 0,\\
&\cdots\cdots\cdots\cdots\cdots\cdots\cdots\cdots\cdots\cdots\\
\gamma^{(k-2)}(0) &= \big(\psi^{(k-2)}(t,\tau_2,\tau_3,\dots,\tau_k)\\
&\qquad - \psi^{(k-2)}(t,\tau_1,\tau_2,\dots,\tau_{k-1})\big)/(\tau_k-\tau_1) = 0,\\
\gamma^{(k-1)}(0) &= \big(\psi^{(k-1)}(t,\tau_2,\tau_3,\dots,\tau_k) - \psi^{(k-1)}(t,\tau_1,\tau_2,\dots,\tau_{k-1})\big)/(\tau_k-\tau_1)\\
&= \big((\tau_2+\tau_3+\dots+\tau_k) - (\tau_1+\tau_2+\dots+\tau_{k-1})\big)/(\tau_k-\tau_1)\\
&= (\tau_k-\tau_1)/(\tau_k-\tau_1) = 1.
\end{aligned}$$

We see that $\gamma(t)$ satisfies the same equation and the same initial conditions as the function $\psi(t,\tau_1,\tau_2,\dots\tau_k)$. By the uniqueness theorem, for $\tau_1 \neq \tau_k$

$$\psi(t,\tau_1,\dots,\tau_k) = \frac{\psi(t,\tau_2,\tau_3,\dots,\tau_k) - \psi(t,\tau_1,\tau_2,\dots,\tau_{k-1})}{\tau_k - \tau_1}.$$

Thus, we have derived one more recurrence relation between the special partial solutions $\psi(t,\tau_1,\tau_2,\dots\tau_k)$. We note that in the obtained relation, the first parameter τ_1 and the last parameter τ_k are distinguished. In fact, since $\psi(t,\tau_1,\tau_2,\dots\tau_k)$ is a symmetric function of parameters, any parameters can be taken as distinguished ones provided that their values do not coincide. For example,

$$\psi(t,\tau_1,\tau_2,\tau_3,\tau_4) = \frac{\psi(t,\tau_1,\tau_3,\tau_4) - \psi(t,\tau_1,\tau_2,\tau_4)}{\tau_3 - \tau_2}.$$

This fact will be often used later in the book. In particular we notice that

$$\psi(t,\tau_1,\tau_2) = \frac{e^{\tau_1 t} - e^{\tau_2 t}}{\tau_1 - \tau_2}$$

and if all τ_j are distinct ($\tau_j \neq \tau_k$), then[4]

$$\psi(t,\tau_1,\tau_2,\ldots,\tau_N) = \frac{\sum\limits_{i=1}^{N} e^{\tau_i t}}{\prod\limits_{\substack{k=1\\k\neq i}}^{N}(\tau_i-\tau_k)}.$$

In the following table we compare the Cauchy problems for the functions $\psi(t,\tau_1)$, $\psi(t,\tau_1,\tau_2),\ldots,\psi(t,\tau_1,\tau_2,\ldots,\tau_N)$ and $\omega_1(t),\omega_2(t),\ldots,\omega_N(t)$.

$\mathcal{P}_1\left(\frac{d}{dt}\right)\psi(t,\tau_1)=0$ $\psi(0,\tau_1)=1$	$\mathcal{P}_N\left(\frac{d}{dt}\right)\omega_1(t)=0$ $\omega_1(0)=1,\ \omega_1'(0)=0,\ldots$ $\omega_1^{(N-1)}(0)=0$
$\mathcal{P}_2\left(\frac{d}{dt}\right)\psi(t,\tau_1,\tau_2)=0$ $\psi(0,\tau_1,\tau_2)=0$ $\psi'(0,\tau_1,\tau_2)=1$	$\mathcal{P}_N\left(\frac{d}{dt}\right)\omega_2(t)=0$ $\omega_2(0)=0,\ \omega_2'(0)=1,\ldots$ $\omega_2^{(N-1)}(0)=0$
$\mathcal{P}_N\left(\frac{d}{dt}\right)\psi(t,\tau_1,\tau_2,\ldots,\tau_N)=0$ $\psi(0,\tau_1,\tau_2,\ldots,\tau_N)=0$ . . . . $\psi^{(N-2)}(0,\tau_1,\tau_2,\ldots,\tau_N)=0$ $\psi^{(N-1)}(0,\tau_1,\tau_2,\ldots,\tau_N)=1$	$\mathcal{P}_N\left(\frac{d}{dt}\right)\omega_N(t)=0$ $\omega_N(0)=0,\ldots,\omega_N^{(N-2)}(0)=0$ $\omega_N^{(N-1)}(0)=1$

One can immediately observe that $\omega_N(t)=\psi(t,\tau_1,\tau_2,\ldots,\tau_N)$. Moreover, since the functions $\psi(t,\tau_1,\tau_2,\ldots,\tau_j)$, $j\leqslant N$, are solutions to the equation

$$\mathcal{P}_N\left(\frac{d}{dt}\right)x(t)=0,$$

the Cauchy problems for the functions $\psi(t,\tau_1,\tau_2,\ldots,\tau_j)$, $1\leqslant j\leqslant N$, can be rewritten as follows:

$$\mathcal{P}_N\left(\frac{d}{dt}\right)\psi(t,\tau_1)=0,$$
$$\psi(0,\tau_1)=1,\ \psi'(0,\tau_1)=\alpha_{11},\ \ldots,\ \psi^{(N-1)}(0,\tau_1)=\alpha_{1,N-1},$$
$$\mathcal{P}_N\left(\frac{d}{dt}\right)\psi(t,\tau_1,\tau_2)=0,\ \psi(0,\tau_1,\tau_2)=0,$$
$$\psi'(0,\tau_1,\tau_2)=1,\ \psi''(0,\tau_1,\tau_2)=\alpha_{22},\ \ldots,\ \psi^{(N-1)}(0,\tau_1,\tau_2)=\alpha_{2,N-1},$$

[4] Compare this formula with the above formula expressing $\psi(t,\tau_1,\tau_2,\ldots,\tau_k)$ as the ratio of two Vandermonde determinants.

$$\mathcal{P}_N\left(\frac{d}{dt}\right)\psi(t,\tau_1,\dots,\tau_{N-1})=0,\ \psi(0,\tau_1,\dots,\tau_{N-1})=0,\ \dots,$$
$$\psi^{(N-2)}(0,\tau_1,\dots,\tau_{N-1})=1,\ \psi^{(N-1)}(0,\tau_1,\dots,\tau_{N-1})=\alpha_{N-1,N-1},$$
$$\mathcal{P}_N\left(\frac{d}{dt}\right)\psi(t,\tau_1,\dots,\tau_N)=0,$$
$$\psi(0,\tau_1,\dots,\tau_N)=0,\ \dots,\ \psi^{(N-1)}(0,\tau_1,\dots,\tau_N)=1,$$

where α_{jk} denote the corresponding values of the derivatives of $\psi(t,\tau_1,\tau_2,\dots,\tau_j)$ at $t=0$. It is easy to describe the connections between the functions $\psi(t,\tau_1,\tau_2,\dots,\tau_j)$ and $\omega_j(t)$, $1\leqslant j\leqslant N$. Indeed,

$$\begin{aligned}&\psi(t,\tau_1,\tau_2,\dots,\tau_N)=\omega_N(t),\\&\psi(t,\tau_1,\tau_2,\dots,\tau_{N-1})=\omega_{N-1}(t)+\alpha_{N-1,N-1}\omega_N(t),\\&\dots\dots\dots\dots\dots\dots\dots\dots\\&\psi(t,\tau_1)=\omega_1(t)+\alpha_{11}\omega_2(t)+\dots+\alpha_{1,N-1}\omega_N(t).\end{aligned}$$

It is possible to indicate the formulas representing the functions $\psi_k(t)$ and $\omega_j(t)$ in terms of an arbitrary fundamental system of solutions to the equation

$$\mathcal{P}_k\left(\frac{d}{dt}\right)x(t)=0.$$

By direct verification, it is easy to prove the equality

$$\psi_k(t)\equiv(\det Y_k(0))^{-1}\det\begin{bmatrix}x_1(t) & \dots & x_k(t)\\ \dots & \dots & \dots\\ x_1^{(k-2)}(t) & \dots & x_k^{(k-2)}(t)\\ x_1^{k-1}(t) & \dots & x_k^{k-1}(t)\end{bmatrix},$$

where $x_1(t),x_2(t),\dots,x_k(t)$ is an arbitrary fundamental system of solutions to the equation $\mathcal{P}_k(\frac{d}{dt})x(t)=0$ and

$$Y_k(0)=\begin{bmatrix}x_1(0) & x_2(0) & \dots & x_k(0)\\ x_1'(0) & x_2'(0) & \dots & x_k'(0)\\ \dots & \dots & \dots & \dots\\ x_1^{(k-1)}(0) & x_2^{(k-1)}(0) & \dots & x_k^{(k-1)}(0)\end{bmatrix}.$$

To conclude the section, we prove that the behavior of a solution to the differential equation

$$\mathcal{P}\left(\frac{d}{dt}\right)\xi(t)=0$$

as $t\to\infty$ can be characterized by the real parts of exponents occurring in the representation

$$\xi(t)=\sum_{j=1}^{M}a_jt^{k_j}e^{(\alpha_j+i\beta_j)t}.$$

Let us prove that the equality

$$\lim_{t\to\infty}|\xi(t)|=0$$

implies that all α_j are negative. Certainly, the converse assertion is also true: the negativity of α_j implies that $|\xi(t)|$ exponentially decreases as $t \to \infty$.

In §2, we proved that any solution $y(t)$ to the homogeneous system (homogeneous vector equation) $y' = Ay$ has the form

$$\begin{aligned} y(t) = e^{tA}c = \Big\{&\psi_1(t)I + \psi_2(t)[A - \tau_1 I] + \dots \\ &+ \psi_N(t)\big[(A-\tau_1 I)(A-\tau_2 I)\cdots(A-\tau_{N-1}I)\big]\Big\}c \\ = {}&\psi_1(t)c_1 + \psi_2(t)c_2 + \dots + \psi_N(t)c_N \end{aligned}$$
$$\big(c_1 = c,\ c_2 = (A-\tau_1 I)c,\ \dots,\ c_N = [A-\tau_1 I][A-\tau_2 I]\cdots[A-\tau_{N-1}I]c\big).$$

Since the functions $\psi_j(t)$ can be represented as linear combinations of the form $t^k e^{\tau_j t} = t^{k_j}\exp[(\alpha_j + i\beta_j)t]$, we can easily prove that each component $y_p(t)$ of the vector-valued function $y(t)$ which is a solution to the homogeneous linear equation $\dfrac{d}{dt}y(t) = Ay(t)$ with a constant matrix A can be represented as a linear combination of the form $t^k e^{(\alpha+i\beta)t}$:

$$y_p(t) = \xi(t) = \sum_{j=1}^{M} a_j t^{k_j} e^{(\alpha_j+i\beta_j)t}.$$

It is obvious that any linear combination $\sum_{p=1}^{N}\gamma_p y_p(t)$ of these components also admits the same representation.

In the following theorem we establish several simple properties of functions $\xi(t)$. These properties will be repeatedly used in the sequel.

THEOREM 1. *If the expression*

$$\xi(t) = \sum_{j=1}^{M} a_j t^{k_j} e^{(\alpha_j+i\beta_j)t}$$

converges to zero as $t \to \infty$, *then there exist constants* $\sigma_0 > 0$, $B_0 > 0, B_1 > 0, \dots, B_m > 0$ *such that for* $t > 0$ *we have*

$$|\xi(t)| < B_0 e^{-\sigma_0 t},\ |\xi'(t)| < B_1 e^{-\sigma_0 t},\ \dots,\ |\xi^{(m)}(t)| < B_m e^{-\sigma_0 t}.$$

Theorem 1 means that the function ξ and its derivatives exponentially decrease as $t \to \infty$. This takes place for the linear combination $\eta(t) = \sum_{p=1}^{N}\gamma_p y_p(t)$, where $y_1, y_2, \dots, y_N$ are components of the vector-valued function $y(t)$ satisfying the $y' = Ay$ and for any solution $\xi(t)$ of an arbitrary homogeneous equation

$$\xi^{(N)} + p_1\xi^{(N-1)} + \dots + p_{N-1}\xi' + p_N\xi = 0$$

with constant coefficients $p_1, p_2, \dots, p_N$ provided that $|\xi(t)| \to 0$ as $t \to \infty$.

Before proving Theorem 1 we establish the following lemma.

LEMMA 1. *If* $\beta_j \neq \beta_k$ *for* $k \neq j$ *and*

$$\xi(t) = \sum_{j=1}^{M} c_j e^{i\beta_j t},$$

then for all t the inequality

$$|\xi(t)| \leqslant \sum_{j=1}^{M} |c_j|$$

holds and there exists a sequence $t_1, t_2, \dots, t_q, \dots$ *such that*

$$\lim_{q\to\infty} t_q = +\infty, \qquad |\xi(t_q)| \geqslant \frac{1}{2}\sqrt{\sum_{j=1}^{M} |c_j|^2}.$$

Proof. The first inequality is obvious:

$$|\xi(t)| = \left|\sum_{j=1}^{M} c_j e^{i\beta_j t}\right| \leqslant \sum_{j=1}^{M} |c_j e^{i\beta_j t}| = \sum_{j=1}^{M} |c_j|.$$

The second assertion can be justified as follows. First, we compute the integral

$$\begin{aligned}\int_T^{2T} |\xi(t)|^2 dt &= \int_T^{2T} \xi(t)\bar{\xi}(t)\, dt = \int_T^{2T} \left[\sum_{j=1}^{M} c_j e^{i\beta_j t} \sum_{k=1}^{M} \bar{c}_k e^{-i\beta_k t}\right] dt \\ &= \int_T^{2T} \left[\sum_{j=1}^{M} c_j\bar{c}_j + \sum_{j\neq k} c_j\bar{c}_k e^{i(\beta_j-\beta_k)t}\right] dt \\ &= T\sum_{j=1}^{M} c_j\bar{c}_j - i\sum_{j\neq k} \frac{c_j\bar{c}_k}{\beta_j-\beta_k}\left[e^{2i(\beta_j-\beta_k)T} - e^{i(\beta_j-\beta_k)T}\right]\end{aligned}$$

and derive the inequality

$$\frac{1}{T}\int_T^{2T} |\xi(t)|^2 dt \geqslant \sum_{j=1}^{M} |c_j|^2 - \frac{1}{T}\frac{2}{\min\limits_{\substack{j,k\\ j\neq k}} |\beta_j-\beta_k|}\left(\sum_{j=1}^{M} |c_j|\right)^2.$$

If T satisfies the condition

$$T \geqslant \frac{8}{3\min\limits_{j\neq k} |\beta_j-\beta_k|}\cdot\left(\sum_{j=1}^{M} |c_j|\right)^2 \left(\sum_{j=1}^{M} |c_j|^2\right)^{-1}$$

then from the previous inequality we get

$$\frac{1}{T}\int_T^{2T} |\xi(t)|^2 dt \geqslant \frac{1}{4}\sum_{j=1}^{M} |c_j|^2.$$

It is clear that the last inequality means that there exists at least one point $t = t^*$ on the segment $[T, 2T]$ at which

$$|\xi(t^*)| \geqslant \frac{1}{2}\sqrt{\sum_{j=1}^{M} |c_j|^2}.$$

We consider the sequence of segments $2^{2^l} \leqslant t \leqslant 2 \cdot 2^{2^l} = 2^{2^l+1}$ whose lengths $T_l = 2^{2^l}$ infinitely increase as t increases. Beginning from some $l = l_1$, the relation

$$2^{2^l} = T_l \geqslant \frac{8}{3 \min\limits_{j \neq k} |\beta_j - \beta_k|} \cdot \left(\sum_{j=1}^{M} |c_j| \right)^2 \left(\sum_{j=1}^{M} |c_j|^2 \right)^{-1}$$

is valid and on the segment $[2^{2^l}, 2^{2^l+1}]$ there exists a point $t = t^* = t_{l-l_1+1}$ at which

$$|\xi(t_{l-l_1+1})| \geqslant \frac{1}{2} \sqrt{\sum_{j=1}^{M} |c_j|^2}.$$

It is obvious that $t_q \geqslant 2^{2^{q+l_1-1}} \geqslant 2^{2^{q-1}}$ and $\lim_{q \to \infty} t_q = \infty$. □

Let

$$\xi(t) = \sum_{j=1}^{M} a_j t^{k_j} e^{(\alpha_j + i\beta_j)t}.$$

We assume that by regrouping the terms, the function $\xi(t)$ can be represented in the form $\xi(t) = \xi_1(t) + \xi_2(t)$, where

$$\xi_1(t) = \sum_{j=1}^{M_0} c_j t^{k_0} e^{(\alpha_0 + i\beta_j)t} = t^{k_0} e^{\alpha_0 t} \sum_{j=1}^{M_0} c_j e^{i\beta_j t},$$

$$\xi_2(t) = \sum_{j=M_0+1}^{M} c_j t^{k_j} e^{(\alpha_j + i\beta_j)t}$$

and the terms that occur in the second sum $\xi_2(t)$ are those for which either $\alpha_j = \alpha_0$ but $k_j < k_0$ or $\alpha_j < \alpha_0$. It is obvious that for sufficiently large t we have

$$|\xi_2(t)| < \frac{1}{4} \sqrt{\sum_{j=1}^{M_0} |c_j|^2}\, t^{k_0} e^{\alpha_0 t}.$$

PROOF OF THEOREM 1. By Lemma 1, there exists an infinitely increasing sequence $t_1, t_2, \dots, t_q, \dots$, such that

$$|\xi_1(t_q)| \geqslant \frac{1}{2} \sqrt{\sum_{j=1}^{M_0} |c_j|^2} \cdot t_q^{k_0} e^{\alpha_0 t_q}.$$

For sufficiently large q,

$$|\xi_2(t_q)| < \frac{1}{4} \sqrt{\sum_{j=1}^{M_0} |c_j|^2} \cdot t_q^{k_0} e^{\alpha_0 t_q}.$$

Consequently,

$$\begin{aligned} |\xi(t_q)| &= |\xi_1(t_q) + \xi_2(t_q)| \geqslant |\xi_1(t_q)| - |\xi_2(t_q)| \\ &> \frac{1}{2} \sqrt{\sum_{j=1}^{M_0} |c_j|^2} \cdot t_q^{k_0} e^{\alpha_0 t_q} - \frac{1}{4} \sqrt{\sum_{j=1}^{M_0} |c_j|^2} \cdot t_q^{k_0} e^{\alpha_0 t_q} = \frac{1}{4} \sqrt{\sum_{j=1}^{M_0} |c_j|^2} \cdot t_q^{k_0} e^{\alpha_0 t_q}. \end{aligned}$$

From the last inequality it follows that if, in addition, $\xi(t) \to 0$ as $t \to \infty$, then $\alpha_j \leqslant \alpha_0 < 0$.

If the function $\xi(t)$ is represented as the linear combination

$$\xi(t) = \sum_{j=1}^{M} a_j t^{k_j} e^{(\alpha_j + i\beta_j)t},$$

then the similar representations hold for the derivatives:

$$\xi'(t) = \sum_{j=1}^{M_1} a_j^{(1)} t^{k_j} e^{(\alpha_j + i\beta_j)t},$$

$$\xi''(t) = \sum_{j=1}^{M_2} a_j^{(2)} t^{k_j} e^{(\alpha_j + i\beta_j)t},$$

$$\dots\dots\dots\dots\dots$$

$$\xi^{(m)}(t) = \sum_{j=1}^{M_m} a_j^{(m)} t^{k_j} e^{(\alpha_j + i\beta_j)t}.$$

In these formulas for the derivatives, the terms $t^{k_j} e^{(\alpha_j + i\beta_j)t}$ have the same exponents $\alpha_j + i\beta_j$ as they have in the formula for the function $\xi(t)$ and the powers t^{k_j} do not exceed the corresponding powers in the monomials with the same exponents in the representation of $\xi(t)$.

From the inequalities $\alpha_j \leqslant \alpha_0 \equiv -2\sigma_0 < 0$ it follows that

$$\begin{aligned} |t^{k_j} e^{(\alpha_j + i\beta_j)t}| &= t^{k_j} e^{\alpha_j t} = t^{k_j} e^{-(\alpha_0 - \alpha_j)t} e^{\alpha_0 t} \leqslant t^{k_j} e^{\alpha_0 t} \\ &\equiv t^{k_j} e^{-2\sigma_0 t} = \frac{k_j!}{\sigma_0^{k_j}} \left[\frac{t^{k_j}}{k_j!} \sigma_0^{k_j} \right] e^{-2\sigma_0 t} \leqslant \frac{k_j!}{\sigma_0^{k_j}} \left(\sum_{k=0}^{\infty} \frac{t^k}{k!} \sigma_0^k \right) e^{-2\sigma_0 t} \\ &= \frac{k_j!}{(\sigma_0^{k_j})} e^{\sigma_0 t} e^{-2\sigma_0 t} = \text{const } e^{-\sigma_0 t}. \end{aligned}$$

Therefore, if $\xi(t) \to 0$ as $t \to \infty$, then $\xi(t), \xi'(t), \dots, \xi^{(m)}(t)$ exponentially decrease:

$$\begin{aligned} &|\xi(t)| < \text{const } e^{-\sigma_0 t}, \\ &|\xi'(t)| < \text{const } e^{-\sigma_0 t}, \\ &\dots\dots\dots\dots \\ &|\xi^{(m)}(t)| < \text{const } e^{-\sigma_0 t} \end{aligned}$$

with some $\sigma_0 > 0$. □

EXERCISE 1. Prove that if

$$\xi(t) = \sum_{j=1}^{M} a_j t^{k_j} e^{(\alpha_j + i\beta_j)t}$$

and

$$\int_0^{\infty} |\xi(t)|^2 dt < \infty$$

then there exist $B_0 > 0$ and $\sigma_0 > 0$ such that

$$|\xi(t)| < B_0 e^{-\sigma_0 t}.$$

COROLLARY 1. *If $\xi(t) = \sum_p a_p y_p(t)$, where $y_p(t)$ are components of the vector-valued solution $y(t)$ to the system $y' = Ay$ and if $\xi(t) \to 0$ as $t \to -\infty$, then there exist constants $\sigma_0 > 0$, $B_0 > 0, B_1 > 0, \dots, B_m > 0$ such that for $t < 0$*

$$\begin{aligned} |\xi(t)| &< B_0 e^{\sigma_0 t} = B_0 e^{-\sigma_0 |t|}, \\ |\xi'(t)| &< B_1 e^{\sigma_0 t} = B_1 e^{-\sigma_0 |t|}, \\ &\dots\dots\dots\dots \\ |\xi^{(m)}(t)| &< B_m e^{\sigma_0 t} = B_m e^{-\sigma_0 |t|}. \end{aligned}$$

PROOF. In fact, Corollary 1 asserts that the function $\xi(t)$, as well as all its derivatives, exponentially decreases. The assertion can be proved by the substitution $\widehat{t} = -t$, which shows that the function $z(\widehat{t}) \equiv y(-\widehat{t}) = y(t)$ satisfies the equation

$$\frac{dz(\widehat{t})}{d\widehat{t}} = -Az(\widehat{t}).$$

Using this substitution, we can represent the linear combination $\xi(t) = \sum_p a_p y_p(t)$ in the form

$$\xi(t) = \zeta(\widehat{t}) = \sum_p a_p z_p(\widehat{t}).$$

The condition that $|\xi(t)| \to 0$ as $t \to -\infty$ takes the form $|\zeta(\widehat{t})| \to 0$ as $\widehat{t} \to +\infty$. By Theorem 1, the exponential decrease estimate

$$|\zeta^{(m)}(\widehat{t})| \leqslant \text{const } e^{-\sigma_0 \widehat{t}}$$

holds for

$$|\zeta^{(m)}(\widehat{t})| = |(-1)^m \xi^{(m)}(t)| = |\xi^{(m)}(t)|, \quad t > 0.$$

This estimate implies that

$$|\zeta^{(m)}(t)| \leqslant \text{const } e^{-\sigma_0 |t|}, \quad t < 0. \quad \square$$

§6. Computation of the matrix exponential by means of reducing matrices to the Jordan form

Matrix exponentials for two-diagonal matrices. The special functions $\psi_i(t)$ as elements of the matrix exponential. An exponential of the Jordan block. An example of discontinuous dependence of the Jordan form on elements of the initial matrix. The use of the block diagonal canonical forms of matrices.

We begin by obtaining explicit formulas for elements of the matrix exponential e^{tA} for a matrix A of the following two-diagonal form:

$$\begin{bmatrix} \tau_1 & 1 & & & \\ & \tau_2 & 1 & & 0 \\ & & \ddots & \ddots & \\ & 0 & & \tau_{N-1} & 1 \\ & & & & \tau_N \end{bmatrix}.$$

On one hand, such matrices can be regarded as simple examples, and on the other hand, as we will see in the sequel, the study of properties of a general matrix exponential can be reduced to the study of such matrices.

We proceed to the construction of the matrix exponential $e^{tA} = Y(t)$, i.e., we will construct a solution $Y(t)$ to the matrix equation

$$\frac{d}{dt}Y(t) = AY(t).$$

Such a solution is a matrix exponential $Y(t) = e^{tA}$ if the initial condition $Y(0) = I$ holds. The elements of the kth column of the matrix $Y(t)$ can be found as those solutions of the system of differential equations

$$\begin{aligned} y'_{1k} &= \tau_1 y_{1k} + y_{2k}, \\ &\cdots\cdots\cdots \\ y'_{N-1\,k} &= \tau_{N-1} y_{N-1\,k} + y_{N\,k}, \\ y'_{Nk} &= \tau_N y_{Nk}, \end{aligned}$$

that satisfy the initial conditions

$$y_{1k}(0) = 0, \ \ldots, \ y_{k-1\,k}(0) = 0, \ y_{kk}(0) = 1, \ \ldots, \ y_{Nk}(0) = 0.$$

A part of these equations, starting with the $(k+1)$th row, can be regarded as an independent system of equations for the unknown functions $y_{k+1\,k}(t), y_{k+2\,k}(t), \ldots, y_{Nk}$ satisfying the zero initial conditions for $t = 0$. It is clear that such equations and the initial conditions are satisfied by the functions

$$y_{k+1\,k}(t) \equiv 0, \ y_{k+2\,k}(t) \equiv 0, \ \ldots, \ y_{Nk}(t) \equiv 0.$$

By the uniqueness theorem, there is no other solution. The fact that $y_k(t) = 0$ for $j > k$ means that $Y(t)$ is an upper triangular matrix

$$Y(t) = \begin{bmatrix} y_{11} & y_{12} & y_{13} & \cdots & y_{1,N-1} & y_{1N} \\ & y_{22} & y_{23} & \cdots & y_{2,N-1} & y_{2N} \\ & & y_{33} & \cdots & y_{3,N-1} & y_{3N} \\ & & & \ddots & \vdots & \vdots \\ & \text{\Huge 0} & & & y_{N-1,N-1} & y_{N-1,N} \\ & & & & & y_{NN} \end{bmatrix}.$$

The nonzero elements of the kth column are found as solutions of the following equations and initial conditions:

$$\begin{aligned} y'_{1k} &= \tau_1 y_{1k} + y_{2k}, \quad y_{1k}(0) = 0, \\ y'_{2k} &= \tau_2 y_{2k} + y_{3k}, \quad y_{2k}(0) = 0, \\ &\cdots\cdots\cdots\cdots\cdots \\ y'_{k-1,k} &= \tau_{k-1} y_{k-1,k} + y_{kk}, \quad y_{k-1,k}(0) = 0, \\ y'_{kk} &= \tau_k y_{kk}, \quad y_{kk}(0) = 1. \end{aligned}$$

Such solutions, written in other notation and with other numbering of parameters, have been studied in §§3–5. In the notation of §5, we have

$$y_{jk}(t) = \psi(t, \tau_j, \tau_{j+1}, \ldots, \tau_k), \quad 1 \leqslant j \leqslant k.$$

In particular, if the matrix A can be represented as a single Jordan block

$$A = \begin{bmatrix} \tau & 1 & & & \\ & \tau & 1 & & \mathbf{0} \\ & & \tau & 1 & \\ & & & \ddots & \ddots \\ & \mathbf{0} & & & \tau & 1 \\ & & & & & \tau \end{bmatrix},$$

then the matrix exponential e^{tA} can be represented as follows:

$$e^{tA} = \begin{bmatrix} e^{\tau t} & \frac{t}{1!}e^{\tau t} & \frac{t^2}{2!}e^{\tau t} & \dots & \frac{t^{N-1}}{(N-1)!}e^{\tau t} \\ & e^{\tau t} & \frac{t}{1!}e^{\tau t} & \dots & \frac{t^{N-2}}{(N-2)!}e^{\tau t} \\ & & e^{\tau t} & \dots & \frac{t^{N-3}}{(N-3)!}e^{\tau t} \\ & \mathbf{0} & & \ddots & \vdots \\ & & & & e^{\tau t} \end{bmatrix};$$

here we use the formula for $\psi(t, \tau, \tau, \dots, \tau)$ from §5.

Any square matrix A can be written in the special canonical form, called the Jordan canonical form. Namely, it is always possible to find a nonsingular matrix T ($\det T \neq 0$) such that

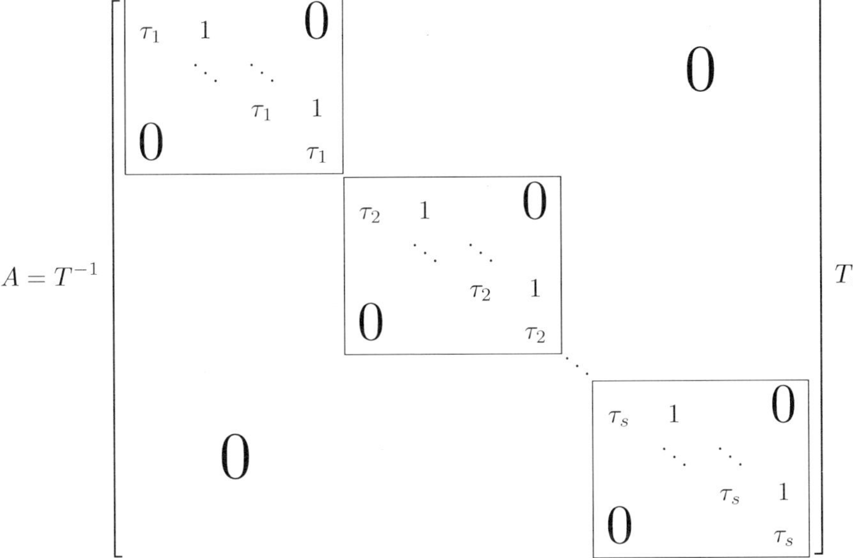

The matrix T realizing the reduction to the canonical form is not unique. Moreover, the canonical form is a discontinuous function of elements of A. To clarify this fact we give an example. Let A be the same matrix as above and let A be written in

terms of T and Jordan blocks. We define the matrix B by the formula

$$B = T^{-1} \begin{bmatrix} 1 & & & & & \\ & 2 & & & \text{\Large 0} & \\ & & 3 & & & \\ & & & \ddots & & \\ & \text{\Large 0} & & & N-1 & \\ & & & & & N \end{bmatrix} T$$

and consider the family of matrices

$$A + \xi B = T^{-1} \begin{bmatrix} \tau_1 + \xi & 1 & & & & \\ & \tau_1 + 2\xi & 1 & & \text{\Large 0} & \\ & & \tau_1 + 3\xi & 1 & & \\ & & & \ddots & \ddots & \\ & \text{\Large 0} & & & \tau_s + (N-1)\xi & 1 \\ & & & & & \tau_s + N\xi \end{bmatrix} T$$

depending on a small parameter ξ. It is clear that for sufficiently small nonzero ξ all eigenvalues of the matrix $A + \xi B$ are distinct. Consequently, for $\xi \neq 0$ (and small enough) the matrix A must be reduced to the diagonal Jordan canonical form:

$$A + \xi B = \widetilde{T}^{-1} \begin{bmatrix} \tau_1 + \xi & & & & & \\ & \tau_1 + 2\xi & & & \text{\Large 0} & \\ & & \tau_1 + 3\xi & & & \\ & & & \ddots & & \\ & \text{\Large 0} & & & \tau_s + (N-1)\xi & \\ & & & & & \tau_s + N\xi \end{bmatrix} \widetilde{T},$$

which shows that the continuous dependence is violated. Furthermore, it is obvious that a similar construction can be realized for any matrix A whose characteristic polynomial has multiple roots. It is always possible to construct a matrix B such that for sufficiently small nonzero ξ all eigenvalues of the matrix $A+\xi B$ are distinct.

Since the Jordan canonical form does not continuously depend on the initial matrix, it is not convenient to use the canonical form in calculations. However, it is widely used in theoretical considerations.

Let $A = T^{-1}A_0T$ and $Y = e^{tA}$. In other words,

$$\frac{d}{dt}Y = AY, \quad Y(0) = I.$$

We transform the equation

$$\frac{d}{dt}Y = T^{-1}A_0TY$$

by multiplying it from the left by T and from the right by T^{-1}:

$$T\left(\frac{d}{dt}Y\right)T^{-1} = TT^{-1}A_0TYT^{-1}.$$

Since T and T^{-1} are constant matrices independent of t, they can be put under the differentiation sign:

$$\frac{d}{dt}(TYT^{-1}) = A_0TYT^{-1}.$$

We introduce the notation

$$Z(t) = TY(t)T^{-1} \qquad \big(Y(t) = T^{-1}Z(t)T\big).$$

We see that

$$\frac{d}{dt}Z(t) = A_0 Z(t), \quad Z(0) = TY(0)T^{-1} = TIT^{-1} = I.$$

In other words,

$$Z(t) = e^{tA_0}, \quad e^{tA} = Y(t) = T^{-1}e^{tA_0}T.$$

We have proved that $e^{tA} = T^{-1}e^{tA_0}T$ if $A = T^{-1}A_0 T$.

We have already established how the matrix exponential for a single Jordan block can be calculated. If the matrix A_0 is block diagonal, i.e.,

$$A_0 = \begin{bmatrix} A_0^{(1)} & & & \mathbf{0} \\ & A_0^{(2)} & & \\ & & \ddots & \\ \mathbf{0} & & & A_0^{(k)} \end{bmatrix},$$

one can try to look for $e^{tA_0} = Y(t)$ as a block diagonal matrix

$$Y(t) = \begin{bmatrix} Y^{(1)}(t) & & & \mathbf{0} \\ & Y^{(2)}(t) & & \\ & & \ddots & \\ \mathbf{0} & & & Y^{(k)}(t) \end{bmatrix},$$

where on the diagonal there are square matrices $Y^{(j)}(t)$ of the same order as the corresponding matrices $A_0^{(j)}$.

It is obvious that the differential equation

$$\frac{d}{dt}Y(t) = A_0 Y(t)$$

can be rewritten as follows:

$$\frac{d}{dt}\begin{bmatrix} Y^{(1)} & & & \mathbf{0} \\ & Y^{(2)} & & \\ & & \ddots & \\ \mathbf{0} & & & Y^{(k)} \end{bmatrix} = \begin{bmatrix} A_0^{(1)} & & & \mathbf{0} \\ & A_0^{(2)} & & \\ & & \ddots & \\ \mathbf{0} & & & A_0^{(k)} \end{bmatrix} \begin{bmatrix} Y^{(1)} & & & \mathbf{0} \\ & Y^{(2)} & & \\ & & \ddots & \\ \mathbf{0} & & & Y^{(k)} \end{bmatrix}.$$

Consequently, it splits into the following independent equations:

$$\frac{d}{dt}Y^{(1)} = A_0^{(1)}Y^{(1)}, \ \frac{d}{dt}Y^{(2)} = A_0^{(2)}Y^{(2)}, \ \ldots, \ \frac{d}{dt}Y^{(k)} = A_0^{(k)}Y^{(k)}.$$

For $Y(0) = I$, it is necessary to require the initial values of $Y^{(j)}$, $j = 1, 2, \ldots, k$, to be the identity matrices.

It is clear that the solution $Y(t) = e^{tA_0}$ can be chosen so that

$$e^{tA_0} = \begin{bmatrix} e^{tA_0^{(1)}} & & & \mathbf{0} \\ & e^{tA_0^{(2)}} & & \\ & & \ddots & \\ \mathbf{0} & & & e^{tA_0^{(k)}} \end{bmatrix}.$$

However, the solution $Y(t)$ is unique (by the uniqueness theorem). We have proved that e^{tA_0} is a block diagonal matrix provided that the matrix A_0 is block diagonal. For example, if

$$A_0 = \begin{bmatrix} \boxed{\tau_1} & & & & & & & \\ & \boxed{\tau_2} & & & & & & \\ & & \tau_3 & 1 & & & & \\ & & & \tau_3 & 1 & & & \\ & & & & \tau_3 & & & \\ & & & & & \boxed{\tau_3} & & \\ & & & & & & \tau_4 & 1 \\ & & & & & & & \tau_4 \end{bmatrix},$$

then

$$e^{tA_0} = \begin{bmatrix} \boxed{e^{\tau_1 t}} & & & & & & & \\ & \boxed{e^{\tau_2 t}} & & & & & & \\ & & e^{\tau_3 t} & te^{\tau_3 t} & \frac{t^2}{2}e^{\tau_3 t} & & & \\ & & 0 & e^{\tau_3 t} & te^{\tau_3 t} & & & \\ & & 0 & 0 & e^{\tau_3 t} & & & \\ & & & & & \boxed{e^{\tau_3 t}} & & \\ & & & & & & e^{\tau_4 t} & te^{\tau_4 t} \\ & & & & & & 0 & e^{\tau_4 t} \end{bmatrix}.$$

A solution to the system $y' = Ay$ can be regarded as a linear combination of the columns of e^{tA}. Therefore, the components of the solution must be represented as linear combinations of the quantities $e^{\tau_j t}, te^{\tau_j t}, t^2e^{\tau_j t}, \ldots$. This is the basis for the following procedure of finding solutions. We look for every component of a solution as a linear combination of the functions $e^{\tau_j t}, te^{\tau_j t}, t^2e^{\tau_j t}, \ldots$. The coefficients of the linear combination are unknown. Substituting this "solution" in the system $y' = Ay$ and comparing the coefficients at the equal functions $t^k e^{\tau_j t}$, we obtain a system of linear algebraic equations for the unknown coefficients.

To find solutions to the system $y' = Ay$ we present a method based on the use of the eigenvectors and adjoined vectors of the matrix A. Let τ_0 be a root of the characteristic equation. Suppose that we have found the eigenvector y_0 and adjoined vectors $y_1, y_2, \ldots, y_r$ corresponding to the root τ_0, so that

$$\begin{aligned} Ay_0 &= \tau_0 y_0, \\ Ay_1 &= \tau_0 y_1 + y_0, \\ Ay_2 &= \tau_0 y_2 + y_1, \\ &\cdots\cdots \\ Ay_r &= \tau_0 y_r + y_{r-1}. \end{aligned}$$

We recall that the vectors $y_0, \ldots, y_r$ are linearly independent. It can be immediately verified that in this case the functions

$$e^{\tau_0 t}y_0,\ e^{\tau_0 t}y_1 + \frac{t}{1!}e^{\tau_0 t}y_0,\ e^{\tau_0 t}y_2 + \frac{t}{1!}e^{\tau_0 t}y_1 + \frac{t^2}{2!}e^{\tau_0 t}y_0,\ \ldots$$

are solutions to the system $y' = Ay$. For example,

$$\begin{aligned}\frac{d}{dt}&\left[e^{\tau_0 t}y_2 + \frac{t}{1!}e^{\tau_0 t}y_1 + \frac{t^2}{2!}e^{\tau_0 t}y_0\right] \\ &= \tau_0 e^{\tau_0 t}y_2 + \left(e^{\tau_0 t}y_1 + \tau_0\frac{t}{1!}e^{\tau_0 t}y_1\right) + \left(\frac{t}{1!}e^{\tau_0 t}y_0 + \tau_0\frac{t^2}{2!}e^{\tau_0 t}y_0\right) \\ &= e^{\tau_0 t}(\tau_0 y_2 + y_1) + \frac{t}{1!}e^{\tau_0 t}(\tau_0 y_1 + y_0) + \frac{t^2}{2!}e^{\tau_0 t}\tau_0 y_0 \\ &= e^{\tau_0 t}Ay_2 + \frac{t}{1!}e^{\tau_0 t}Ay_1 + \frac{t^2}{2!}e^{\tau_0 t}Ay_0 \\ &= A\left[e^{\tau_0 t}y_2 + \frac{t}{1!}e^{\tau_0 t}y_1 + \frac{t^2}{2!}e^{\tau_0 t}y_0\right].\end{aligned}$$

The initial values of these solutions at $t = 0$ are linearly independent because they coincide with the linearly independent vectors $y_0, y_1, \ldots$.

We will be able to construct a complete system of linearly independent solutions to a given system if we solve the algebraic problem of finding a basis consisting of eigenvectors and adjoined vectors of the matrix A. The basis contains only eigenvectors if the matrix has no multiple eigenvalues.

§7. The definition of a Green matrix. Existence and uniqueness

The definition of a Green matrix. The construction of a Green matrix by reducing a matrix A to some special form. The absence of nontrivial bounded solutions to the equation $y' = Ay$ under the assumption that the matrix A has no purely imaginary eigenvalues. The uniqueness of a Green matrix.

In §2, we studied matrix exponentials, which play an important role in our exposition because they are very convenient in the study of solutions to equations with constant coefficients. Green matrices play a similar role in the theory of differential equations. We begin to study Green matrices in this section.

DEFINITION 1. A matrix-valued function $G_0(t, A) \equiv G_0(t)$ defined for all $t \neq 0$, continuous and continuously differentiable for $t \neq 0$, is called a Green matrix if the following conditions are satisfied:

(a) for $t \neq 0$ the following differential equation holds:

$$\frac{d}{dt}G_0(t) = AG_0(t);$$

(b) for $t = 0$ the matrix-valued function $G_0(t)$ has a discontinuity of the first kind

$$G_0(+0) - G_0(-0) = I;$$

(c) the norm $\|G(t)\|$ is bounded for $-\infty < t < \infty$.

We show that if a matrix A has no purely imaginary eigenvalues, then there exists a unique Green matrix $G_0(t, A)$.

THEOREM 1 (existence). *If a matrix A has no purely imaginary eigenvalues, then there exists a Green matrix $G_0(t, A)$.*

PROOF. Let A have no eigenvalues on the imaginary axis. Reducing the matrix A to some special form, we construct a Green matrix $G_0(t, A)$. In linear algebra it is proved that under the assumption of Theorem 1 it is possible to find a nonsingular matrix T such that

$$A = T^{-1} \begin{bmatrix} B & 0 \\ 0 & C \end{bmatrix} T, \tag{1}$$

where all eigenvalues of the matrix B have strictly negative real parts and all eigenvalues of the matrix C have positive real parts. The possibility of the representation (1) follows, e.g., from the theory of the Jordan canonical form.[5]

We define the matrix $G_0(t, A)$ by the formula

$$G_0(t, A) = \begin{cases} T^{-1} \begin{bmatrix} e^{tB} & 0 \\ 0 & 0 \end{bmatrix} T & \text{for } t > 0, \\ T^{-1} \begin{bmatrix} 0 & 0 \\ 0 & -e^{tC} \end{bmatrix} T & \text{for } t < 0 \end{cases}$$

and verify that $G_0(t, A)$ satisfies all the requirements from Definition 1.

The matrix $G_0(t, A) \equiv G_0(t)$ is defined for all $t \neq 0$, is continuous, and is continuously differentiable for $t \neq 0$. Let us verify Condition (a) from Definition 1 for $t > 0$. Indeed,

$$\begin{aligned} \frac{d}{dt} G_0(t) &= T^{-1} \begin{bmatrix} \frac{d}{dt} e^{tB} & 0 \\ 0 & 0 \end{bmatrix} T = T^{-1} \begin{bmatrix} Be^{tB} & 0 \\ 0 & 0 \end{bmatrix} T = T^{-1} \begin{bmatrix} B & 0 \\ 0 & C \end{bmatrix} \begin{bmatrix} e^{tB} & 0 \\ 0 & 0 \end{bmatrix} T \\ &= T^{-1} \begin{bmatrix} B & 0 \\ 0 & C \end{bmatrix} T T^{-1} \begin{bmatrix} e^{tB} & 0 \\ 0 & 0 \end{bmatrix} T = A G_0(t), \quad t > 0. \end{aligned}$$

The case $t < 0$ can be verified in a similar way.

To check Condition (b) from Definition 1, we compute the limits of $G_0(t)$ from the left and from the right:

$$G_0(+0) = T^{-1} \begin{bmatrix} I & 0 \\ 0 & 0 \end{bmatrix} T,$$

$$G_0(-0) = T^{-1} \begin{bmatrix} 0 & 0 \\ 0 & -I \end{bmatrix} T,$$

which implies the equality $G_0(+0) - G_0(-0) = I$.

It remains to show that the norm $\|G_0(t)\|$ is bounded for $-\infty < t < \infty$. We show that it is bounded for $t \geqslant 0$. For $t \geqslant 0$ we have $\|G_0(t)\| \leqslant \|T^{-1}\| \, \|e^{tB}\| \, \|T\|$. Let the matrix B be of order k. We assume that all eigenvalues of the matrix B satisfy the inequality $\operatorname{Re} \lambda_j(B) \leqslant -\delta < 0$. As was proved in §3, under this assumption,

$$\|e^{tB}\| \leqslant \gamma(k) \left(\frac{\|B\|}{\delta} \right)^{k-1} e^{-\frac{\delta}{2} t} \quad (t \geqslant 0).$$

Hence $\|e^{tB}\|$ (thereby, $\|G_0(t)\|$) is bounded for $t \geqslant 0$.

---

[5] We note that B or C in the representation (1) is absent if real parts of the eigenvalues of the matrix A are either all positive or all negative. It is obvious that such a representation of the matrix A is not unique.

The boundedness of $\|G_0(t)\|$ for $t \leqslant 0$ is proved in a similar way. Namely, for $t \leqslant 0$ we have

$$\|G_0(t)\| \leqslant \|T^{-1}\| \, \|e^{tC}\| \, \|T\|.$$

Let the matrix C be of order m and let $\operatorname{Re}\lambda_j(C) \geqslant \delta > 0$. Then

$$\operatorname{Re}\lambda_i(-C) \leqslant -\delta < 0,$$
$$\|e^{tC}\| = \|e^{(-t)(-C)}\| \leqslant \gamma(m) \left(\frac{\|C\|}{\delta}\right)^{m-1} e^{\frac{\delta}{2}t} \quad (t \leqslant 0).$$

Hence $\|G_0(t)\|$ is also bounded for $t \leqslant 0$. □

To establish the uniqueness of a Green matrix (in particular, the independence of the way A is reduced to the special form) we need some auxiliary lemmas.

LEMMA 1. *Let $u(t)$, $-\infty < t < \infty$, be a bounded ($\|u(t)\| \leqslant h$) solution to the vector equation (the system of order k)*

$$\frac{du(t)}{dt} = Bu(t)$$

and let all eigenvalues $\tau_1, \tau_2, \ldots, \tau_k$ of the matrix B lie in the left half-plane: $\operatorname{Re}\tau_j \leqslant -\delta < 0$. *Then $u(t) \equiv 0$.*

PROOF. We use the equality $u(0) = e^{tB}u(-t)$ and the inequalities ($t > 0$)

$$\|u(0)\| \leqslant \|e^{tB}\| \, \|u(-t)\| \leqslant \|e^{tB}\|h \leqslant \gamma(k) \left(\frac{\|B\|}{\delta}\right)^{k-1} e^{-\frac{\delta}{2}t}h.$$

Since the last relation is valid for any $t > 0$, we have

$$\|u(0)\| \leqslant h\gamma(k) \left(\frac{\|B\|}{\delta}\right)^{k-1} \inf_{t>0} e^{-\frac{\delta}{2}t} = 0,$$
$$u(0) = 0, \quad u(t) = e^{tB}u(0) = 0. \quad \square$$

LEMMA 2. *Let $v(t)$, $-\infty < t < \infty$, be a bounded ($\|v(t)\| \leqslant h$) solution to the vector equation (the system of order m)*

$$\frac{dv(t)}{dt} = Cv(t),$$

and let all eigenvalues $\tau_1, \tau_2, \ldots, \tau_m$ of the matrix C lie in the right half-plane $\operatorname{Re}\tau_j \geqslant \delta > 0$. *Then $v(t) \equiv 0$.*

PROOF. It suffices to note that $v(-t)$ satisfies the equation

$$\frac{d}{dt}v(-t) = -Cv(-t),$$

where all eigenvalues of the matrix $-C$ lie in the left half-plane. Lemma 1 implies the equality $v(-t) = 0$, which, in turn, implies $v(t) \equiv 0$. □

THEOREM 2. *Let $y(t)$ be a bounded ($\|y(t)\| \leqslant h$) solution to the vector equation (the system of order N)*

$$\frac{dy(t)}{dt} = Ay(t), \tag{2}$$

which is defined for all t ($-\infty < t < \infty$), with the matrix A has no purely imaginary eigenvalues ($|\operatorname{Re} \tau_j| \geqslant \delta > 0$, $j = 1, 2, \dots, N$). Then $y(t) \equiv 0$.

PROOF. By using a nonsingular transformation, we reduce the matrix A to the block form (1), where the real parts of the eigenvalues of the matrix B are negative and the real parts of the eigenvalues of the matrix C are positive. If $y(t)$ is a solution to (2), then

$$\frac{d}{dt}Ty = \begin{bmatrix} B & 0 \\ 0 & C \end{bmatrix} Ty, \qquad \|Ty\| \leqslant \|T\|\,\|y\| \leqslant \|T\|h.$$

It is convenient to write $Ty = z$ as a composite vector $Ty = z = \begin{pmatrix} u \\ v \end{pmatrix}$ so that

$$\frac{d}{dt}u(t) = Bu(t), \quad \frac{d}{dt}v(t) = Cv(t).$$

It is clear that

$$\|u(t)\| \leqslant \|z(t)\| \leqslant \|T\|h, \quad \|v(t)\| \leqslant \|z(t)\| \leqslant \|T\|h.$$

By Lemmas 1 and 2, $u(t) \equiv 0$ and $v(t) \equiv 0$. Consequently, $z(t) \equiv 0$. It is obvious that

$$y(t) = T^{-1}Ty(t) = T^{-1}z(t) = 0. \quad \square$$

COROLLARY 1. *Any bounded ($\|Y(t)\| \leqslant H$) and defined for all t, $-\infty < t < \infty$, matrix-valued solution $Y(t)$ to the equation*

$$\frac{d}{dt}Y(t) = AY(t)$$

vanishes ($Y(t) = 0$) if the matrix A has no purely imaginary eigenvalues.

THEOREM 3 (uniqueness). *If a matrix A has no purely imaginary eigenvalues, then a Green matrix is unique.*

PROOF. Let $G_0^{(1)}(t)$ and $G_0^{(2)}(t)$ be two Green matrices. Then their difference $\widehat{G}_0(t) = G_0^{(2)}(t) - G_0^{(1)}(t)$ is continuous for all t ($-\infty < t < \infty$) and bounded. For $t < 0$ and $t > 0$ the matrix satisfies the equation

$$\frac{d}{dt}\widehat{G}_0(t) = A\widehat{G}_0(t),$$

which must be satisfied at the point $t = 0$ by continuity (give an accurate proof of this fact). By Corollary 1, we conclude that $\widehat{G}_0(t) \equiv 0$, i.e., $G_0^{(1)}(t) \equiv G_0^{(2)}(t)$. $\square$

§8. The polynomial representation of the Green matrix. Green functions

The special functions $g_i(t)$, their properties, estimates, and differential equations. The representation of the Green matrix in the form of a polynomial in A. An estimate of the Green matrix. The Green function for a single equation of an arbitrary order.

In §3, we obtained the polynomial representation of the matrix exponential e^{tA}. A similar representation can be derived for the Green matrix $G_0(t)$. To do this, we construct special functions $g(t,\tau_1,\tau_2,\dots,\tau_k)$ which are similar to the functions $\psi(t,\tau_1,\tau_2,\dots,\tau_k)$ appearing in the polynomial representation of e^{tA} (cf. (9) in §3). The functions $g(t,\tau_1,\tau_2,\dots,\tau_k)$ can be defined by solutions of differential equations satisfying some additional conditions or by integral representations.

We recall the integral representations (cf. §3)

$$\psi(t,\tau_1) = e^{\tau_1 t},$$

$$(1)\qquad \begin{aligned}\psi(t,\tau_1,\tau_2,\dots,\tau_k) &= \int_0^t \psi(s,\tau_1,\tau_2,\dots,\tau_{k-1})e^{\tau_k(t-s)}ds\\ &= \int_0^t \psi(s,\tau_1,\tau_2,\dots,\tau_{k-1})\psi(t-s,\tau_k)\,ds.\end{aligned}$$

By analogy with $\psi(t,\tau_1,\tau_2,\dots,\tau_k)$, we define the functions $g(t,\tau_1,\tau_2,\dots,\tau_k)$ by integral representations. We assume (this restriction is essential) that there are no purely imaginary numbers among $\tau_1,\tau_2,\dots,\tau_k$ ($\operatorname{Re}\tau_j \neq 0$). We set

$$g(t,\tau_1) = \begin{cases} \text{if } \operatorname{Re}\tau_1 > 0: & \begin{cases} 0 & \text{for } t>0,\\ -e^{\tau_1 t} & \text{for } t<0;\end{cases}\\ \text{if } \operatorname{Re}\tau_1 < 0: & \begin{cases} e^{\tau_1 t} & \text{for } t>0,\\ 0 & \text{for } t<0;\end{cases}\end{cases}$$

$$g(t,\tau_1,\tau_2,\dots,\tau_k) = \begin{cases} \int_{-\infty}^{t} g(s,\tau_1,\tau_2,\dots,\tau_{k-1})e^{\tau_k(t-s)}ds & \text{if } \operatorname{Re}\tau_k<0,\\ -\int_{t}^{\infty} g(s,\tau_1,\tau_2,\dots,\tau_{k-1})e^{\tau_k(t-s)}ds & \text{if } \operatorname{Re}\tau_k>0,\end{cases}$$

or, which is equivalent,

$$(2)\qquad g(t,\tau_1,\tau_2,\dots,\tau_k) = \int_{-\infty}^{\infty} g(s,\tau_1,\tau_2,\dots,\tau_{k-1})g(t-s,\tau_k)\,ds.$$

We note that if all parameters τ_j lie in the left half-plane ($\operatorname{Re}\tau_j < 0$ for $j = 1,2,\dots,k$), then

$$g(t,\tau_1,\tau_2,\dots,\tau_k) = \begin{cases} \psi(t,\tau_1,\tau_2,\dots,\tau_k), & t>0,\\ 0, & t<0.\end{cases}$$

Introduce the notation $g_k(t) = g(t,\tau_1,\tau_2,\dots,\tau_k)$. Let $|\operatorname{Re}\tau_j| \geqslant \delta$, $j - 1,2,\dots$. It is obvious that $g_1(t)$ is a piecewise continuous function with a single discontinuity point of the first kind $t = 0$ at which the jump is equal to 1:

$$[g_1(0)] \equiv g_1(+0) - g_1(-0) = 1.$$

LEMMA 1. *The functions* $g_2(t), g_3(t), \dots, g_k(t), \dots$ *are continuous and the following estimates hold:*

$$|g_j(t)| < \frac{\Omega(j)}{\delta^{j-1}} e^{-\frac{\delta}{2}|t|}, \quad j = 1, 2, \dots, \tag{3}$$

where $\Omega(j)$ *is a constant depending only on* j.

PROOF. We proceed by induction on j. It is obvious that $g_1(t)$ satisfies the estimate

$$|g_1(t)| \leqslant e^{-\delta|t|} < e^{-\frac{\delta}{2}|t|}.$$

Note that for $\Omega(j)$ we can take, e.g., $\Omega(j) = 3^{j-1}$. If it is true for g_{j-1} , $j \geqslant 2$, then $g_j(t)$ uniformly converge on any segment $|t| \leqslant T$ for any $\tau_1, \tau_2, \dots, \tau_j$ such that $|\tau_j| \leqslant M$, $|\operatorname{Re}\tau_j| \geqslant \delta$. Consequently, $g_j(t) = g_j(t, \tau_1, \tau_2, \dots, \tau_j)$ continuously depend on $t, \tau_1, \tau_2, \dots, \tau_j$.

Assume that for some $j \geqslant 2$ we have established the estimate (3), i.e.,

$$|g_{j-1}(t)| < \frac{\Omega(j-1)}{\delta^{j-2}} e^{-\frac{\delta}{2}|t|}$$

(for $j = 1$ this fact is known: $\Omega(1) = 1$). To prove the estimate (3) for $g_j(t)$ we begin with the case $\operatorname{Re}\tau_j \geqslant \delta > 0$. The case $\operatorname{Re}\tau_j \leqslant -\delta < 0$ can be handled in the same way.

For $\operatorname{Re}\tau_j \geqslant \delta$ we have

$$|g_j(t)| = \left| \int_t^{\infty} g_{j-1}(s) e^{\tau_j(t-s)} ds \right| \leqslant \int_{-\infty}^{0} |g_{j-1}(t-\xi)| e^{\delta\xi} d\xi.$$

Let $t \geqslant 0$. Then $t - \xi \geqslant 0$ for $0 \geqslant \xi > -\infty$. Therefore,

$$\begin{aligned} |g_j(t)| &\leqslant \int_{-\infty}^{0} |g_{j-1}(t-\xi)| e^{\delta\xi} d\xi \leqslant \frac{\Omega(j-1)}{\delta^{j-2}} \int_{-\infty}^{0} e^{-\frac{\delta}{2}(t-\xi)} e^{\delta\xi} d\xi \\ &= \frac{2\Omega(j-1)}{3\delta^{j-1}} e^{-\frac{\delta}{2}t} = \frac{2\Omega(j-1)}{3\delta^{j-1}} e^{-\frac{\delta}{2}|t|}. \end{aligned}$$

Let $t < 0$. Then

$$\begin{aligned} |g_j(t)| &\leqslant \int_{-\infty}^{0} |g_{j-1}(t-\xi)| e^{\delta\xi} d\xi \leqslant \frac{\Omega(j-1)}{\delta^{j-2}} \int_{-\infty}^{0} e^{-\frac{\delta}{2}|t-\xi|} e^{\delta\xi} d\xi \\ &= \frac{\Omega(j-1)}{\delta^{j-2}} \int_{-\infty}^{t} e^{\frac{\delta}{2}(\xi-t)} e^{\delta\xi} d\xi + \frac{\Omega(j-1)}{\delta^{j-2}} \int_{t}^{0} e^{\frac{\delta}{2}(t-\xi)} e^{\delta\xi} d\xi \\ &= \frac{2\Omega(j-1)}{3\delta^{j-1}} e^{\delta t} + \frac{2\Omega(j-1)}{\delta^{j-1}} e^{\frac{\delta}{2}t} \left(1 - e^{\frac{\delta}{2}t}\right) \\ &\leqslant \frac{3\Omega(j-1)}{\delta^{j-1}} e^{\frac{\delta}{2}t} = \frac{3\Omega(j-1)}{\delta^{j-1}} e^{-\frac{\delta}{2}|t|}. \end{aligned}$$

Thus, for any t the estimate (3) holds if $\operatorname{Re}\tau_j \geqslant \delta > 0$. For $\Omega(j)$ we can take

$$\Omega(j) = 3\Omega(j-1) = \max\Big\{\frac{2}{3}\Omega(j-1),\, 3\Omega(j-1)\Big\}.$$

For $\operatorname{Re}\tau_j \leqslant -\delta$ we derive

$$|g_j(t)| = \left|\int_{-\infty}^{t} g_{j-1}(s)e^{\tau_j(t-s)}\,ds\right| \leqslant \int_0^{\infty} |g_{j-1}(t-\xi)|e^{-\delta\xi}\,d\xi.$$

Let $t \geqslant 0$. Then

$$\begin{aligned} |g_j(t)| &\leqslant \int_0^{\infty} |g_{j-1}(t-\xi)|e^{-\delta\xi}\,d\xi \leqslant \frac{\Omega(j-1)}{\delta^{j-2}}\int_0^{\infty} e^{-\frac{\delta}{2}(t-\xi)}e^{-\delta\xi}\,d\xi \\ &= \frac{2\Omega(j-1)}{3\delta^{j-1}}e^{\frac{\delta}{2}t} = \frac{2\Omega(j-1)}{3\delta^{j-1}}e^{-\frac{\delta}{2}|t|}. \end{aligned}$$

Let $t < 0$. Then

$$\begin{aligned} |g_j(t)| &\leqslant \int_0^{\infty} |g_{j-1}(t-\xi)|e^{-\delta\xi}\,d\xi \leqslant \frac{\Omega(j-1)}{\delta^{j-2}}\int_0^{\infty} e^{-\frac{\delta}{2}|\xi-t|}e^{-\delta\xi}\,d\xi \qquad \square \\ &= \frac{\Omega(j-1)}{\delta^{j-2}}\int_t^{\infty} e^{\frac{\delta}{2}(t-\xi)}e^{-\delta\xi}\,d\xi + \frac{\Omega(j-1)}{\delta^{j-2}}\int_0^{t} e^{\frac{\delta}{2}(\xi-t)}e^{-\delta\xi}\,d\xi \\ &= \frac{2\Omega(j-1)}{\delta^{j-1}}e^{\frac{\delta}{2}t}\left(1-e^{-\frac{\delta}{2}t}\right) + \frac{2\Omega(j-1)}{3\delta^{j-1}}e^{-\delta t} \\ &\leqslant \frac{\Omega(j)}{\delta^{j-1}}e^{-\frac{\delta}{2}t} = \frac{\Omega(j)}{\delta^{j-1}}e^{-\frac{\delta}{2}|t|}. \end{aligned}$$

Differentiating the integrals occurring in the definition of $g_1(t), g_2(t), g_3(t), \dots$, it is easy to verify that for $t \neq 0$ these functions satisfy the following differential equations:

$$\frac{dg_1}{dt} = \tau_1 g_1, \qquad \frac{dg_j}{dt} = \tau_j g_j + g_{j-1} \quad (j \geqslant 2).$$

Note that these equations differ from the equations for the functions $\psi_1(t), \psi_2(t), \dots$ only by the notation of the unknown functions. It is known that if $|\operatorname{Re}\tau_j| \geqslant \delta > 0$, then for $t \neq 0$ these equations have a solution such that

$$|g_j(t)| \leqslant \frac{\Omega(j)}{\delta^{j-1}}e^{-\frac{\delta}{2}|t|}$$

and

$$\begin{aligned} [g_1(0)] &= g_1(+0) - g_1(-0) = 1, \\ [g_2(0)] &= g_2(+0) - g_2(-0) = 0, \\ &\dots\dots\dots\dots \\ [g_k(0)] &= g_k(+0) - g_k(-0) = 0, \\ &\dots\dots\dots\dots \end{aligned}$$

The equations for $g_j(t)$, which are valid for $t \neq 0$, can be rewritten as a matrix equation for the vector with the components $g_1(t), g_2(t), \ldots$:

$$\frac{d}{dt}\begin{pmatrix} g_1(t) \\ g_2(t) \\ \\ \vdots \\ \\ g_N(t) \end{pmatrix} = \begin{bmatrix} \tau_1 & & & & 0 \\ 1 & \tau_2 & & & \\ & 1 & \tau_3 & & \\ & & \ddots & \ddots & \\ 0 & & & 1 & \tau_N \end{bmatrix}\begin{pmatrix} g_1(t) \\ g_2(t) \\ \\ \vdots \\ \\ g_N(t) \end{pmatrix} \equiv Ag(t).$$

In the last equation, the matrix of coefficients has the eigenvalues $\tau_1, \tau_2, \ldots, \tau_N$ which are assumed to satisfy the conditions $|\operatorname{Re}\tau_j| \geqslant \delta > 0$.

We suppose that the equations for $g_i(t)$ and the boundary conditions

$$g_i(+0) - g_i(-0) = 0, \quad i \geqslant 2,$$
$$g_1(+0) - g_1(-0) = 1$$

admit two bounded for $-\infty < t < \infty$ solutions $g^{[1]}(t)$ and $g^{[2]}(t)$. The difference $\widetilde{g}(t) = g^{[1]}(t) - g^{[2]}(t)$ is continuous at $t = 0$. At the point $t = 0$, the left and right derivatives

$$\frac{d}{dt}[g^{[1]} - g^{[2]}]\bigg|_{+0} = A[g^{[1]}(+0) - g^{[2]}(+0)],$$
$$\frac{d}{dt}[g^{[1]} - g^{[2]}]\bigg|_{-0} = A[g^{[1]}(-0) - g^{[2]}(-0)]$$

coincide. Therefore, the difference $\widetilde{g}(t) = g^{[1]}(t) - g^{[2]}(t)$ satisfies the equation

$$\frac{d}{dt}\widetilde{g} = A\widetilde{g}(t)$$

on the entire line. The difference is bounded and, by Theorem 2 in §7, vanishes, i.e., $g^{[1]}(t) = g^{[2]}(t)$. We have proved the following assertion.

PROPOSITION 1. *The problem of finding bounded functions $g_j(t)$, $j = 1, 2, \ldots, N$, such that*

$$\begin{aligned} &\frac{dg_1}{dt} = \tau_1 g_1, \quad t \neq 0, \\ &\frac{dg_i}{dt} = \tau_i g_i + g_{i-1} \quad (i \geqslant 2), \quad t \neq 0, \end{aligned} \tag{4}$$

$$g_1(+0) - g_1(-0) = 1, \quad g_i(+0) - g_i(-0) = 0, \quad i \geqslant 2 \tag{5}$$

has a unique solution provided that there are no purely imaginary numbers among τ_j. If $|\operatorname{Re}\tau_j| \geqslant \delta > 0$, then the solution satisfies the estimate (3).

By analogy with the formula (cf. (9) in §3)

$$e^{tA} = \psi_1(t)I + \psi_2(t)[A - \tau_1 I] + \cdots + \psi_N(t)[A - \tau_1 I][A - \tau_2 I]\cdots[A - \tau_{N-1} I] \tag{6}$$

we construct the matrix polynomial

$$G_0(t) = g_1(t)I + g_2(t)[A - \tau_1 I] + \cdots + g_N(t)[A - \tau_1 I][A - \tau_2 I]\cdots[A - \tau_{N-1} I] \tag{7}$$

provided that the eigenvalues $\tau_1, \tau_2, \ldots, \tau_N$ of the matrix A do not lie on the imaginary axis ($|\operatorname{Re}\tau_j| \geqslant \delta$). It is easy to see that for $t \neq 0$ the matrix $G_0(t)$ is a continuous matrix-valued function of t. Moreover, it is infinitely differentiable in t and continuously depends on the elements and eigenvalues of the matrix A. In the theory of functions of complex variables it is proved, using the Rouché theorem, that the set of the roots depends continuously on the coefficients of the polynomial. Consequently, the set of the eigenvalues depends continuously on the elements of the matrix. Hence $G_0(t) \equiv G_0(t, A)$ is a continuous function of A.

At $t = 0$, the dependence of $G_0(t, A)$ on t is discontinuous:

$$G_0(+0, A) - G_0(-0, A) = g_1(+0)I - g_1(-0)I = I,$$

which follows from (5) and the properties of $g_j(t)$.

In §3, the proof of the fact that the polynomial

$$\mathcal{Q}(t, A) = \psi_1(t)I + \psi_2(t)[A - \tau_1 I] + \cdots + \psi_N(t)[A - \tau_1 I][A - \tau_2 I] \cdots [A - \tau_{N-1} I] \tag{8}$$

satisfies the differential equation

$$\frac{d}{dt}\mathcal{Q}(t, A) = A\mathcal{Q}(t, A)$$

was based only on the differential equations for $\psi_j(t)$. But for $t \neq 0$ the same equations are satisfied by the functions $g_j(t)$ occurring in the expression (7) for $G_0(t) = G_0(t, A)$ in the same way as the functions $\psi_j(t)$ occurring in the expression (8) for $\mathcal{Q}(t, A)$. Therefore,

$$\frac{d}{dt}G_0(t) = AG_0(t), \quad t \neq 0.$$

The matrix $G_0(t)$ is bounded if $|\operatorname{Re} \tau_j| \geqslant \delta > 0$ and decreases as $|t| \to \infty$. This fact follows from inequalities which are similar to those for the matrix exponential:

$$\begin{aligned}
\|G_0(t)\| &\leqslant |g_1(t)| + |g_2(t)|\,\|A - \tau_1 I\| + |g_3(t)|\,\|A - \tau_1 I\|\,\|A - \tau_2 I\| + \ldots \\
&\quad + |g_N(t)|\,\|A - \tau_1 I\|\,\|A - \tau_2 I\| \cdots \|A - \tau_{N-1} I\| \\
&\leqslant |g_1(t)| + |g_2(t)| \cdot 2\|A\| + |g_3(t)|(2\|A\|)^2 + \cdots + |g_N(t)|(2\|A\|)^{N-1} \\
&\leqslant e^{-\frac{\delta}{2}|t|} + \frac{\Omega(2)}{\delta} e^{-\frac{\delta}{2}|t|}(2\|A\|) + \frac{\Omega(3)}{\delta^2} e^{-\frac{\delta}{2}|t|}(2\|A\|)^2 \\
&\quad + \cdots + \frac{\Omega(N)}{\delta^{N-1}} e^{-\frac{\delta}{2}|t|}(2\|A\|)^{N-1}.
\end{aligned}$$

Since $\delta \leqslant |\operatorname{Re} \tau_j| \leqslant |\tau_j| \leqslant \|A\|$,

$$1 \leqslant \frac{\|A\|}{\delta} \leqslant \left(\frac{\|A\|}{\delta}\right)^2 \leqslant \cdots \leqslant \left(\frac{\|A\|}{\delta}\right)^{N-1},$$

we can obtain a simpler but less precise inequality for $G_0(t)$ as follows:

$$\|G_0(t)\| \leqslant \gamma(N) \left(\frac{\|A\|}{\delta}\right)^{N-1} e^{-\delta|t|/2}, \tag{9}$$

where $\gamma(N)$ is a constant depending only on the order N of the matrix A. A similar estimate (cf. (11) in §3) was obtained for the matrix exponential e^{tA}. Certainly, the constant $\gamma(N)$ was defined differently in these estimates, we use for it the same notation. Choosing the maximal value among two values of $\gamma(N)$ (in (9) and (11), §3), we can use it in both estimates. Thus, no significance should be attached to the difference of these constants; moreover, the values of factors play no role in what follows. Since for $t \neq 0$ the components $g_{kj}(t)$ of the kth column of the matrix $G_0(t)$ satisfy the homogeneous system

$$\frac{d}{dt}\begin{pmatrix} g_{k1}(t) \\ g_{k2}(t) \\ \vdots \\ g_{kN}(t) \end{pmatrix} = A \begin{pmatrix} g_{k1}(t) \\ g_{k2}(t) \\ \vdots \\ g_{kN}(t) \end{pmatrix}$$

and converge to zero as $t \to \pm\infty$, from Theorem 1 and Corollary 1 in §5 it follows that the components $g_{kj}(t)$ and their derivatives exponentially decrease as $t \to \pm\infty$:

$$\left| \frac{d^m}{dt^m} g_{kj}(t) \right| < B_{mkj} e^{-\sigma_0 |t|},$$

where B_{mkj} and σ_0 are some positive constants. Therefore,

$$\left\| \frac{d^m}{dt^m} G_0(t) \right\| \leqslant B_m e^{-\sigma_0 |t|}.$$

Thus, to solve the nonhomogeneous system

$$\frac{d}{dt} y(t) = Ay(t) + f(t), \tag{10}$$

we use the Green matrix, whereas to solve the single nonhomogeneous differential equation of the Nth order

$$\mathcal{P}_N\left(\frac{d}{dt}\right) x(t) = q(t), \tag{11}$$

where

$$\mathcal{P}_N\left(\frac{d}{dt}\right) = \frac{d^N}{dt^N} + p_1 \frac{d^{N-1}}{dt^{N-1}} + \cdots + p_{N-1}\frac{d}{dt} + p_N,$$

we will use the Green function $g(t)$ which can be constructed if the characteristic equation $\mathcal{P}_N(\tau) = 0$ has no purely imaginary solutions.

DEFINITION 1. A function $g(t)$ is called a Green function if it satisfies the following conditions:

(a) $|g(t)| \to 0, |g'(t)| \to 0, \dots, |g^{(N-1)}(t)| \to 0$ as $|t| \to \infty$;

(b) $g(t)$ is a solution to the homogeneous equation

$$\mathcal{P}_N\left(\frac{d}{dt}\right) g(t) = 0, \quad t \neq 0;$$

(c) $g(t), g'(t), \dots, g^{(N-2)}(t)$ are continuous at $t = 0$, whereas $g^{(N-1)}(t)$ has the discontinuity of the first kind at $t = 0$ and

$$g^{(N-1)}(+0) - g^{(N-1)}(-0) = 1.$$

We show that a Green function exists and is unique under the additional assumption on the roots τ_j of the characteristic polynomial $\mathcal{P}_N(\tau)$.

THEOREM 1. *If* $\operatorname{Re}\tau_j \neq 0$, $j = 1, 2, \dots, N$, *then a Green function* $g(t)$ *exists and is unique.*

PROOF. At the beginning of the section, under the assumption that $\operatorname{Re}\tau_j \neq 0$ for all $j = 1, 2, \dots, N$, we constructed the functions $g_1(t), g_2(t), \dots, g_N(t)$ that are uniquely defined by (3)–(5). Introducing the vector $z(t)$ with the components $g_1(t), g_2(t), \dots, g_N(t)$, we rewrite (3)–(5) as follows:

$$z(+0) - z(-0) = \begin{pmatrix} g_1(+0) \\ g_2(+0) \\ \vdots \\ g_N(+0) \end{pmatrix} - \begin{pmatrix} g_1(-0) \\ g_2(-0) \\ \vdots \\ g_N(-0) \end{pmatrix} = \begin{pmatrix} 1 \\ 0 \\ \vdots \\ 0 \end{pmatrix}, \tag{12}$$

$$\|z(t)\| \to \infty \text{ as } t \to \pm\infty, \tag{13}$$

$$\frac{d}{dt} z(t) = \begin{bmatrix} \tau_1 & & & & & \\ 1 & \tau_2 & & & \mathbf{0} & \\ & 1 & \tau_3 & & & \\ & & \ddots & \ddots & & \\ & \mathbf{0} & & 1 & \tau_{N-1} & \\ & & & & 1 & \tau_N \end{bmatrix} z(t). \tag{14}$$

For $z(t)$ we can take the first column of the Green matrix $G_0(t, A)$ constructed for the system $y' = Ay$, where

$$A = \begin{bmatrix} \tau_1 & & & & & \\ 1 & \tau_2 & & & \mathbf{0} & \\ & 1 & \tau_3 & & & \\ & & \ddots & \ddots & & \\ & \mathbf{0} & & 1 & \tau_{N-1} & \\ & & & & 1 & \tau_N \end{bmatrix}.$$

As was shown, the components of the Green matrix $G_0(t, A)$ and all their derivatives exponentially decrease as $t \to \pm\infty$. Therefore, the functions $g_j(t)$, which form the first column of $G_0(t, A)$, satisfy the estimates

$$\left| \frac{d^m}{dt^m} g_j(t) \right| < B_{mj} e^{-\sigma_0 |t|}$$

with some $B_{mj} > 0$ and $\sigma_0 > 0$. Using the equalities

$$\left(\frac{d}{dt} - \tau_1\right) g_1 = 0,$$

$$\left(\frac{d}{dt} - \tau_2\right) g_2 = g_1, \quad \left(\frac{d}{dt} - \tau_1\right)\left(\frac{d}{dt} - \tau_2\right) g_2(t) = 0,$$

$$\dots\dots\dots\dots\dots\dots\dots\dots\dots\dots$$

$$\left(\frac{d}{dt} - \tau_N\right) g_N = g_{N-1}, \quad \left(\frac{d}{dt} - \tau_1\right)\left(\frac{d}{dt} - \tau_2\right) \cdots \left(\frac{d}{dt} - \tau_N\right) g_N(t) = 0$$

for $t \neq 0$, we establish that for $t \neq 0$ the function $g_N(t)$ is a solution to the equation

$$\mathcal{P}_N\left(\frac{d}{dt}\right) g_N(t) \equiv \left(\frac{d}{dt} - \tau_1\right)\left(\frac{d}{dt} - \tau_2\right) \cdots \left(\frac{d}{dt} - \tau_N\right) g_N(t) = 0.$$

The continuity of the functions $g_N(t), g_{N-1}(t), \dots, g_2(t)$ at $t=0$ and the equalities

$$\begin{aligned}
&\left(\frac{d}{dt}-\tau_N\right) g_N = g_{N-1}, \quad \frac{d}{dt} g_N = \tau_N g_{N-1},\\
&\left(\frac{d}{dt}-\tau_{N-1}\right)\left(\frac{d}{dt}-\tau_N\right) g_N = g_{N-2},\\
&\frac{d^2}{dt^2} g_N = (\tau_N+\tau_{N-1})\frac{d}{dt} g_N - \tau_N\tau_{N-1} g_N + g_{N-2},\\
&\dots\dots\dots\dots\dots\dots\dots\\
&\left(\frac{d}{dt}-\tau_3\right)\left(\frac{d}{dt}-\tau_4\right)\cdots\left(\frac{d}{dt}-\tau_N\right) g_N = g_2,\\
&\frac{d^{N-2}}{dt^{N-2}} g_N = (\tau_3+\tau_4+\dots+\tau_N)\frac{d^{N-3}}{dt^{N-3}} g_N + \dots\\
&\qquad\qquad\qquad + (-1)^{N-3}\tau_3\tau_4\cdots\tau_N g_N + g_2
\end{aligned}$$

imply the continuity of the derivatives

$$\frac{d}{dt} g_N, \frac{d^2}{dt^2} g_N, \dots, \frac{d^{N-2}}{dt^{N-2}} g_N.$$

The equality

$$\left(\frac{d}{dt}-\tau_2\right)\left(\frac{d}{dt}-\tau_3\right)\cdots\left(\frac{d}{dt}-\tau_N\right) g_N = g_1$$

rewritten in the form

$$\frac{d^{N-1}}{dt^{N-1}} g_N(t) = (\tau_2+\tau_3+\dots+\tau_N)\frac{d^{N-2}}{dt^{N-2}} g_N(t) + \dots + (-1)^N \tau_2\tau_3\cdots\tau_N g_N(t) + g_1(t),$$

together with the condition $g_1(+0)-g_1(-0)=1$, implies that

$$\left[\frac{d^{N-1}}{dt^{N-1}} g_N(t)\right]_{t=0} = g_N^{(N-1)}(+0) - g_N^{(N-1)}(-0) = 1.$$

It is clear that by setting $g(t)=g_N(t)$, all Conditions (a)–(c) from Definition 1 of a Green function are satisfied. The existence of a Green function is proved.

To prove the uniqueness of a Green function, we assume the contrary, i.e., that we can find two different Green functions $g^{[1]}(t)$ and $g^{[2]}(t)$. Construct the functions $g_j^{[p]}(t)$, $p=1,2$; $j=1,2,\dots,N$, by the following formulas ($t\neq 0$):

$$\begin{aligned}
&g_N^{[1]} = g^{[1]}, \quad g_N^{[2]} = g^{[2]},\\
&g_{N-1}^{[1]} = \left(\frac{d}{dt}-\tau_N\right) g_N^{[1]}, \quad g_{N-1}^{[2]} = \left(\frac{d}{dt}-\tau_N\right) g_N^{[2]},\\
&\dots\dots\dots\dots\dots\dots\dots\\
&g_{k-1}^{[1]} = \left(\frac{d}{dt}-\tau_k\right) g_k^{[1]}, \quad g_{k-1}^{[2]} = \left(\frac{d}{dt}-\tau_k\right) g_k^{[2]},\\
&\dots\dots\dots\dots\dots\dots\dots\\
&g_1^{[1]} = \left(\frac{d}{dt}-\tau_2\right) g_2^{[1]}, \quad g_1^{[2]} = \left(\frac{d}{dt}-\tau_2\right) g_2^{[2]}.
\end{aligned}$$

It is obvious that $g_j^{[1]}$ and $g_j^{[2]}$ are linear combinations of

$$g^{[p]}(t), \frac{d}{dt}g^{[p]}(t), \ldots, \frac{d^{N-j}}{dt^{N-j}}g^{[p]}(t)$$

of the following form:

$$g_j^{[p]}(t) = \frac{d^{N-j}}{dt^{N-j}}g^{[p]}(t) + \left\{ \begin{array}{l} \text{a linear combination of the derivatives of} \\ g^{[p]}(t) \text{ of order at most } N-j-1. \end{array} \right.$$

Therefore, $g_N^{[p]}(t), g_{N-1}^{[p]}(t), \ldots, g_2^{[p]}(t)$ are continuous at $t = 0$ and $g_1^{[p]}(t)$ has discontinuity at $t = 0$:

$$g_1^{[p]}(+0) - g_1^{[p]}(-0) = \left[\frac{d^{N-1}}{dt^{N-1}}g^{[p]}(t)\right]_{t=+0} - \left[\frac{d^{N-1}}{dt^{N-1}}g^{[p]}(t)\right]_{t=-0} = 1.$$

The functions $g_j^{[p]}(t)$, $p = 1, 2$; $j = 1, 2, \ldots, N$, converge to zero as $t \to \pm\infty$ because the first $N-1$ derivatives of $g^{[p]}(t)$, from which $g_j^{[p]}$ can be expressed as linear combinations, possess the same property. Furthermore, it is easy to see that for $t \neq 0$

$$\begin{aligned} g_1^{[p]}(t) &= \left(\frac{d}{dt} - \tau_2\right)\left(\frac{d}{dt} - \tau_3\right)\cdots\left(\frac{d}{dt} - \tau_N\right)g^{[p]}(t), \\ \left(\frac{d}{dt} - \tau_1\right)g_1^{[p]}(t) &= \left(\frac{d}{dt} - \tau_1\right)\left(\frac{d}{dt} - \tau_2\right)\cdots\left(\frac{d}{dt} - \tau_N\right)g^{[p]}(t) \\ &\equiv \mathcal{P}_N\left(\frac{d}{dt}\right)g^{[p]}(t) = 0. \end{aligned}$$

Thus, if $g^{[1]}(t)$ and $g^{[2]}(t)$ are two functions subject to Conditions (a)–(c) from Definition 1, then, starting with $g^{[1]}(t)$ and $g^{[2]}(t)$, it is possible to construct two vector-valued functions

$$z^{[1]}(t) = \begin{pmatrix} g_1^{[1]}(t) \\ g_2^{[1]}(t) \\ \vdots \\ g_N^{[1]}(t) \end{pmatrix}, \quad z^{[2]}(t) = \begin{pmatrix} g_1^{[2]}(t) \\ g_2^{[2]}(t) \\ \vdots \\ g_N^{[2]}(t) \end{pmatrix}$$

such that their components $g_j^{[p]}(t)$ satisfy the following conditions:

$$\begin{gathered} \frac{d}{dt}g_1^{[p]} = \tau_1 g_1^{[p]}, \quad t \neq 0, \\ \frac{d}{dt}g_j^{[p]} = \tau_j g_{j-1}^{[p]}, \quad j \geqslant 2, \quad t \neq 0, \\ g_1^{[p]}(+0) - g_1^{[p]}(-0) = 1, \quad g_j^{[p]}(+0) - g_j^{[p]}(-0) = 0, \quad j \geqslant 2, \\ |g_j^{[p]}(t)| \to 0 \text{ as } |t| \to \infty, \quad j = 1, 2, \ldots, N. \end{gathered}$$

We know that the functions $g_1(t), g_2(t), \ldots, g_N(t)$, which were used in the construction of the Green matrix, are uniquely determined by these conditions. They are unique because there are no purely imaginary numbers among $\tau_1, \tau_2, \ldots, \tau_N$.

Hence $g_j^{[1]}(t) = g_j^{[2]}(t) = g_j(t)$, $j = 1, 2, \ldots, N$. Consequently, the Green functions coincide, because $g^{[1]}(t) = g_N^{[1]}(t)$ and $g^{[2]}(t) = g_N^{[2]}(t)$ by construction. $\square$

REMARK 1. It is seen from the proof of Theorem 1 that Condition (a) in Definition 1 can be weakened. It suffices to require that $|g(t)| \to 0$ as $t \to \pm\infty$. Then from the differential equation $\mathcal{P}_N\left(\frac{d}{dt}\right)g(t) = 0$ for $t \neq 0$ we obtain $|g'(t)| \to 0, |g''(t)| \to 0, \ldots, |g^{(N-1)}(t)| \to 0$ as $t \to \pm\infty$, which is included in Condition (a).

§9. Nonhomogeneous linear equations

The vector Cauchy problem, the representation of a solution and estimates. Bounded solutions on the whole line $-\infty < t < \infty$ and the representation by means of Green matrices. The case of a single equation of an arbitrary order. The existence of an exponentially decreasing solution to the equation $x' = Ax + \varphi(t)$, where $\varphi(t)$ exponentially decreases. The notion of a generalized solution.

Up to now we studied equations of the form

$$\frac{d}{dt}y(t) - Ay(t) = 0$$

or

$$x^{(n)} + p_1 x^{(n-1)} + \cdots + p_{n-1}x' + p_n x = 0$$

which are called homogeneous equations. Nonhomogeneous equations (equations with the right-hand side)

$$\frac{d}{dt}y(t) - Ay(t) = f(t),$$

$$x^{(n)} + p_1 x^{(n-1)} + \cdots + p_{n-1}x' + p_n x = q(t)$$

play an important role in many applications. Beginning with this section, the nonhomogeneous equations will receive much attention in our consideration.

We consider the nonhomogeneous equation

$$\frac{d}{dt}y(t) = Ay(t) + f(t) \tag{1}$$

and the Cauchy problem for (1) with the initial condition

$$y(t_0) = b. \tag{2}$$

The right-hand side $f(t)$ in (1) is assumed to be piecewise continuous with a finite number of discontinuities. To construct solutions to the system of nonhomogeneous equations, we can use the matrix exponential and the Green matrix.

The use of matrix exponentials.

PROPOSITION 1. *A solution $y(t)$ to the Cauchy problem* (1), (2) *exists, is unique, can be represented by the Cauchy formula*

$$y(t) = e^{(t-t_0)A}b + \int_{t_0}^{t} e^{(t-s)A} f(s)\, ds, \tag{3}$$

and satisfies the estimate

$$\|y(t)\| \leqslant e^{|t-t_0|\,\|A\|}\|y(t_0)\| + \frac{e^{|t-t_0|\,\|A\|}-1}{\|A\|}\max_{s\in[t_0,t]}\|f(s)\|. \tag{4}$$

On the right-hand side of (3) the first term $e^{(t-t_0)A}b$ is a general form of a solution to the homogeneous equation $y'(t) = Ay(t)$ and the second term is some partial solution to the nonhomogeneous equation (1). It is easy to see that all solutions to the nonhomogeneous equation (1) can be obtained by the formula (3). Indeed, if $y(t)$ is a solution to (3), then, by setting $y(t_0) = b$, we obtain the Cauchy problem which $y(t)$ solves.

The use of Green matrices.

PROPOSITION 2. *If the matrix A has no purely imaginary eigenvalues (so that $|\operatorname{Re}\tau_j(A)| \geqslant \delta$) and $f(t)$ is bounded for $-\infty < t < \infty$ ($\|f(t)\| \leqslant F$), then there exists a unique bounded ($-\infty < t < \infty$) solution $y(t)$ to the equation (1) such that*

$$\|y(t)\| \leqslant \frac{4F}{\delta}\gamma(N)\left(\frac{\|A\|}{\delta}\right)^{N-1}. \tag{5}$$

The solution $y(t)$ can be determined by the formula

$$y(t) = \int_{-\infty}^{\infty} G_0(t-s)f(s)\,ds, \tag{6}$$

where $G_0(t)$ is the Green matrix (see the construction of $G_0(t)$ in §7, 8).

At the end of this section, we will show how to solve nonhomogeneous equations of higher order.

PROOF OF PROPOSITION 1. We assume that $f(t)$ is continuous for simplicity. We consider the function $y(t)$ defined for $t_0 \leqslant t \leqslant t_1$ by the formula (3) and compute its derivative with respect to t:

$$\begin{aligned}\frac{d}{dt}y(t) &= \frac{d}{dt}[e^{(t-t_0)A}\,b] + \frac{d}{dt}\left[\int_{t_0}^{t} e^{(t-s)A}f(s)\,ds\right]\\ &= Ae^{(t-t_0)A}\,b + \left[f(t) + \int_{t_0}^{t} Ae^{(t-s)A}f(s)\,ds\right]\\ &= A\left[e^{(t-t_0)A}\,b + \int_{t_0}^{t} e^{(t-s)A}f(s)\,ds\right] + f(t) = Ay(t) + f(t).\end{aligned}$$

We see that $y(t)$ is a solution to the nonhomogeneous equation (1). It is obvious that $y(t)$ satisfies the initial condition (2). Indeed,

$$y(t_0) = e^{(t_0-t_0)A}\,b + \int_{t_0}^{t_0} e^{(t_0-s)A}f(s)\,ds = Ib + 0 = b.$$

It is easy to show that the differentiation is valid here and, thereby, the function $y(t)$ is continuous and continuously differentiable on the segment $t_0 \leqslant t \leqslant t_1$, i.e., $y(t)$ is a solution to the Cauchy problem (1), (2).

To prove the uniqueness of a solution to the problem (1), (2), we assume the contrary. Let $y^{[1]}(t)$ and $y^{[2]}(t)$ be continuous and continuously differentiable vector-valued functions such that

$$\frac{d}{dt}y^{[1]} = Ay^{[1]} + f(t), \quad y^{[1]}(t_0) = b,$$
$$\frac{d}{dt}y^{[2]} = Ay^{[2]} + f(t), \quad y^{[2]}(t_0) = b.$$

Denoting $z(t) = y^{[1]}(t) - y^{[2]}(t)$, we easily verify that the difference $z(t)$ satisfies the homogeneous equation and the zero initial condition

$$\frac{d}{dt}z(t) = Az(t), \quad z(t_0) = 0.$$

Consequently, $y^{[1]}(t) - y^{[2]}(t) = z(t) = 0$.

We assumed that the initial condition (2) is imposed at the left endpoint $t = t_0$ of the segment $[t_0, t_1]$ on which the right-hand side $f(t)$ of the equation (1) is defined. However, this assumption is not essential.

Let $f(t)$ be defined for $t_1 \leqslant t \leqslant t_2$. Let $t_0 \in [t_1, t_2]$. The Cauchy formula (3) defines the continuous function $y(t)$ on the segment $[t_1, t_2]$. It is easy to verify that the above proof of the fact that $y(t)$ satisfies the equation (1), which is based on the direct differentiation, does not use the fact that t_0 is the left endpoint of the segment of definition of $f(t)$. This assumption is also not essential in the proof of the uniqueness of a solution.

Starting from the formula

$$y(t) = e^{(t-t_0)A}\, y(t_0) + \int_{t_0}^{t} e^{(t-s)A} f(s)\, ds, \tag{7}$$

we derive the estimate (4) for the norm of a solution to the Cauchy problem. It is obvious that

$$\|y(t)\| \leqslant \|e^{(t-t_0)A}\|\, \|y(t_0)\| + \left| \int_{t_0}^{t} \|e^{(t-s)A}\|\, \|f(s)\|\, ds \right|.$$

Recalling that

$$\|e^{(t-t_0)A}\| \leqslant e^{|t-t_0|\,\|A\|}, \quad \|e^{(t-s)A}\| \leqslant e^{|t-s|\,\|A\|},$$

we obtain

$$\|y(t)\| \leqslant e^{|t-t_0|\,\|A\|}\, \|y(t_0)\| + \left| \int_{t_0}^{t} e^{|t-s|\,\|A\|}\, ds \right| \max_{s\in[t_0,t]} \|f(s)\|.$$

Computing the integral on the right-hand side of the last inequality, we arrive at the required estimate (4).

This formula cannot be used only in a single exceptional case $\|A\| = 0$, i.e., $A = 0$. In this case, the elements of the matrix of coefficients A vanish and the equation (1) takes the simplest form

$$\frac{d}{dt}y(t) = f(t).$$

It is evident that

$$y(t) = y(t_0) + \int_{t_0}^{t} f(s)\,ds, \quad e^{t\|A\|} = e^{t\cdot 0} = 1,$$

$$\|y(t)\| \leqslant \|y(t_0)\| + |t - t_0| \max_{s\in[t_0,t]} \|f(s)\| = e^{|t-t_0|\,\|A\|}\,\|y(t_0)\| + |t-t_0| \max_{s\in[t_0,t]} \|f(s)\|.$$

Hence, if $\|A\| = 0$, then the coefficient

$$\frac{e^{|t-t_0|\,\|A\|} - 1}{\|A\|}$$

at the second term on the right-hand side of the estimate (4) must be replaced with the coefficient $|t - t_0|$. The replacement is natural because

$$\lim_{\|A\|\to 0} \frac{e^{|t-t_0|\,\|A\|} - 1}{\|A\|} = |t - t_0|.$$

The assumption $\|A\| \neq 0$ is not usually mentioned, regarding that the expression $(e^{|t-t_0|\,\|A\|} - 1)/\|A\|$ is always extended by continuity for $\|A\| = 0$. □

Considering the Cauchy problems

$$\frac{d}{dt}y^{[1]}(t) = Ay^{[1]}(t) + f^{[1]}(t), \quad y^{[1]}(t_0) = b^{[1]},$$
$$\frac{d}{dt}y^{[2]}(t) = Ay^{[2]}(t) + f^{[2]}(t), \quad y^{[2]}(t_0) = b^{[2]},$$

we see that

$$\frac{d}{dt}(y^{[1]}(t) - y^{[2]}(t)) = A(y^{[1]}(t) - y^{[2]}(t)) + (f^{[1]}(t) - f^{[2]}(t)),$$
$$y^{[1]}(t_0) - y^{[2]}(t_0) = b^{[1]} - b^{[2]}.$$

In view of (4), we obtain the inequality

$$\|y^{[1]}(t) - y^{[2]}(t)\| \leqslant e^{|t-t_0|\,\|A\|}\|b^{[1]} - b^{[2]}\| + \frac{e^{|t-t_0|\,\|A\|} - 1}{\|A\|} \max_{t\in[t_1,t_2]} \|f^{[1]}(t) - f^{[2]}\|$$

which will be referred to as the theorem on continuous dependence of solutions to the Cauchy problem on initial data and right-hand sides.

PROOF OF PROPOSITION 2. Let $f(t)$ be continuous and bounded for all $-\infty < t < \infty$ ($\|f(t)\| \leqslant F$). Let the matrix A have no purely imaginary eigenvalues ($|\operatorname{Re}\tau_j(A)| \geqslant \delta$) and let $G_0(t)$ be the Green matrix. Consider the integral (cf. (6))

$$y_0(t) = \int_{-\infty}^{\infty} G_0(t-s)f(s)\,ds \equiv \int_{-\infty}^{t} G_0(t-s)f(s)\,ds + \int_{t}^{\infty} G_0(t-s)f(s)\,ds. \tag{8}$$

The inequalities

$$\|f(t)\| \leqslant F, \quad \|G_0(t-s)\| \leqslant \gamma(N)\left(\frac{\|A\|}{\delta}\right)^{N-1} e^{-\frac{\delta}{2}|t-s|}$$

imply the inequalities ($t \neq s$)

$$\|G_0(t-s)f(s)\| \leqslant F\,\gamma(N)\left(\frac{\|A\|}{\delta}\right)^{N-1} e^{-\frac{\delta}{2}|t-s|},$$

$$\left\|\frac{d}{dt}G_0(t-s)f(s)\right\| \leqslant \|AG_0(t-s)f(s)\| \leqslant F\gamma(N)\frac{\|A\|^N}{\delta^{N-1}}e^{-\frac{\delta}{2}|t-s|}.$$

In view of the last inequalities both integrals on the right-hand side of (8) uniformly converge. Moreover, to compute $\dfrac{d}{dt}y(t)$ we can formally differentiate these integrals. Hence we obtain the estimate (5):

$$\begin{aligned}
\|y(t)\| &\leqslant \int_{-\infty}^{t} \|G_0(t-s)f(s)\|\,ds + \int_{t}^{\infty} \|G_0(t-s)f(s)\|\,ds \\
&\leqslant F\,\gamma(N)\left(\frac{\|A\|}{\delta}\right)^{N-1}\left[\int_{-\infty}^{t} e^{-\frac{\delta}{2}|t-s|}ds + \int_{t}^{\infty} e^{-\frac{\delta}{2}|t-s|}ds\right] \\
&= F\gamma(N)\left(\frac{\|A\|}{\delta}\right)^{N-1}\cdot\frac{4}{\delta}.
\end{aligned}$$

Differentiating the integrals in (8), we compute the derivative

$$\begin{aligned}
\frac{d}{dt}y(t) &= \frac{d}{dt}\int_{-\infty}^{t} G_0(t-s)f(s)\,ds + \frac{d}{dt}\int_{t}^{\infty} G_0(t-s)f(s)\,ds \\
&= \left[G_0(+0)f(t) + \int_{-\infty}^{t} \frac{d}{dt}G_0(t-s)f(s)\,ds\right] \\
&\quad + \left[-G_0(-0)f(t) + \int_{t}^{\infty} \frac{d}{dt}G_0(t-s)f(s)\,ds\right] \\
&= [G_0(+0) - G_0(-0)]f(0) + \int_{-\infty}^{\infty} AG_0(t-s)f(s)\,ds
\end{aligned}$$

$$= f(t) + A \int_{-\infty}^{\infty} G_0(t-s) f(s)\, ds = f(t) + Ay(t).$$

We have proved that the vector-valued function $y(t)$ represented by the formula (6) is a solution to the nonhomogeneous equation (1).

We note that $f(t)$ was assumed to be continuous. However, it is not hard to prove that if a bounded function $f(t)$ is discontinuous at some point $t = t_0$ and is continuous at all other points, then the function $y(t)$ represented by the formula (6) is bounded and continuous at $t = t_0$, differentiable for $t \neq t_0$, and satisfies the equation (1). To prove this assertion it suffices to verify that the computation of the derivative $y'(t)$ remains valid for $t \neq t_0$ in the case of a discontinuous function $f(t)$. In particular, for $t \neq 0$ the function

$$w(t) = \int_{0}^{\infty} G_0(t-s) f(s)\, ds$$

is a solution to the equation

$$\frac{d}{dt} w(t) = Aw(t) + \begin{cases} f(t) & \text{for } t > 0, \\ 0 & \text{for } t < 0. \end{cases}$$

We pass to the proof of the uniqueness of a bounded solution to the equation (1). If $y^{[1]}(t)$ and $y^{[2]}(t)$ are two solutions, i.e.,

$$\frac{d}{dt} y^{[1]}(t) = Ay^{[1]}(t) + f(t),$$
$$\frac{d}{dt} y^{[2]}(t) = Ay^{[2]}(t) + f(t),$$

then the difference $y^{[1]}(t) - y^{[2]}(t)$ is a bounded solution to the homogeneous equation

$$\frac{d}{dt}[y^{[1]}(t) - y^{[2]}(t)] = A[y^{[1]}(t) - y^{[2]}(t)]$$

and, consequently, must vanish. □

We consider the nonhomogeneous equation of the nth order

$$x^{(n)} + p_1 x^{(n-1)} + \cdots + p_{n-1} x' + p_n x = q(t). \tag{9}$$

It can be rewritten as a vector equation of the first order

$$\frac{d}{dt}\begin{pmatrix} x \\ x' \\ \vdots \\ x^{(n-1)} \\ x^{(n)} \end{pmatrix} - \begin{bmatrix} 0 & 1 & 0 & \dots & 0 & 0 \\ 0 & 0 & 1 & \dots & 0 & 0 \\ \dots & \dots & \dots & \dots & \dots & \dots \\ 0 & 0 & 0 & \dots & 0 & 1 \\ -p_n & -p_{n-1} & -p_{n-2} & \dots & -p_2 & -p_1 \end{bmatrix} \begin{pmatrix} x \\ x' \\ \vdots \\ x^{(n-1)} \\ x^{(n)} \end{pmatrix} = \begin{pmatrix} 0 \\ 0 \\ \vdots \\ 0 \\ q(t) \end{pmatrix}.$$

Denote by A the matrix of coefficients:

$$A = \begin{bmatrix} 0 & 1 & 0 & \dots & 0 & 0 \\ 0 & 0 & 1 & \dots & 0 & 0 \\ \dots & \dots & \dots & \dots & \dots & \dots \\ 0 & 0 & 0 & \dots & 0 & 1 \\ -p_n & -p_{n-1} & -p_{n-2} & \dots & -p_2 & -p_1 \end{bmatrix}.$$

The characteristic equation

$$\det[\tau I - A] = \tau^n + p_1\tau^{n-1} + \cdots + p_{n-1}\tau + p_n = 0$$

has roots $\tau_1, \tau_2, \ldots, \tau_n$ satisfying the condition $|\operatorname{Re}\tau_j| \geqslant \delta > 0$. This condition implies the existence of the Green matrix

$$G(t) = \begin{bmatrix} g_{11}(t) & g_{12}(t) & \ldots & g_{1n}(t) \\ g_{21}(t) & g_{22}(t) & \ldots & g_{2n}(t) \\ \cdots & \cdots & \cdots & \cdots \\ g_{n1}(t) & g_{n2}(t) & \ldots & g_{nn}(t) \end{bmatrix}.$$

By definition,

$$\begin{aligned} &\|G(t)\| \to 0 \quad \text{as } t \to \pm\infty, \\ &\frac{d}{dt}G(t) = AG(t) \quad \text{for } t \neq 0, \\ &G(t+0) - G(t-0) = I. \end{aligned}$$

In view of the estimate for the Green matrix (cf. (9) in §8), every element $g_{ik}(t)$ of the matrix $G(t)$ satisfies the following estimate:

$$|g_{jk}(t)| \leqslant \|G(t)\| \leqslant \gamma(n)\left(\frac{\|A\|}{\delta}\right)^{n-1} e^{-|t|\delta/2}. \tag{10}$$

A bounded solution $x(t)$ to the equation (9) (if $q(t)$ is continuous and bounded) exists, is unique, and can be represented as the integral

$$\begin{pmatrix} x(t) \\ x'(t) \\ \vdots \\ x^{(n-2)}(t) \\ x^{(n-1)}(t) \end{pmatrix} = \int_{-\infty}^{\infty} G(t-s) \begin{pmatrix} 0 \\ 0 \\ \vdots \\ 0 \\ q(s) \end{pmatrix} ds.$$

We write out this integral representation term-by-term and find that

$$\begin{aligned} x(t) &= \int_{-\infty}^{\infty} g_{1n}(t-s)q(s)\,ds, \\ x'(t) &= \int_{-\infty}^{\infty} g_{2n}(t-s)q(s)\,ds, \\ &\cdots\cdots\cdots\cdots \\ x^{(n-1)}(t) &= \int_{-\infty}^{\infty} g_{nn}(t-s)q(s)\,ds. \end{aligned} \tag{11}$$

It is notable that there is a simple connection between the functions $g_{kn}(t-s)$ in the representation (11) and the Green function $g(t)$ constructed in §8. Indeed, from the matrix equation

$$\frac{d}{dt}G(t) = AG(t), \quad t \neq 0,$$

it follows that for $t \neq 0$ the last column of the matrix $G(t)$ consisting of $g_{kn}(t)$ satisfies the equations

$$\frac{d}{dt}\begin{pmatrix} g_{1n} \\ g_{2n} \\ \vdots \\ g_{n-1,n} \\ g_{nn} \end{pmatrix} = \begin{bmatrix} 0 & 1 & 0 & \dots & 0 & 0 \\ 0 & 0 & 1 & \dots & 0 & 0 \\ \dots & \dots & \dots & \dots & \dots & \dots \\ 0 & 0 & 0 & \dots & 0 & 1 \\ -p_n & -p_{n-1} & -p_{n-2} & \dots & -p_2 & -p_1 \end{bmatrix} \begin{pmatrix} g_{1n} \\ g_{2n} \\ \vdots \\ g_{n-1,n} \\ g_{nn} \end{pmatrix}$$

which imply the relations ($t \neq 0$)

$$\frac{d^n}{dt^n} g_{1n} + p_1 \frac{d^{n-1}}{dt^{n-1}} g_{1n} + \dots + p_{n-1}\frac{d}{dt} g_{1n} + p_n g_{1n} = 0,$$
$$g_{kn} = \frac{d^{k-1} g_{1n}}{dt^{k-1}}, \quad k = 2, 3, \dots, n.$$

The fact that the last columns in the relation $G(+0) - G(-0) = I$ are equal is equivalent to the conditions ($k = 1, 2, \dots, n-1$)

$$\text{(12)} \qquad \begin{aligned} \left.\frac{d^{k-1}}{dt^{k-1}} g_{1n}\right|_{t=+0} - \left.\frac{d^{k-1}}{dt^{k-1}} g_{1n}\right|_{t=-0} &= 0. \\ \left.\frac{d^{n-1}}{dt^{n-1}} g_{1n}\right|_{t=+0} - \left.\frac{d^{n-1}}{dt^{n-1}} g_{1n}\right|_{t=-0} &= 1. \end{aligned}$$

We see that the relations (12) imposed on g_{1n} literally coincide with the conditions that uniquely determine the Green function $g(t)$. Therefore,

$$g_{1n}(t) = g(t),\ g_{2n}(t) = g'(t),\ g_{3n}(t) = g''(t),\ \dots,\ g_{nn}(t) = g^{(n-1)}(t).$$

In the considered case, the norm $\|A\|$ is completely defined by the coefficients p_j in the equation (9) and, consequently, by the roots of the corresponding characteristic equation. Therefore, the inequalities (10) lead to the conclusion that for $t \neq 0$

$$\begin{gathered} |g(t)| \leqslant \Omega \exp\{-|t|\delta/2\}, \\ |g^{(k)}(t)| \leqslant \Omega \exp\{-|t|\delta/2\}, \quad k = 1, 2, \dots, n-1, \\ \Omega \equiv \Omega(n, \tau_1, \tau_2, \dots, \tau_n). \end{gathered}$$

We have proved that for a bounded continuous function $q(t)$ a bounded solution $x(t)$ to the equation (9) exists, is unique, and can be represented by the formula

$$\text{(13)} \qquad x(t) = \int_{-\infty}^{\infty} g(t-s) q(s)\, ds.$$

The derivatives $x'(t), x''(t), \dots, x^{(n-1)}(t)$ of the solution $x(t)$ also admit the integral representations

$$\text{(14)} \qquad x^{(k)}(t) = \int_{-\infty}^{\infty} g^{(k)}(t-s) q(s)\, ds.$$

The integrals in (13) and (14) uniformly converge because of the inequalities for $g(t), g'(t), \dots, g^{(n-1)}(t)$.

It is easy to derive the estimate

$$|x(t)| \leqslant \int_{-\infty}^{\infty} |g(t-s)|\,|q(s)|\,ds \leqslant \sup_t |q(t)| \int_{-\infty}^{\infty} \Omega e^{-\delta|t-s|/2} ds$$
$$= 2\sup_t |q(t)|\Omega \int_0^{\infty} e^{-\delta\xi/2} d\xi = \delta^{-1} \sup_t |q(t)|\Omega,$$

which implies the inequality

$$\sup_t |x(t)| \leqslant \sup_t |q(t)|\Omega/\delta. \tag{15}$$

EXERCISE 1. Using the representation

$$x(t) = \int_{-\infty}^{\infty} g(t-s)q(s)\,ds \equiv \int_{-\infty}^{t} g(t-s)q(s)\,ds + \int_{t}^{\infty} g(t-s)q(s)\,ds$$

and the properties of $g(t)$, verify by direct substitution that $x(t)$ is a bounded solution to the equation (9).

Now we study one property of the nonhomogeneous equation

$$\frac{dx}{dt} = Ax + \varphi(t). \tag{16}$$

This property is of great importance for our further needs.

PROPOSITION 3. *If $\varphi(t)$ exponentially decreases as $t \to \infty$, i.e.,*

$$\|\varphi(t)\| \leqslant Ke^{-\sigma_0 t} \quad (\sigma_0 > 0), \tag{17}$$

then there is at least one solution $x(t)$ to the equation (16) *that also decreases exponentially as $t \to \infty$:*

$$\|x(t)\| \leqslant Me^{-\sigma t} \tag{18}$$

(for σ we can take any number such that $0 < \sigma < \sigma_0$).

REMARK 1. We emphasize that no assumptions about properties of the matrix of coefficients A and location of the eigenvalues of A are required. We also emphasize that an exponentially decreasing solution is not necessarily unique.

PROOF OF PROPOSITION 3. Substituting $x(t) = e^{-\sigma_* t} y(t)$, $0 < \sigma_* < \sigma_0$, in equation (16), we see that $y(t)$ satisfies the equation

$$\frac{dy(t)}{dt} = (A + \sigma_* I)y(t) + f(t),$$

where $f(t) = e^{\sigma_* t}\varphi(t)$. We choose $\sigma_* \in (0, \sigma_0)$ so that the matrix $A_\sigma \equiv A + \sigma_* I$ has no purely imaginary eigenvalues ($|\operatorname{Re}\lambda_j(A_\sigma)| \geqslant \delta > 0$). It is clear that at most a finite number of the values of σ_* on the interval $(0, \sigma_0)$ are inadmissible. We note that δ depends on σ_*. Since $0 < \sigma_* < \sigma_0$, we have $\|f(t)\| \leqslant K$ for $t \geqslant 0$. Therefore, for $t > 0$ the function

$$y(t) = \int_0^{\infty} G_0(t-s, A + \sigma_* I) f(s)\,ds$$

is a bounded solution to the equation

$$\frac{dy(t)}{dt} = A_{\sigma_*} y(t) + f(t), \quad t > 0,$$

and the estimate $\|y(t)\| \leqslant M$ holds. Hence for $t > 0$ the function $x(t) = e^{-\sigma_* t} y(t)$ satisfies equation (16) and the estimate $\|x(t)\| \leqslant e^{-\sigma_* t} M$. For σ_* we can take any number in the interval $(0, \sigma_0)$ except perhaps a finite number of values. Consequently, the estimate (18) is valid for any $\sigma \in (0, \sigma_0)$. But it is necessary to keep in mind that M depends on σ. □

To conclude the section, we present arguments leading to a notion which plays an important role in the theory of differential equations with ordinary derivatives and partial derivatives as well. The notion is concerned with generalized solutions to equations. We will be able only to give a brief exposition of ideas connected with this notion, discussing the simplest example of nonhomogeneous equations with constant coefficients.

We again apply the Cauchy formula (cf. (3))

$$y(t) = e^{(t-t_0)A} a + \int_{t_0}^{t} e^{(t-s)A} f(s)\, ds \tag{19}$$

which represents a solution $y(t)$ to the Cauchy problem

$$\frac{dy}{dt} = Ay + f(t), \quad y(t_0) = a. \tag{20}$$

We seek $y(t)$ on the segment $t_0 \leqslant t \leqslant t_1$ on which the right-hand side $f(t)$ is known. We will use the Cauchy formula (19) to clarify questions leading to the notion of generalized solutions to differential equations.

In many cases, we need to consider a sequence of solutions $y_j(t)$, $j = 1, 2, \ldots$, to the Cauchy problems

$$\frac{dy_j}{dt} = Ay_j + f_j(t), \quad y_j(t_0) = a_j, \qquad j = 1, 2, \ldots. \tag{21}$$

We assume that the sequence of the initial data a_j converges to some vector a_0 and the sequence of the right-hand sides f_j converges to some vector-valued function f_0 as $j \to \infty$. The question arises: may we assert that the sequence of the solutions $y_j(t)$ converges to the solution $y_0(t)$ to the Cauchy problem

$$\frac{dy_0}{dt} = Ay_0 + f_0(t), \quad y_0(t_0) = a_0? \tag{22}$$

To answer this question it is necessary to specify what we mean by "convergence." Consider several cases.

If the sequence of continuous functions $f_j(t)$ satisfies the Cauchy criterion which yields the uniform convergence of $f_j(t)$, then $f_j(t)$ uniformly converges to a continuous function $f_0(t)$. By the theorem about the continuous dependence of solutions on the initial data and right-hand sides, the inequalities

$$\|f_j(t) - f_0(t)\| \leqslant \varepsilon_j, \quad \|a_j - a_0\| \leqslant \varepsilon_j \quad \left(\lim_{j\to\infty} \varepsilon_j = 0, \ t_0 \leqslant t \leqslant t_1\right)$$

imply that $y_j(t)$ converge uniformly to $y_0(t)$. The differentiability of $y_0(t)$ can be established on the basis of the Cauchy formula in view of continuity if $f_0(t)$. We note that the estimate

$$\begin{aligned}\|y_j(t)-y_0(t)\| &\leqslant e^{(t_1-t_0)\|A\|}\,\|a_j-a_0\| \\ &\quad+\frac{e^{(t_1-t_0)\|A\|}-1}{\|A\|}\max_{t_0\leqslant t\leqslant t_1}\|f_j(t)-f_0(t)\|\end{aligned}$$

can be established with the help of the Cauchy formula.

The situation is more complicated if the convergence of $f_j(t)$ to $f_0(t)$ is not uniform. We will consider the convergence in the Hilbert space of square integrable functions. In the sense of this convergence, we may only assume that

$$\sqrt{\int\limits_{t_0}^{t_1}\|f_j(t)-f_0(t)\|^2dt}\leqslant\varepsilon_j,\quad \|a_j-a_0\|\leqslant\varepsilon_j\quad(\lim_{j\to\infty}\varepsilon_j=0).$$

Even if all $f_j(t)$ are continuous, we cannot regard $f_0(t)$ as a continuous function. We give an example illustrating this assertion.

EXAMPLE 1. Consider scalar functions

$$f_j(t)=\begin{cases}0, & t_0\leqslant t\leqslant\dfrac{t_0+t_1}{2}-\dfrac{t_1-t_0}{2j},\\ \dfrac{1}{2}+\dfrac{j}{t_1-t_0}\Big(t-\dfrac{t_0+t_1}{2}\Big), & \dfrac{t_0+t_1}{2}-\dfrac{t_1-t_0}{2j}\leqslant t\leqslant\dfrac{t_0+t_1}{2}+\dfrac{t_1-t_0}{2j},\\ 1, & \dfrac{t_0+t_1}{2}+\dfrac{t_1-t_0}{2j}\leqslant t\leqslant t_1,\end{cases}$$

$$f_0(t)=\begin{cases}0, & t_0\leqslant t<\dfrac{t_0+t_1}{2},\\ 1, & \dfrac{t_0+t_1}{2}\leqslant t\leqslant t_1.\end{cases}$$

It is clear that $f_1(t),f_2(t),f_3(t),\dots$ are continuous while $f_0(t)$ is discontinuous. At the same time,

$$|f_j(t)-f_0(t)|=\begin{cases}0, & \Big|t-\dfrac{t_0+t_1}{2}\Big|\geqslant\dfrac{t_1-t_0}{2j},\\ \dfrac{1}{2}+\dfrac{j}{t_1-t_0}\Big(t-\dfrac{t_0+t_1}{2}\Big), & \dfrac{t_0+t_1}{2}-\dfrac{t_1-t_0}{2j}\leqslant t\leqslant\dfrac{t_0+t_1}{2},\\ \dfrac{1}{2}-\dfrac{j}{t_1-t_0}\Big(t-\dfrac{t_0+t_1}{2}\Big), & \dfrac{t_0+t_1}{2}\leqslant t\leqslant\dfrac{t_0+t_1}{2}+\dfrac{t_1-t_0}{2j},\end{cases}$$

$$\begin{aligned}\int\limits_{t_0}^{t_1}|f_j(t)-f_0(t)|^2dt&=\int\limits_{\frac{t_0+t_1}{2}-\frac{t_1-t_0}{2j}}^{\frac{t_0+t_1}{2}}\frac{1}{4(t_1-t_0)^2}\big[2jt+(t_1-t_0)-j(t_0+t_1)\big]^2dt\\ &+\int\limits_{\frac{t_0+t_1}{2}}^{\frac{t_0+t_1}{2}+\frac{t_1-t_0}{2j}}\frac{1}{4(t_1-t_0)^2}\big[2jt-(t_1-t_0)-j(t_0+t_1)\big]^2dt=\frac{t_1-t_0}{12j}.\end{aligned}$$

In other words, the continuous functions $f_j(t)$ converge to the discontinuous function $f_0(t)$ in the Hilbert space norm. In this example, the function $f_0(t)$ has a single discontinuity point of the first kind. In general, the limiting function can have considerably more complicated character of discontinuity.

Example 1 shows that if the space of continuous vector-valued functions $f(t)$, $t_0 \leqslant t \leqslant t_1$, is equipped with the Hilbert space norm

$$\|f(t)\|_{L_2} = \sqrt{\int_{t_0}^{t_1} \|f(t)\|^2 dt},$$

the obtained normed space of continuous vector-valued functions $f(t)$ is not closed with respect to the convergence in the norm $\|\cdot\|_{L_2}$. A sequence of continuous functions $f_1(t), f_2(t), \dots$ satisfying the Cauchy criterion

$$\|f_j(t) - f_k(t)\|_{L_2} < \varepsilon_{\min\{j,k\}}, \quad \lim_{j\to\infty} \varepsilon_j = 0$$

may have no continuous limiting function $f_0(t)$ such that

$$\|f_j(t) - f_0(t)\|_{L_2} \leqslant \varepsilon_j.$$

Usually for the Hilbert space L_2 one takes the completion of the space of continuous functions with respect to the norm $\|\cdot\|_{L_2}$. The completion procedure consists in adding to the set of continuous functions all functions that can be regarded as limit (in the $\|\cdot\|_{L_2}$-norm) of sequences of continuous functions $f_j(t)$ satisfying the Cauchy criterion. Such a completion is similar to the standard procedure from mathematical analysis which is used in the definition of irrational numbers using sequences of rational numbers satisfying the Cauchy criterion. The converging sequence can be understood as a sequence of more and more exact approximations to an irrational number. We refer the reader to [**15, 18**] for the accurate presentation of the theory of completion of a normed space.

We now explain what must be regarded as a solution to the Cauchy problem if the vector-valued function $f(t)$ on the right-hand side of the differential equation belongs to the space L_2 but is not continuous. Let $\{f_j(t)\}$ be a sequence of continuous vector-valued functions. Assume that it is a Cauchy sequence in the $\|\cdot\|_{L_2}$-norm and has the limit $f_0(t)$, i.e.,

$$\sqrt{\int_{t_0}^{t_1} \|f_j(t) - f_k(t)\|^2 dt} \leqslant \varepsilon_{\min\{j,k\}},$$

$$\sqrt{\int_{t_0}^{t_1} \|f_j(t) - f_0(t)\|^2 dt} \leqslant \varepsilon_j \quad (\lim_{j\to\infty} \varepsilon_j = 0).$$

We also assume that the sequence of vectors $\{a_j\}$ converges (in the norm of the Euclidean space) to a vector a_0:

$$\|a_j - a_k\| \leqslant \varepsilon_{\min\{j,k\}}, \quad \|a_j - a_0\| \leqslant \varepsilon_j.$$

We construct a sequence of solutions $y_j(t)$ to the Cauchy problems (21) and show that the sequence of continuous vector-valued functions $\{y_j(t)\}$ continuously converges on $[t_0, t_1]$. It is natural to consider the limit (a continuous vector-valued function $y_0(t)$) of the sequence $\{y_j(t)\}$ as a "solution" to the Cauchy problem (22). We write "solution" in quotes because we cannot hope that there exists the derivative $\dfrac{dy_0(t)}{dt}$.

Let us show rigorously that the sequence $\{y_j(t)\}$ satisfies the Cauchy criterion which provides the uniform convergence of this sequence. Indeed, the equalities

$$\frac{dy_j(t)}{dt} = Ay_j(t) + f_j(t), \quad y_j(t_0) = a_j$$
$$\frac{dy_k(t)}{dt} = Ay_k(t) + f_k(t), \quad y_k(t_0) = a_k$$

imply that the differences $y_j(t) - y_k(t)$ are solutions to the problems

$$\frac{d}{dt}[y_j(t) - y_k(t)] = A[y_j(t) - y_k(t)] + [f_j(t) - f_k(t)],$$
$$[y_j(t_0) - y_k(t_0)] = a_j - a_k$$

and, by the Cauchy formula, admit the representation

$$y_j(t) - y_k(t) = e^{(t-t_0)A}\,[a_j - a_k] + \int_{t_0}^{t} e^{(t-s)A}\,[f_j(s) - f_k(s)]ds$$

and satisfy the estimates

$$\|y_j(t) - y_k(t)\| \leqslant e^{(t_1-t_0)\|A\|}\|a_j - a_k\| + e^{(t_1-t_0)\|A\|}\int_{t_0}^{t_1}\|f_j(s) - f_k(s)\|ds.$$

By the Cauchy–Schwarz–Bunyakovskii inequality,

$$\int_{t_0}^{t_1}\|f_j(s) - f_k(s)\|ds \leqslant \sqrt{\int_{t_0}^{t_1} 1^2 ds}\sqrt{\int_{t_0}^{t_1}\|f_j(s) - f_k(s)\|^2 ds}$$
$$\leqslant \sqrt{t_1 - t_0}\sqrt{\int_{t_0}^{t_1}\|f_j(s) - f_k(s)\|^2 ds}.$$

Therefore,

$$\|y_j(t) - y_k(t)\| \leqslant e^{(t_1-t_0)\|A\|}\left(\|a_j - a_k\| + \sqrt{t_1 - t_0}\sqrt{\int_{t_0}^{t_1}\|f_j(s) - f_k(s)\|^2 ds}\right)$$
$$\leqslant e^{(t_1-t_0)\|A\|}(1 + \sqrt{t_1 - t_0})\varepsilon_{\min\{j,k\}}.$$

Hence the sequence $\{y_j(t)\}$ uniformly converges, which implies the continuity of $y_0(t)$.

The integral

$$\int_{t_0}^{t_1} f(t)g(t)\,dt$$

makes sense for $f \in L_2$ and $g \in L_2$. (We again refer the reader to [**15, 18**] where this assertion is explained in detail.) Therefore, for any vector-valued function $f_0 \in L_2$ and any vector a_0 we can define the vector-valued function

$$\widehat{y}_0(t) = e^{(t-t_0)A}\,a_0 + \int_{t_0}^{t} e^{(t-s)A}\,f_0(s)\,ds.$$

If

$$\sqrt{\int_{t_0}^{t_1} \|f_0(s) - f_j(s)\|^2 ds} \leqslant \varepsilon_j, \quad \|a_0 - a_j\| \leqslant \varepsilon_j$$

and

$$y_j(t) = e^{(t-t_0)A}\,a_j + \int_{t_0}^{t} e^{(t-s)A}\,f_j(s)\,ds,$$

then the difference $y_j(t) - \widehat{y}_0(t)$ admits the representation

$$y_j(t) - \widehat{y}_0(t) = e^{(t-t_0)A}\,[a_j - a_0] + \int_{t_0}^{t} e^{(t-s)A}\,[f_j(s) - f_0(s)]\,ds$$

and satisfies the estimate

$$\|y_j(t) - \widehat{y}_0(t)\| \leqslant e^{(t_1-t_0)\|A\|}\big[1 + \sqrt{t_1 - t_0}\,\big]\varepsilon_j$$

which follows from the representation.

It is obvious that $y_j(t)$ uniformly converge to $\widehat{y}_0(t)$ as $j \to \infty$. Consequently, $\widehat{y}_0(t)$ coincides with the "solution" $y_0(t)$ to the Cauchy problem (22) with the right-hand side $f_0(t)$.

DEFINITION 1. The above introduced "solution" $y_0(t)$ is usually called a generalized solution to the Cauchy problem (22). In other words, by a generalized solution to the Cauchy problem (22) on $t_0 \leqslant t \leqslant t_1$ we mean a continuous vector-valued function

$$y_0(t) = e^{(t-t_0)A}\,a_0 + \int_{t_0}^{t} e^{(t-s)A}\,f_0(s)\,ds. \tag{23}$$

In fact, the formula (23) coincides with the Cauchy formula (19), but (23) can be applied to discontinuous functions $f_0(t)$. (It suffices that $f_0 \in L_2$.) Moreover, the vector-valued function $y_0(t)$ is not necessarily differentiable. Consequently, it cannot be regarded as a classical solution to the differential equation.

The detailed study of generalized solutions is an important part of the current theory of differential equations, but it does not answer the purpose of this book. Therefore, we restricted ourselves by the above remarks. Hereinafter, we will use generalized solutions in the construction of optimal control theory.

§10. The representation of the matrix exponential and of the Green matrix by contour integrals

Auxiliary polynomials computed from the characteristic polynomial. The representation of the special functions $\omega_j(t)$ by contour integrals. The contour integral for the matrix exponential. The representation of the Green matrix by a complex integral along the imaginary axis. The use of the Plancherel theorem for deducing an integral identity connected with the Green matrix.

We begin by considering the characteristic polynomial $\pi_N(\tau) = \det[\tau I - A]$ of a matrix A, which we write as follows:

(1) $$\pi_N(\tau) = \tau^N + p_1\tau^{N-1} + \cdots + p_{N-1}\tau + p_N.$$

We define the polynomials $\pi_{N-1}(\tau), \pi_{N-2}(\tau), \ldots, \pi_1(\tau), \pi_0(\tau)$ by the formulas

(2) $$\begin{aligned} &\pi_{N-1}(\tau) = \tau^{N-1} + p_1\tau^{N-2} + \cdots + p_{N-2}\tau + p_{N-1}, \\ &\pi_{N-2}(\tau) = \tau^{N-2} + p_1\tau^{N-3} + \cdots + p_{N-2}, \\ &\quad\cdots\cdots\cdots\cdots\cdots\cdots \\ &\pi_1(\tau) = \tau + p_1, \\ &\pi_0(\tau) = 1 \end{aligned}$$

and verify that they are connected by the recurrence relations

(3) $$\begin{aligned} &\pi_N(\tau) - \tau\pi_{N-1}(\tau) = p_N, \\ &\pi_{N-1}(\tau) - \tau\pi_{N-2}(\tau) = p_{N-1}, \\ &\quad\cdots\cdots\cdots\cdots \\ &\pi_2(\tau) - \tau\pi_1(\tau) = p_2, \\ &\pi_1(\tau) - \tau\pi_0(\tau) = p_1. \end{aligned}$$

On the complex plane, we choose a smooth contour Γ surrounding a simply connected domain that contains the roots $\tau_1, \tau_2, \ldots, \tau_N$ of the polynomial $\pi_N(\tau)$. We define $\omega_1, \omega_2, \ldots, \omega_N$ by the integrals

(4) $$\omega_j(t) = \frac{1}{2\pi i}\oint_\Gamma \frac{\pi_{N-j}(\tau)}{\pi_N(\tau)} e^{\tau t}\, d\tau$$

and study their properties.

First, we compute $\omega_1(0)$:

(5) $$\omega_1(0) = \frac{1}{2\pi i}\oint_\Gamma \frac{\pi_{N-1}(\tau)}{\pi_N(\tau)}\, d\tau.$$

By setting $\tau = Re^{i\varphi}$, $d\tau = iRe^{i\varphi}d\varphi$, the contour of integration in this integral can be replaced by a circle of any sufficiently large radius R so that the value of the integral does not change. We obtain

(6)
$$\omega_1(0) = \frac{1}{2\pi}\int_0^{2\pi} \frac{\pi_N(Re^{i\varphi}) - p_N}{\pi_N(Re^{i\varphi})}\frac{Re^{i\varphi}\, d\varphi}{Re^{i\varphi}} = 1 - \frac{p_N}{2\pi}\int_0^{2\pi}\frac{d\varphi}{\pi_N(Re^{i\varphi})} = 1 + O\left(\frac{1}{R^N}\right).$$

It is obvious that $\omega_1(0) = 1$ since the value $\omega_1(0)$ is independent of the choice of R.

Let us compute

$$\omega_k(0) = \frac{1}{2\pi i}\oint_\Gamma \frac{\pi_{N-k}(\tau)}{\pi_N(\tau)}\,d\tau, \quad k = 2, 3, \ldots, N. \tag{7}$$

In this case, it is also convenient to replace Γ by a circle $\tau = Re^{i\varphi}$ of a sufficiently large radius R such that the circle contains all roots of $\pi_N(\tau)$. Then we can write the integral representation in the form

$$\omega_k(0) = \frac{1}{2\pi}\int_0^{2\pi} \frac{R\pi_{N-k}(Re^{i\varphi})\cdot e^{i\varphi}}{\pi_N(Re^{i\varphi})}\,d\varphi \tag{8}$$

and conclude that $\omega_k(0) = O\left(\frac{1}{R^{k-1}}\right)$. Since R can be taken as large as necessary, we have $\omega_k(0) = 0$ for $k \geqslant 2$. Thus,

$$\omega_1(0) = 1, \quad \omega_2(0) = \omega_3(0) = \cdots = \omega_N(0) = 0. \tag{9}$$

Differentiating $\omega_j(t)$ with respect to t:

$$\begin{aligned}
\frac{d}{dt}\omega_j(t) &= \frac{d}{dt}\frac{1}{2\pi i}\oint_\Gamma \frac{\pi_{N-j}(\tau)}{\pi_N(\tau)}e^{\tau t}\,d\tau = \frac{1}{2\pi i}\oint_\Gamma \frac{\tau\pi_{N-j}(\tau)}{\pi_N(\tau)}e^{\tau t}\,d\tau \\
&= \frac{1}{2\pi i}\oint_\Gamma \frac{\pi_{N-j+1}(\tau) - p_{N-j+1}}{\pi_N(\tau)}e^{\tau t}\,d\tau = \frac{1}{2\pi i}\oint_\Gamma \frac{\pi_{N-j+1}(\tau)}{\pi_N(\tau)}e^{\tau t}\,d\tau \\
&\quad - \frac{p_{N-j+1}}{2\pi i}\oint_\Gamma \frac{\pi_0(\tau)}{\pi_N(\tau)}e^{\tau t}\,d\tau = \begin{cases} 0 - p_N\omega_N(t) & (j = 1) \\ \omega_{j-1}(t) - p_{N-j+1}\omega_N(t) & (j > 1) \end{cases}
\end{aligned}$$

we see that the functions $\omega_1, \omega_2, \ldots, \omega_N$ form a solution to the Cauchy problem

$$\begin{aligned}
\frac{d\omega_1}{dt} &= -p_N\omega_N, \quad \omega_1(0) = 1, \\
\frac{d\omega_2}{dt} &= \omega_1 - p_{N-1}\omega_N, \quad \omega_2(0) = 0, \\
&\cdots\cdots\cdots\cdots \\
\frac{d\omega_N}{dt} &= \omega_{N-1} - p_1\omega_N, \quad \omega_N(0) = 0.
\end{aligned} \tag{10}$$

We recall that the functions $\omega_1, \omega_2, \ldots, \omega_N$ were used in the polynomial representation of the matrix exponential (cf. §4)

$$e^{tA} = \omega_1(t)I + \omega_2(t)A + \cdots + \omega_N(t)A^{N-1}. \tag{11}$$

Substituting (4) in (11), we obtain

$$e^{tA} = \frac{1}{2\pi i}\oint_\Gamma \frac{1}{\pi_N(\tau)}\left[\pi_{N-1}(\tau)I + \pi_{N-2}(\tau)A + \cdots + \pi_0(\tau)A^{N-1}\right]e^{\tau t}\,d\tau. \tag{12}$$

Multiplying the matrix polynomial under the integral sign by $\tau I - A$ and transforming the product with the help of the recurrence relations (3), we find

$$\begin{aligned}
&\frac{1}{\pi_N(\tau)}[\tau I - A]\big[\pi_{N-1}(\tau)I + \pi_{N-2}(\tau)A + \dots + \pi_0(\tau)A^{N-1}\big] \\
&\quad = \frac{1}{\pi_N(\tau)}\Big\{\tau\pi_{N-1}(\tau)I + [\tau\pi_{N-2}(\tau) - \pi_{N-1}(\tau)]A + \dots \\
&\qquad\qquad + [\tau\pi_0(\tau) - \pi_1(\tau)]A^{N-1} - \pi_0(\tau)A^N\Big\} \\
&\quad = \frac{1}{\pi_N(\tau)}\Big\{[\pi_N(\tau) - p_N]I - p_{N-1}A - p_{N-2}A^2 - \dots - p_1A^{N-1} - A^N\Big\} \\
&\quad = I - \frac{1}{\pi_N(\tau)}\Big[A^N + p_1A^{N-1} + p_2A^{N-2} + \dots + p_{N-1}A + p_NI\Big] = I.
\end{aligned}$$

Thus, using the fact that a matrix satisfies its characteristic equation, we have proved the equality

$$(\tau I - A)\frac{1}{\pi_N(\tau)}\big[\pi_{N-1}(\tau)I + \pi_{N-2}(\tau)A + \dots + \pi_0(\tau)A^{N-1}\big] = I,$$

i.e.,

$$\frac{1}{\pi_N(\tau)}\big[\pi_{N-1}(\tau)I + \pi_{N-2}(\tau)A + \dots + \pi_0(\tau)A^{N-1}\big] = [\tau I - A]^{-1}. \tag{13}$$

As a result, we have justified the integral representation for the matrix exponential

$$e^{tA} = \frac{1}{2\pi i}\oint_\Gamma e^{\tau t}[\tau I - A]^{-1}d\tau, \tag{14}$$

where the integral is taken over any contour surrounding a simply connected domain containing all eigenvalues of A.

The representation (14) is often used. We need it to obtain a new formula for the Green matrix $G_0(t, A)$ (cf. (§7–9).

We begin with auxiliary constructions. Suppose that all eigenvalues of a matrix B lie in the left half-plane ($\operatorname{Re}\tau_j(B) \leqslant -\delta$, $\delta > 0$). We compute e^{tB} for $t > 0$. We compose the contour Γ from the half-circle $\tau = Re^{i\varphi}$ ($\pi/2 \leqslant \varphi \leqslant 3\pi/2$) of radius $> 2\|B\|$ located in the left half-plane and the segment on the imaginary axis $\tau = i\omega$, $-R \leqslant \omega \leqslant R$, which closes the half-circle. It is obvious that all eigenvalues $\tau_j(B)$ lie inside domain surrounded by Γ ($|\tau_j(B)| \leqslant \|B\| < 2\|B\|$). Therefore, for $t > 0$ we have

$$e^{tB} = \frac{1}{2\pi}\int_{-R}^{R} e^{i\omega t}[i\omega I - B]^{-1}d\omega + \frac{1}{2\pi}\int_{\frac{\pi}{2}}^{\frac{3\pi}{2}} e^{R(\cos\varphi + i\sin\varphi)t}[Re^{i\varphi}I - B]^{-1}Re^{i\varphi}d\varphi. \tag{15}$$

The second integral on the right-hand side of (15) is estimated as follows:

$$\left\| \frac{1}{2\pi} \int_{\frac{\pi}{2}}^{\frac{3\pi}{2}} e^{R(\cos\varphi + i\sin\varphi)t} [Re^{i\varphi} I - B]^{-1} Re^{i\varphi} d\varphi \right\|$$

$$\leqslant \frac{1}{2\pi} \frac{R}{R - \|B\|} 2 \int_0^{\frac{\pi}{2}} e^{-Rt\sin\psi} d\psi$$

$$= \frac{R}{\pi[R - \|B\|]} \left[\int_0^{\frac{\pi}{4}} \frac{e^{-Rt\sin\psi}\cos\psi}{\cos\psi} d\psi + \int_{\frac{\pi}{4}}^{\frac{\pi}{2}} e^{-Rt\sin\psi} d\psi \right]$$

$$\leqslant \frac{R}{\pi[R - \|B\|]} \left[\int_0^{\frac{\pi}{4}} \frac{e^{-Rt\sin\psi} d\sin\psi}{\sqrt{2}/2} + \int_{\frac{\pi}{4}}^{\frac{\pi}{2}} e^{-Rt\frac{\sqrt{2}}{2}} d\psi \right]$$

$$= \frac{R}{\pi[R - \|B\|]} \left[\frac{\sqrt{2}}{Rt} \left(1 - e^{-\frac{Rt}{\sqrt{2}}}\right) + \frac{\pi}{4} e^{-\frac{Rt}{\sqrt{2}}} \right] = O\Big(\frac{1}{Rt}\Big).$$

We have proved that if $\operatorname{Re}\tau_j(B) \leqslant -\delta < 0$ for all j, then

$$e^{tB} = \lim_{R\to\infty} \frac{1}{2\pi} \int_{-R}^{R} e^{i\omega t} [i\omega I - B]^{-1} d\omega$$

for $t > 0$, or, more exactly,

$$e^{tB} = \frac{1}{2\pi} \int_{-R}^{R} e^{i\omega t} [i\omega I - B]^{-1} d\omega + O\Big(\frac{1}{Rt}\Big). \tag{16}$$

Consider the same matrix whose eigenvalues lie in the left half-plane and a contour Γ located in the right half-plane or such that a part of Γ lies on the imaginary axis. Then the integral

$$\frac{1}{2\pi i} \oint_{\Gamma'} e^{\tau t} [\tau I - B]^{-1} d\tau$$

vanishes. Indeed, every element of the matrix under the integral sign is a regular analytic function in the right half-plane (and on the imaginary axis). But the integral of such a function over a closed contour vanishes. We choose the contour Γ' consisting of the half-circle $\tau = Re^{i\varphi}$ $(-\pi/2 \leqslant \varphi \leqslant \pi/2,\ R \geqslant 2\|B\|)$ located in the right half-plane and the segment on the imaginary axis $\tau = i\omega$ $(-R \geqslant \omega \geqslant -R)$ which closes the half-circle.

For $t < 0$ we see that

$$\frac{1}{2\pi} \int_{-\frac{\pi}{2}}^{+\frac{\pi}{2}} e^{R(\cos\varphi + i\sin\varphi)t} [Re^{i\varphi} I - B]^{-1} Re^{i\varphi} d\varphi - \frac{1}{2\pi} \int_{-R}^{+R} e^{i\omega t} [i\omega I - B]^{-1} d\omega = 0. \tag{17}$$

We estimate the first term on the left-hand side of (17) by the method similar to the method we just used in the case $t > 0$. We again arrive at the conclusion that this term is estimated by $O\left(\frac{1}{R|t|}\right)$. Therefore, for $t < 0$ we obtain

$$0 = \frac{1}{2\pi} \int_{-R}^{+R} e^{i\omega t}[i\omega I - B]^{-1} d\omega + O\left(\frac{1}{R|t|}\right).$$

Recalling (cf. §7) that

$$G_0(t, B) = \begin{cases} e^{tB} & \text{for } t > 0, \\ 0 & \text{for } t < 0 \end{cases}$$

if all eigenvalues of the matrix B lie in the left half-plane, we see that

$$G_0(t, B) = \frac{1}{2\pi} \int_{-R}^{+R} e^{i\omega t}[i\omega I - B]^{-1} d\omega + O\left(\frac{1}{R|t|}\right), \quad t \neq 0. \tag{18}$$

If instead of the matrix B we take a matrix C whose spectrum lies in the strictly right half-plane, then for the Green matrix $G_0(t, C)$ we obtain the formula

$$G_0(t, C) = \begin{cases} 0 & \text{for } t > 0, \\ -e^{tC} & \text{for } t < 0. \end{cases}$$

In this case, for $t \neq 0$ we again have the representation (18) with B replaced with C. We recommend the reader to prove this assertion accurately. The proof follows the same scheme as in the case $\operatorname{Re} \tau_j(B) < 0$ and consists in the study of the matrix integral

$$\oint_{\Gamma} e^{\tau t}[\tau I - C]^{-1} d\tau$$

which is taken over a contour consisting of half-circles of radius R and center at the origin and the segments on the imaginary axis which close these half-circles.

If A has no purely imaginary eigenvalues, then

$$A = T^{-1} \begin{bmatrix} B & 0 \\ 0 & C \end{bmatrix} T,$$

where T is some nonsingular matrix T, all eigenvalues of B lie in the left half-plane and all eigenvalues of C lie in the right half-plane. Moreover,

$$\tau I - A = T^{-1} \begin{bmatrix} \tau I_k - B & 0 \\ 0 & \tau I_{N-k} - C \end{bmatrix} T,$$

$$\frac{1}{2\pi}\int_{-R}^{+R} e^{i\omega t}[i\omega I - A]^{-1}d\omega$$

$$= T^{-1}\begin{bmatrix} \frac{1}{2\pi}\int_{-R}^{+R} e^{i\omega t}[i\omega I_k - B]^{-1}d\omega & 0 \\ 0 & \frac{1}{2\pi}\int_{-R}^{+R} e^{i\omega t}[i\omega I_{N-k} - C]^{-1}d\omega \end{bmatrix} T$$

$$+ O\Big(\frac{1}{R|t|}\Big) = \begin{cases} T^{-1}\begin{bmatrix} e^{tB} & 0 \\ 0 & 0 \end{bmatrix} T + O\big(\frac{1}{R|t|}\big) & \text{for } t > 0 \\ T^{-1}\begin{bmatrix} 0 & 0 \\ 0 & -e^{tC} \end{bmatrix} T + O\big(\frac{1}{R|t|}\big) & \text{for } t < 0 \end{cases}$$

$$= G(t, A) + O\Big(\frac{1}{R|t|}\Big).$$

Hence

$$G(t, A) = \lim_{R\to\infty} \frac{1}{2\pi}\int_{-R}^{+R} e^{i\omega t}[i\omega I - A]^{-1}d\omega.$$

This representation of the Green matrix $G(t, A)$ is often used in various applications.

In detailed courses of mathematical analysis, in connection with the so-called Fourier transform the following theorem is established.

THEOREM 1 (Plancherel). *Let $\varphi(\omega)$ be a square integrable function, i.e.,*

$$\int_{-\infty}^{+\infty} |\varphi(\omega)|^2 d\omega < \infty.$$

Then the sequence of functions

$$\widehat{\varphi}_k(t) = \frac{1}{\sqrt{2\pi}}\int_{-R_k}^{+R_k} e^{i\omega t}\varphi(\omega)\, d\omega, \tag{19}$$

where $R_k \to \infty$ as $k \to \infty$, form a Cauchy sequence in the sense of mean convergence, i.e.,

$$\int_{-\infty}^{\infty} |\widehat{\varphi}_k(t) - \widehat{\varphi}_l(t)|^2 dt < \varepsilon(\min\{k, l\}), \quad \varepsilon(k) \to 0 \text{ as } k \to \infty.$$

Therefore, there exists a square integrable function $\widehat{\varphi}(t)$ which is the limit of the sequence (19).

DEFINITION 1. The function $\widehat{\varphi}(t)$ in Theorem 1 is called the Fourier transform of the function $\varphi(\omega)$.

In addition, the Parseval equality

$$\int_{-\infty}^{\infty} |\widehat{\varphi}(t)|^2 dt = \int_{-\infty}^{\infty} |\varphi(\omega)|^2 d\omega$$

holds. If $\widehat{\psi}(t)$ is the Fourier transform of $\psi(\omega)$, then

$$\int_{-\infty}^{\infty} |\widehat{\psi}(t)|^2 dt = \int_{-\infty}^{\infty} |\psi(\omega)|^2 d\omega,$$

$$\int_{-\infty}^{\infty} \widehat{\varphi}(t)\overline{\widehat{\psi}}(t)\, dt = \int_{-\infty}^{\infty} \varphi(\omega)\overline{\psi}(\omega)\, d\omega.$$

The last equality is also called the Parseval equality.

Denoting the elements of the matrix $G(t, A)$ by $G_{jk}(t)$ and the elements of the matrix $D(\omega) = [i\omega I - A]^{-1}$ by $D_{jk}(\omega)$, we can interpret the equality

$$G_{jk}(t) = \lim_{R\to\infty} \frac{1}{2\pi} \int_{-R}^{+R} e^{i\omega t} D_{jk}(\omega)\, d\omega$$

as follows: $\sqrt{2\pi}G_{jk}(t)$ is the Fourier transform of $D_{jk}(\omega)$. By the Parseval equality, we have

$$2\pi \int_{-\infty}^{\infty} \overline{G}_{jl}(t)\, G_{jk}(t)\, dt = \int_{-\infty}^{\infty} \overline{D}_{jl}(\omega)\, D_{jk}(\omega)\, d\omega$$

$$\int_{-\infty}^{\infty} \left(\sum_{j=1}^{N} \overline{G}_{jl}(t)\, G_{jk}(t)\right) dt = \frac{1}{2\pi} \int_{-\infty}^{\infty} \left(\sum_{j=1}^{N} \overline{D}_{jl}(\omega)\, D_{jk}(\omega)\right) d\omega.$$

In other words, the last equality implies the matrix equality

$$\int_{-\infty}^{\infty} G^*(t)\, G(t)\, dt = \frac{1}{2\pi} \int_{-\infty}^{\infty} D^*(\omega)\, D(\omega)\, d\omega = \frac{1}{2\pi} \int_{-\infty}^{\infty} [A^* + i\omega I]^{-1}[A - i\omega I]^{-1} d\omega$$

which will play an important role in what follows.

§11. Boundary-value problem on a segment

The formulation of the boundary-value problem and the condition of the unique solvability. The representation of a solution by means of the Green matrix and some conditions. The existence and uniqueness of the Green matrix.

In this section, we study the problem of finding a solution $y(t)$ to the nonhomogeneous equation

$$\frac{d}{dt}y(t) = Ay(t) + f(t), \quad t_L \leqslant t \leqslant t_R \qquad (t_L \leqslant t_R) \tag{1}$$

with the boundary conditions

$$Ly(t_L) = l, \quad Ry(t_R) = r. \tag{2}$$

We assume that each of the matrices L, R has N columns (N is the number of components of the vector $y(t)$). The number of rows of the matrix L can differ from that of the matrix R. We denote the number of rows of the matrix L by k_L and the number of rows of the matrix R by k_R.

We try to clarify the conditions on L, l, R, r under which the boundary-value problem (1), (2) has a unique solution. Let $Y(t)$ be a fundamental matrix of solutions to the homogeneous equation

$$\frac{d}{dt}y(t) = Ay(t), \tag{3}$$

i.e.,

$$\frac{d}{dt}Y(t) = AY(t), \quad \det Y(t) \neq 0. \tag{4}$$

We know that

$$Y(t) = e^{(t-t_L)A}\,Y(t_L), \quad \det Y(t_L) \neq 0.$$

Consider the partial solution

$$y_0(t) = \int_{t_L}^{t} e^{(t-s)A}\,f(s)$$

to the nonhomogeneous equation (1) such that $y_0(t_L) = 0$. Any solution to the nonhomogeneous equation (1) can be represented in the form $y(t) = Y(t)a + y_0(t)$ and satisfies the condition $y(t_L) = Y(t_L)a$. It satisfies the boundary conditions (2) if

$$\begin{aligned} LY(t_L)a &= -Ly_0(t_L) + l \equiv l, \\ RY(t_R)a &= -Ry_0(t_R) + r \equiv r - R\int_{t_L}^{t_R} e^{(t_R-s)A}\,f(s)\,ds, \end{aligned}$$

i.e.,

$$\begin{bmatrix} LY(t_L) \\ RY(t_R) \end{bmatrix} a \equiv \begin{bmatrix} l \\ r - R\int_{t_L}^{t_R} e^{(t_R-s)A}\,f(s)\,ds \end{bmatrix}. \tag{5}$$

Thus, we reduce the question on the solvability condition for the boundary-value problem (1), (2) to that for a system of linear algebraic equations (5).

A necessary and sufficient condition for the system (5) to be solvable is that the rank of the augmented matrix

$$\begin{bmatrix} LY(t_L) & l \\ RY(t_R) & r - R\int_{t_L}^{t_R} e^{(t_R-s)A}\,f(s)\,ds \end{bmatrix} \tag{6}$$

coincides with the rank of the matrix of coefficients

$$\begin{bmatrix} LY(t_L) \\ RY(t_R) \end{bmatrix} \tag{7}$$

to the system (5). We impose one more condition in the statement of the boundary-value problem (1), (2). Namely, we require that the problem (1), (2) be solvable (in addition, the solution be unique) not only for a fixed, but for any l, r, $f(s)$.

Now we show that the matrix of coefficients (7) must be a square matrix. Let the rank of the matrix of coefficients (7) be equal to k, $k \leqslant N$. The augmented matrix has $N+1$ columns. Therefore, assuming that the number of rows (equal to

$k_L + k_R$) of (6) is greater than N, we always can choose l, r, $f(s)$ so that the last column is not a linear combination of the remaining columns and, consequently, the rank of the matrix (6) is equal to $k+1$. We see that the assumption $k_L + k_R > N$ contradicts the requirement that the system is solvable for any l, r, $f(s)$. Hence, the inequality $k_L + k_R \leqslant N$ must be valid. Since, by assumption, the number of rows of the matrix of coefficients (7) is less than N, there exists a nontrivial linear combination of the columns which is equal to zero. In this case, adding the vector composed of the coefficients of such a linear combination to some solution to the system (5), we again obtain a solution to the system (5). Consequently, the assumption $k_L + k_R < N$ contradicts the requirement that the system (5) has a unique solution. Thus,

$$k_L + k_R = N, \tag{8}$$

i.e., the matrix of coefficients (7) must be a square matrix.

For the unique solvability of a system of linear algebraic equations (5) with a square matrix of coefficients (7) it is necessary and sufficient that the determinant of (7) differs from zero:

$$\det \begin{bmatrix} LY(t_L) \\ RY(t_R) \end{bmatrix} \neq 0. \tag{9}$$

It may appear that the obtained condition depends on the choice of a fundamental system of solutions $Y(t)$. Let us show that this is not the case. We have $Y(t_R) = e^{(t_R - t_L)A} Y(t_L)$ and $\det Y(t_L) \neq 0$. Therefore, if the relation

$$\begin{aligned} \det \begin{bmatrix} LY(t_L) \\ RY(t_R) \end{bmatrix} &= \det \begin{bmatrix} LY(t_L) \\ Re^{(t_R-t_L)A} Y(t_L) \end{bmatrix} = \det \left\{ \begin{bmatrix} L \\ Re^{(t_R-t_L)A} \end{bmatrix} Y(t_L) \right\} \\ &= \det Y(t_L) \det \begin{bmatrix} L \\ Re^{(t_R-t_L)A} \end{bmatrix} \neq 0 \end{aligned}$$

is valid for some fundamental matrix, then it remains valid for any fundamental matrix. In particular, setting $Y(t) = e^{(t-t_L)A}$ ($Y(t_L) = I$, $\det Y(t_L) = 1$), we conclude that the boundary-value problem (1), (2) is uniquely solvable if

$$\det \begin{bmatrix} L \\ Re^{(t_R-t_L)A} \end{bmatrix} \neq 0. \tag{10}$$

Thus, we have proved the following theorem.

THEOREM 1. *The boundary-value problem* (1), (2) *has a unique solution* $y(t)$ *for any* l, r, $f(t)$ *if and only if the conditions* (8) *and* (10) *hold.*

EXAMPLE 1. Let $t_L = 0$, $t_R = 1$, and

$$A = \begin{bmatrix} 0 & -p \\ p & 0 \end{bmatrix}.$$

Consider the system

$$\frac{d}{dt}\begin{pmatrix} y_1 \\ y_2 \end{pmatrix} - \begin{bmatrix} 0 & -p \\ p & 0 \end{bmatrix} \begin{pmatrix} y_1 \\ y_2 \end{pmatrix} = \begin{pmatrix} f_1(t) \\ f_2(t) \end{pmatrix}$$

and the boundary conditions

$$y_1(0) = l, \quad y_2(1) = r,$$

i.e., $L = (1, 0)$, $R = (0, 1)$. Since

$$\exp\left(t\begin{bmatrix} 0 & -p \\ p & 0 \end{bmatrix}\right) = \begin{bmatrix} \cos pt & -\sin pt \\ \sin pt & \cos pt \end{bmatrix},$$

$$Re^{tA} \equiv \exp\left(t_R\begin{bmatrix} 0 & -p \\ p & 0 \end{bmatrix}\right) = (0 \quad 1)\begin{bmatrix} \cos p & -\sin p \\ \sin p & \cos p \end{bmatrix} = (\sin p \quad \cos p),$$

$$\det\begin{bmatrix} L \\ Re^{tA} \end{bmatrix} = \begin{bmatrix} 1 & 0 \\ \sin p & \cos p \end{bmatrix} = \cos p,$$

the problem is uniquely solvable for any l, r if $\cos p \neq 0$, i.e., if $p \neq (n + 1/2)\pi$, where n is an integer.

EXAMPLE 2. Let

$$A = \begin{bmatrix} p & 0 & 0 \\ 0 & 0 & 0 \\ 0 & 0 & -p \end{bmatrix}.$$

Consider the system

$$\frac{d}{dt}\begin{pmatrix} y_1 \\ y_2 \\ y_3 \end{pmatrix} = \begin{bmatrix} p & 0 & 0 \\ 0 & 0 & 0 \\ 0 & 0 & -p \end{bmatrix}\begin{pmatrix} y_1 \\ y_2 \\ y_3 \end{pmatrix} = \begin{pmatrix} f_1(t) \\ f_2(t) \\ f_3(t) \end{pmatrix}, \qquad 0 \leqslant t \leqslant 1,$$

and the boundary conditions (2) at $t = t_L = 0$ and $t = t_R = 1$ with

$$L = \begin{bmatrix} 1 & 0 & -1 \\ 1 & 1 & 0 \end{bmatrix}, \qquad R = (1 \quad -p \quad 1),$$

the right-hand side l in the left boundary condition $Ly(t_L) = l$ is a two-dimensional vector with components l_1 and l_2, whereas the right-hand side r in the right boundary condition $Ry(t_R) = r$ has only one component which will be denoted by r. In this example,

$$e^{tA} = \begin{bmatrix} e^{pt} & 0 & 0 \\ 0 & 1 & 0 \\ 0 & 0 & e^{-pt} \end{bmatrix},$$

$$Re^{tA} = (e^{pt} - pe^{-pt}),$$

$$\det\begin{bmatrix} L \\ Re^{A} \end{bmatrix} = \det\begin{bmatrix} 1 & 0 & -1 \\ 1 & 1 & 0 \\ e^p & -p & e^{-p} \end{bmatrix} = p + e^p + e^{-p} = p + 2\operatorname{ch} p.$$

It is easy to show that the equation $p + 2\operatorname{ch} p = 0$ has no real roots. Therefore, the boundary-value problem is uniquely solvable for any l_1, l_2, r, and any continuous vector-valued function $f(t)$ independently of number p.

We turn to the construction of the theory of boundary-value problems on the segment $t_L \leqslant t \leqslant t_R$. The conditions (8), (10) are assumed to hold. Choosing s in the interval $t_L < s < t_R$, we define $y_0(t, s)$ by the equality

$$y_0(t, s) = \begin{cases} 0 & \text{for } t_L \leqslant t < s, \\ e^{(t-s)A} f_0 & \text{for } s \leqslant t < t_R, \end{cases}$$

where f_0 is a constant vector. It is obvious that for $t \neq s$ the vector-valued function $y_0(t,s)$ is continuously differentiable; moreover,

$$\frac{d}{dt} y_0(t,s) = Ay_0(t,s) \quad (t \neq s).$$

For $t = s$ each of the components of the vector $y_0(t,s)$ has a discontinuity of the first kind, hence

$$y_0(s+0,s) - y_0(s-0,s) = f_0.$$

It is not difficult to verify that any vector-valued function of the form

$$y(t,s) = e^{(t-t_L)A}\, a + y_0(t,s)$$

possesses the same properties:

$$\frac{d}{dt} y(t,s) = Ay(t,s) \text{ for } t \neq s,$$
$$y(s+0,s) - y(s-0,s) = f_0.$$

Take a so as to satisfy the boundary conditions:

$$Ly(t_L,s) = l, \quad Ry(t_R,s) = r.$$

Since

$$y(t_L,s) = a,$$
$$y(t_R,s) = e^{(t_R-t_L)A}\, a + y_0(t_R,s),$$

the vector a can be obtained as a solution to the system

$$\begin{bmatrix} L \\ Re^{(t_R-t_L)A} \end{bmatrix} a = \begin{pmatrix} l \\ r - Ry_0(t_R,s) \end{pmatrix}.$$

The vector a is uniquely found from this system of linear algebraic equations because its determinant is assumed to be not zero. Thus, we have proved the following lemma.

LEMMA 1. *Under the conditions* (8) *and* (10), *for any* $t_L < s < t_R$ *and vectors* l, r, f_0 *of dimensions* k_L, k_R, N *respectively there exists a vector-valued function* $y(t,s)$ *that is continuously differentiable in* t *on each of the intervals* $t_L \leqslant t < s$, $s < t \leqslant t_R$ *and satisfies the homogeneous differential equation*

$$\frac{d}{dt} y(t) = Ay(t) \tag{11}$$

on this interval. Moreover, for $t = s$ *the function* $y(t,s)$ *has a discontinuity of the first kind*

$$y(s+0,s) - y(s-0,s) = f_0 \tag{12}$$

and at $t = t_L$ *and* $t = t_R$ *satisfies the boundary conditions*

$$Ly(t_L,s) = l, \quad Ry(t_R,s) = r. \tag{13}$$

REMARK 1. It is not difficult to complement the statement of Lemma 1 by the assertion that for $t \neq s$ the vector-valued function $y(t,s)$ is a continuous function of s provided that f_0 is independent of s. We strongly recommend the reader to justify accurately this additional assertion because we will use it hereinafter.

We apply Lemma 1 to construct Green matrices that will be used for the representation of a solution to the boundary-value problem for arbitrary $f(t)$, l, r.

DEFINITION 1. By Green matrices of the boundary-value problem (1), (2) we mean

(a) the $N \times N$-matrix $G(t,s)$ such that

$$\begin{cases} G(s+0,s) - G(s-0,s) = I_N, \\ \dfrac{d}{dt}G(t,s) = AG(t,s) \quad (t \neq s), \\ LG(t_L,s) = 0, \quad RG(t_R,s) = 0; \end{cases} \tag{14}$$

(b) the $N \times k_L$-matrix $G_L(t)$ such that

$$\begin{cases} \dfrac{d}{dt}G_L(t) = AG_L(t), \\ LG_L(t_L) = I_{k_L}, \quad RG_L(t_R) = 0; \end{cases} \tag{15}$$

(c) the $N \times k_R$-matrix $G_R(t)$ such that

$$\begin{cases} \dfrac{d}{dt}G_R(t) = AG_R(t), \\ LG_R(t_L) = 0, \quad RG_R(t_R) = I_{k_R}. \end{cases} \tag{16}$$

It is easy to show that the Green matrices $G(t,s)$, $G_L(t)$, and $G_R(t)$ are uniquely defined by the conditions (14), (15), and (16) respectively.

We begin with the matrix $G(t,s)$ consisting of N columns, each being an N-dimensional vector. For $t_L < t < s$ and $s < t < t_R$ the pth column $g_p(t,s)$ satisfies the differential equation

$$\frac{d}{dt}g_p(t,s) = Ag_p(t,s)$$

and the boundary conditions

$$Lg_p(t_L,s) = 0, \quad Rg_p(t_R,s) = 0,$$

$$g_p(s+0,s) - g_p(s-0,s) = \begin{pmatrix} 0 \\ \vdots \\ 0 \\ 1 \\ 0 \\ \vdots \\ 0 \end{pmatrix} \begin{matrix} \\ \\ \\ \leftarrow \text{the } p\text{th row.} \\ \\ \\ \\ \end{matrix} \tag{17}$$

We know that under the assumption (10) the vector-valued function $g_p(t,s)$ is unique.

Having constructed $g_1(t,s), g_2(t,s), \dots, g_N(t,s)$, we thereby construct all columns of $G(t,s)$. Thus, $G(t,s)$ is unique. In view of Remark 1, we can prove that

$G(t, s)$ is continuous with respect to both variables for $t \neq s$.[6] The existence and uniqueness of the matrices $G_L(t)$ and $G_R(t)$ can be justified in the same way.

Each of the columns of these matrices is a solution to the boundary-value problem which is uniquely solvable. For example, the qth column $g_{L,q}(t)$ of the matrix $G_L(t)$ satisfies the following conditions:

$$\frac{d}{dt} g_{L,q}(t) = A g_{L,q}(t),$$

$$L g_{L,q}(t_L) = \left.\begin{pmatrix} 0 \\ \vdots \\ 0 \\ 1 \\ 0 \\ \vdots \\ 0 \end{pmatrix} \begin{matrix} \\ \\ \\ \leftarrow \text{the } q\text{th component} \\ \\ \\ \\ \end{matrix} \right\} k_L \text{components},$$

$$R g_{L,q}(t_R) = \left.\begin{pmatrix} 0 \\ 0 \\ \vdots \\ 0 \\ \vdots \\ 0 \end{pmatrix} \right\} k_R \text{ components}.$$

At this point we complete the justification of the existence and uniqueness of the Green matrices $G(t, s)$, $G_L(t)$, and $G_R(t)$.

REMARK 2. The matrix $G(t, s)$ defined on the square $t_L < t < t_R$, $t_L < s < t_R$, depends continuously on t and on s if $t \neq s$, i.e., the elements of $G(t, s)$ may have discontinuity only on the diagonal $t = s$ of this square.

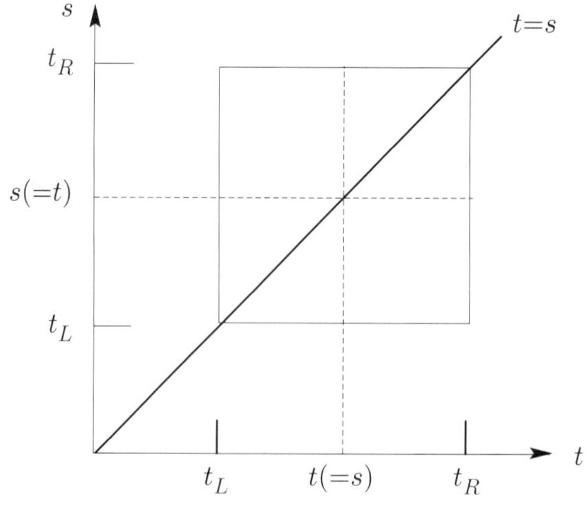

FIGURE 1

[6] In fact, $G(t, s)$ is even continuous for $t \geqslant s$ if we set that $G(s, s) = G(s + 0, s)$. Similarly, $G(t, s)$ is continuous for $t \leqslant s$ under the assumption that $G(s, s) = G(s - 0, s)$.

means that if we watch the change of $G(t,s)$ fixing s and increasing t, i.e., moving along the horizontal line to the right, then we see that passing through the diagonal $t = s$, the matrix has a step discontinuity of size I_N.

Figure 1 shows that the limit from the left $\lim\limits_{\substack{t\to s\\ t<s}} G(t,s) = G(s-0,s)$ is equal to the limit from above $\lim\limits_{\substack{s\to t\\ s>t}} G(t,s) = G(t,t+0)$ on the diagonal. Similarly, the limit from the right $\lim\limits_{\substack{t\to s\\ t>s}} G(t,s) = G(s+0,s)$ is equal to the limit from below $\lim\limits_{\substack{s\to t\\ s<t}} G(t,s) = G(t,t-0)$ on the diagonal. This fact allows us to rewrite the equality (17) for the matrix $G(t,s)$ in the following equivalent form:

$$G(t,t-0) - G(t,t+0) = I_N. \tag{18}$$

Now, we can derive a formula in which the Green matrices are used in order to represent a solution.

We show that the function

$$y(t) = G_L(t)l + \int_{t_L}^{t_R} G(t,s)f(s)\,ds + G_R(t)r \tag{19}$$

is a solution to the boundary-value problem (1), (2). We compute the derivative $(t_L < t < t_R)$

$$\begin{aligned}
\frac{d}{dt}y(t) &= \left[\frac{d}{dt}G_L(t)\right] l + \frac{d}{dt}\left\{\int_{t_L}^{t} G(t,s)f(s)\,ds + \int_{t}^{t_R} G(t,s)f(s)\,ds\right\} + \left[\frac{d}{dt}G_R(t)\right] r \\
&= AG_L(t)l + \left\{\int_{t_L}^{t} \left[\frac{d}{dt}G(t,s)\right] f(s)\,ds + G(t,t-0)f(t)\right. \\
&\qquad\qquad \left. + \int_{t}^{t_R} \left[\frac{d}{dt}G(t,s)\right] f(s)\,ds - G(t,t+0)f(t)\right\} + AG_R(t)r \\
&= AG_L(t)l + A\int_{t_L}^{t} G(t,s)f(s)\,ds + A\int_{t}^{t_R} G(t,s)f(s)\,ds + AG_R(t)r \\
&\quad + \big[G(t,t-0) - G(t,t+0)\big]f(t) \\
&= A\left\{G_L(t)l + \int_{t_L}^{t_R} G(t,s)f(s)\,ds + G_R(t)r\right\} + f(t) = Ay(t) + f(t).
\end{aligned}$$

Hence $y(t)$ is a solution to the differential equation (1).

We show that $y(t)$ satisfies the boundary conditions (2). Indeed, for $t_L < t < s < t_R$ we have

$$Ly(t) = L\left\{G_L(t)l + \int_{t_L}^{t_R} G(t,s)f(s)\,ds + G_R(t)r\right\}$$
$$= [LG_L(t)]l + \int_{t_L}^{t_R} [LG(t,s)]f(s)\,ds + [LG_R(t)]r.$$

Using the continuity of $y(t)$, $G_L(t)$, and $G_R(t)$ for $t_L \leqslant t \leqslant t_R$ and the equalities

$$LG_L(t_L) = I_{k_L}, \quad LG(t_L, s) = 0, \quad LG_R(t_L) = 0,$$

we find that

$$Ly(t_L) = [LG_L(t_L)]l + \int_{t_L}^{t_R} [LG(t_L,s)]f(s)\,ds + [LG_R(t_L)]r = I_{k_L}l = l.$$

The right boundary condition $Ry(t_R) = r$ can be verified in a similar way. Therefore, the following theorem has been proved.

THEOREM 2. *Under the assumptions of Theorem* 1, *a unique solution* $y(t)$ *to the boundary-value problem* (1), (2) *is represented in the form* (19).

§12. The closeness estimates for solutions to the boundary-value problems with close coefficients and close right-hand sides

An estimate for a solution to the boundary-value problem by means of the estimates for Green matrices. A new criterion for the unique solvability of the boundary-value problem. The stability of a solution under perturbations of coefficients, boundary conditions, and right-hand sides.

In this section, we deal with the boundary-value problem

$$\frac{d}{dt}y(t) = Ay(t) + f(t), \quad 0 < t < T, \tag{1}$$
$$Ly(0) = \varphi, \quad Ry(T) = \psi. \tag{2}$$

We assume that the necessary conditions hold under which the problem (1), (2) is well posed (cf. §11). In particular, the following conditions (cf. (10) in §11) must be valid:

$$\Delta = \det\begin{bmatrix} L \\ Re^{TA} \end{bmatrix} \neq 0.$$

Certainly, the total number of rows of the matrices L and R must coincide with the number of components of the vector-valued function $y(t)$. Under this assumption, the solution $y(t)$ to the boundary-value problem (1), (2) can be represented in terms of the Green matrices $G(t,s)$, $G_L(t)$, and $G_R(t)$ by the formula

$$y(t) = G_L(t)\varphi + \int_0^T G(t,s)f(s)\,ds + G_R(t)\psi. \tag{3}$$

If

$$\|G_L\| \leqslant K, \quad \|G(t,s)\| \leqslant K, \quad G_R(t)\| \leqslant K, \tag{4}$$

then the solution $y(t)$ satisfies the estimate

$$\|y(t)\| \leqslant K(\|\varphi\| + \|\psi\|) + KT \max_s \|f(s)\|. \tag{5}$$

To characterize properties of the boundary-value problem (1), (2), it is convenient to use the constant K appearing in the estimates (4) for the Green functions instead of the determinant Δ. The constant K can be used in the study of the stability of a solution under small variations of parameters occurring in the matrices A, L, and R and in the right-hand sides $f(t)$, φ, ψ.

EXAMPLE 1.

$$A = \begin{bmatrix} a & 0 & 0 \\ 0 & 0 & 0 \\ 0 & 0 & -a \end{bmatrix}, \quad L = (1 \quad 0 \ -1), \quad R = \begin{bmatrix} 1 & 0 & 0 \\ 1 & -1 & 0 \end{bmatrix}, \quad T = 1.$$

It is required to find a solution to the system

$$\begin{aligned} \frac{dy_1}{dt} &= ay_1 + f_1(t) \\ \frac{dy_2}{dt} &= f_2(t) \\ \frac{dy_3}{dt} &= -ay_3 + f_3(t) \end{aligned}$$

with the boundary conditions

$$y_1(0) - y_3(0) = \varphi, \quad y_1(1) = \psi_1, \quad y_1(1) - y_2(1) = \psi_2.$$

For this problem we have

$$Re^{TA} = \begin{bmatrix} 1 & 0 & 0 \\ 1 & -1 & 0 \end{bmatrix} \cdot \begin{bmatrix} e^a & 0 & 0 \\ 0 & 1 & 0 \\ 0 & 0 & e^{-a} \end{bmatrix} = \begin{bmatrix} e^a & 0 & 0 \\ e^a & -1 & 0 \end{bmatrix}$$

$$\Delta = \det \begin{bmatrix} L \\ Re^{TA} \end{bmatrix} = \det \begin{bmatrix} 1 & 0 & -1 \\ e^a & 0 & 0 \\ e^a & -1 & 0 \end{bmatrix} = e^a \neq 0.$$

The Green functions are represented by the formulas

$$G(t,s) = \begin{cases} \begin{bmatrix} -e^{(t-s)a} & 0 & 0 \\ 0 & -1 & 0 \\ -e^{-(t+s)a} & 0 & 0 \end{bmatrix}, & t < s, \\ \begin{bmatrix} 0 & 0 & 0 \\ 0 & 0 & 0 \\ -e^{-(t+s)a} & 0 & e^{-(t-s)a} \end{bmatrix}, & t > s, \end{cases}$$

$$G_L(t) = \begin{bmatrix} 0 \\ 0 \\ -e^{-ta} \end{bmatrix},$$

$$G_R(t) = \begin{bmatrix} e^{(t-1)a} & 0 \\ 1 & -1 \\ e^{-(t+1)a} & 0 \end{bmatrix}.$$

For $a > 0$ all nonzero elements of the Green matrices do not exceed 1 by module. Since the norm of a matrix is less than the square root of the sum of squares of its elements, we have $(a > 0)$

$$\|G(t,s)\| \leqslant \sqrt{3}, \qquad \|G_L(t)\| \leqslant 1, \qquad \|G_R(t)\| \leqslant 2,$$

i.e., we can take $K = 2$.

We consider simultaneously two boundary-value problems of type (1), (2) with close date. Namely, let $y(t)$, $0 \leqslant t \leqslant T$, be a solution to the problem (1), (2) with A, L, R and let $\widetilde{y}(t)$ be a solution to the boundary-value problem

$$\frac{d}{dt}\widetilde{y}(t) = \widetilde{A}\widetilde{y}(t) + f(t) \tag{6}$$

$$\widetilde{L}\widetilde{y}(0) = \widetilde{\varphi}, \quad \widetilde{R}\widetilde{y}(T) = \widetilde{\psi}. \tag{7}$$

The matrices $\widetilde{L}$ and $\widetilde{R}$ are assumed to have the same number of rows and columns as the matrices L and R respectively. We mean that the problems (1), (2) and (6), (7) are close in the following sense:

$$\begin{aligned} &\|\widetilde{A} - A\| \leqslant \varepsilon, \quad \|\widetilde{f} - f\| \leqslant \varepsilon, \quad \|\widetilde{\varphi} - \varphi\| \leqslant \varepsilon, \\ &\|\widetilde{\psi} - \psi\| \leqslant \varepsilon, \quad \|\widetilde{L} - L\| \leqslant \varepsilon, \quad \|\widetilde{R} - R\| \leqslant \varepsilon, \end{aligned} \tag{8}$$

where ε is a sufficiently small number. Throughout the section, $f(t)$ and $\widetilde{f}(t)$ are continuous vector-valued functions.

We assume that the problem (1), (2) is uniquely solvable for any right-hand sides $f(t)$, φ, ψ and the solution $y(t)$ satisfies the estimate (5). Using the equation (1), we easily establish the following inequalities for the derivative $y'(t)$:

$$\begin{aligned} &\|y'(t)\| \leqslant \|A\|\,\|y(t)\| + \|f(t)\|, \\ &\|y'(t)\| \leqslant \|A\|K\big[\|\varphi\| + \|\psi\|\big] + \big[1 + \|A\|KT\big] \max_{0\leqslant t\leqslant T} \|f(t)\|. \end{aligned}$$

It turns out that the solvability of the problem (1), (2) implies the solvability of the problem (6), (7) if ε is not too large. Let us prove this assertion.

Starting from solution $y(t)$ to the problem (1), (2), we define the vectors φ_1, ψ_1 and the vector-valued function $f_1(t)$ by the formulas

$$\begin{aligned} \varphi_1 &= (L - \widetilde{L})y(0) + \widetilde{\varphi} - \varphi, \\ \psi_1 &= (R - \widetilde{R})y(T) + \widetilde{\psi} - \psi, \\ f_1(t) &= (\widetilde{A} - A)y(t) + \widetilde{f}(t) - f(t). \end{aligned}$$

The following estimates hold:

$$\|\varphi_1\| \leqslant \varepsilon F, \quad \|\psi_1\| \leqslant \varepsilon F, \quad \|f_1(t)\| \leqslant \varepsilon F,$$

where $F = 1 + K\big[\|\varphi\| + \|\psi\| + T\max\limits_{0\leqslant t\leqslant T}\|f(t)\|\big]$. Now we can construct $y_1(t)$ as the solution to the boundary-value problem

$$\frac{d}{dt}y_1(t) = Ay_1(t) + f_1(t),$$

$$Ly_1(0) = \varphi_1, \quad Ry_1(T) = \psi_1$$

which differs from the problem (1), (2) by the right-hand side $f_1(t)$. The solution $y_1(t)$ exists, is unique, and satisfies the estimate (5). Having constructed $y_1(t)$, we define the vectors φ_2, ψ_2 and the vector-valued function $f_2(t)$ by the formulas

$$\begin{aligned}\varphi_2 &= (L - \widetilde{L})y_1(0),\\ \psi_2 &= (R - \widetilde{R})y_1(T),\\ f_2(t) &= (\widetilde{A} - A)y_1(t).\end{aligned}$$

We can assert that

$$\|\varphi_2\| \leqslant \varepsilon^2K(2+T)F, \quad \|\psi_2\| \leqslant \varepsilon^2K(2+T)F, \quad \|f_2(t)\| \leqslant \varepsilon^2K(2+T)F.$$

A solution $y_2(t)$ to the boundary-value problem

$$\frac{d}{dt}y_2(t) = Ay_2(t) + f_2(t),$$

$$Ly_2(0) = \varphi_2, \quad Ry_2(t) = \psi_2$$

exists, is unique, and satisfies the estimate

$$\|y_2(t)\| \leqslant [\varepsilon K(2+T)]^2F.$$

We describe the process of successive definitions of continuous and continuously differentiable vector-valued functions $y_2(t), y_3(t), \dots, y_j(t), \dots$. We assume that we already have y_{j-1} such that

$$\|y_{j-1}(t)\| \leqslant [\varepsilon K(2+T)]^{j-1}F.$$

We set

$$\varphi_j = (L - \widetilde{L})y_{j-1}(0), \quad \psi_j = (R - \widetilde{R})y_{j-1}(T), \quad f_j(t) = (\widetilde{A} - A)y_{j-1}(t)$$

and verify that

$$\begin{aligned}\|\varphi_j\| &\leqslant \varepsilon[\varepsilon K(2+T)]^{j-1}F,\\ \|\psi_j\| &\leqslant \varepsilon[\varepsilon K(2+T)]^{j-1}F,\\ \|f_j(t)\| &\leqslant \varepsilon[\varepsilon K(2+T)]^{j-1}F.\end{aligned}$$

The function $y_j(t)$ is found as a solution to the boundary-value problem

$$\frac{d}{dt}y_j(t) = Ay_j(t) + f_j(t), \tag{9}$$

$$Ly_j(0) = \varphi_j, \quad Ry(T) = \psi_j. \tag{10}$$

A solution $y_j(t)$ to the last problem exists and is unique. Indeed, this problem differs from the problem (1), (2) only by right-hand sides (φ_j instead of φ, $f_j(t)$

instead of $f(t)$ and ψ_j instead of ψ). But, by assumption, the problem (1), (2) is uniquely solvable for any right-hand sides. The solution $y_j(t)$ satisfies the estimate

$$\begin{aligned}\|y_j(t)\| &\leqslant K\big(\|\varphi_j\| + \|\psi_j\| + T \max_{0\leqslant t\leqslant T} \|f_j(t)\|\big)\\ &\leqslant K(2+T)\varepsilon[\varepsilon K(2+T)]^{j-1}F = [\varepsilon K(2+T)]^j F.\end{aligned}$$

We note that the vector-valued function $y_j(t)$ is a solution to the differential equation, is differentiable, and its derivative can be estimated as follows:

$$\begin{aligned}\left\|\frac{dy_j}{dt}\right\| &\leqslant \|A\|\,\|y_j(t)\| + \|f_j(t)\| \leqslant \|A\|[\varepsilon K(2+T)]^j F + \varepsilon[\varepsilon K(2+T)]^{j-1}F\\ &= \left[\|A\| + \frac{1}{K(2+T)}\right][\varepsilon K(2+T)]^j F.\end{aligned}$$

If

$$\varepsilon K(2+T) < 1, \tag{11}$$

then from the above estimates it follows that the series

$$\widetilde{y}(t) = y(t) + \sum_{j=1}^{\infty} y_j(t), \qquad y'(t) + \sum_{j=1}^{\infty} y_j'(t)$$

converge uniformly on $[0,T]$. Consequently, $\widetilde{y}(t)$, being the sum of the first series, is a continuous and continuously differentiable vector-valued function.

It is easy to check that

$$\begin{aligned}L\widetilde{y}(0) &= Ly(0) + \varphi_1 + \varphi_2 + \varphi_3 + \dots\\ &= Ly(0) + \big[(L-\widetilde{L})y(0) + (\widetilde{\varphi} - \varphi)\big] + \sum_{j=1}^{\infty}(L-\widetilde{L})y_j(0)\\ &= Ly(0) + (L-\widetilde{L})\left[y(0) + \sum_{j=1}^{\infty} y_j(0)\right] + \widetilde{\varphi} - \varphi\\ &= Ly(0) + (L-\widetilde{L})\widetilde{y}(0) + \widetilde{\varphi} - \varphi = \varphi + (L-\widetilde{L})\widetilde{y}(0) + \widetilde{\varphi} - \varphi\\ &= (L-\widetilde{L})\widetilde{y}(0) + \widetilde{\varphi}\end{aligned}$$

so that $\widetilde{L}\widetilde{y}(0) = \widetilde{\varphi}$. Similarly, it is easy to check the validity of the right boundary condition $\widetilde{R}y(T) = \widetilde{\psi}$. Furthermore,

$$\begin{aligned}\frac{d}{dt}\widetilde{y}(t) - A\widetilde{y}(t) &= \left(\frac{d}{dt}y(t) - Ay(t)\right) + \sum_{j=1}^{\infty}\left(\frac{dy_j}{dt} - Ay_j(t)\right) = f(t) + \sum_{j=1}^{\infty} f_j(t)\\ &= f(t) + (\widetilde{A} - A)y(t) + \widetilde{f}(t) - f(t) + \sum_{j=1}^{\infty}(\widetilde{A} - A)y_j(t)\\ &= \widetilde{f}(t) + (\widetilde{A} - A)\left[y(t) + \sum_{j=1}^{\infty} y_j(t)\right] = \widetilde{f}(t) + (\widetilde{A} - A)\widetilde{y}(t).\end{aligned}$$

Having

$$\frac{d}{dt}\widetilde{y} - A\widetilde{y} = \widetilde{f}(t) + (\widetilde{A} - A)\widetilde{y}(t),$$

we have shown that $\widetilde{y}(t)$ is a solution to the problem (6), (7) under the assumptions (8), (11). Since the boundary-value problem (1), (2) is solvable for any right-hand sides $f(t)$, φ, ψ, the problem (6), (7) is also solvable for any $\widetilde{f}(t)$, $\widetilde{\varphi}$, $\widetilde{\psi}$ provided that

$$\|A-\widetilde{A}\| \leqslant \varepsilon, \quad \|R-\widetilde{R}\| \leqslant \varepsilon, \quad \|L-\widetilde{L}\| \leqslant \varepsilon, \quad \varepsilon K(2+T)<1.$$

We rewrite the problem (6), (7) in the form

$$\begin{aligned} L\widetilde{y}(0) &= (L-\widetilde{L})\widetilde{y}(0)+\widetilde{\varphi}, \\ \frac{d}{dt}\widetilde{y} &= Ay(t)+\big[(\widetilde{A}-A)\widetilde{y}(t)+\widetilde{f}(t)\big], \\ R\widetilde{y}(T) &= (R-\widetilde{R})\widetilde{y}(T)+\widetilde{\psi} \end{aligned} \tag{12}$$

and consider it as the problem (1), (2) with the right-hand sides $(L-\widetilde{L})\widetilde{y}(0)+\widetilde{\varphi}$, $(\widetilde{A}-A)\widetilde{y}+\widetilde{f}$, and $(R-\widetilde{R})\widetilde{y}(T)+\widetilde{\psi}$. For $\widetilde{y}(t)$ we obtain the inequalities

$$\begin{aligned} \|\widetilde{y}(t)\| \leqslant K\Big[&\|L-\widetilde{L}\|\,\|\widetilde{y}(0)\|+\|\widetilde{\varphi}\|+\|R-\widetilde{R}\|\,\|\widetilde{y}(T)\| \\ &+\|\widetilde{\psi}\|+T\max_{0\leqslant t\leqslant T}\big(\|\widetilde{A}-A\|\,\|\widetilde{y}(t)\|+\|\widetilde{f}(t)\|\big)\Big], \end{aligned}$$

$$\max_{0\leqslant t\leqslant T}\|\widetilde{y}(t)\| \leqslant \varepsilon K(2+T)\max_{0\leqslant t\leqslant T}\|\widetilde{y}(t)\|+K\Big[\|\widetilde{\varphi}\|+\|\widetilde{\psi}\|+T\max_{0\leqslant t\leqslant T}\|\widetilde{f}(t)\|\Big],$$

$$\|\widetilde{y}(t)\| \leqslant \frac{K}{1-\varepsilon K(2+T)}\Big[\|\widetilde{\varphi}\|+\|\widetilde{\psi}\|+T\max_{0\leqslant t\leqslant T}\|\widetilde{f}(t)\|\Big].$$

If $\widetilde{\varphi}=0$, $\widetilde{\psi}=0$, and $\widetilde{f}(t)=0$, then $\widetilde{y}(t)=0$ is a solution to the problem (6), (7), i.e., the latter is uniquely solvable.

Now we can compare the solutions to the problems (1), (2) and (6), (7). Since

$$\begin{gathered} \frac{dy}{dt}=Ay+f(t), \quad \frac{d\widetilde{y}}{dt}=\widetilde{A}\widetilde{y}+\widetilde{f}(t), \\ Ly(0)=\varphi, \quad Ry(T)=\psi, \quad \widetilde{L}\widetilde{y}(0)=\widetilde{\varphi}, \quad \widetilde{R}\widetilde{y}(T)=\widetilde{\psi}, \end{gathered}$$

the difference $y(t)-\widetilde{y}(t)$ satisfies the equation

$$\frac{d}{dt}(y-\widetilde{y})=A(y-\widetilde{y})+(A-\widetilde{A})\widetilde{y}+f(t)-\widetilde{f}(t) \tag{13}$$

and the boundary conditions

$$\begin{aligned} L[y(0)-\widetilde{y}(0)] &= \varphi-\widetilde{\varphi}+(\widetilde{L}-L)\widetilde{y}(0), \\ R[y(T)-\widetilde{y}(T)] &= \psi-\widetilde{\psi}+(\widetilde{R}-R)\widetilde{y}(T). \end{aligned} \tag{14}$$

In addition,

$$\begin{aligned} \|(A-\widetilde{A})\widetilde{y}(t)+f(t)-\widetilde{f}(t)\| &\leqslant \varepsilon\|\widetilde{y}(t)\|+\varepsilon \\ &\leqslant \varepsilon+\frac{\varepsilon K}{1-\varepsilon K(2+T)}\Big[\|\varphi\|+\|\psi\|+T\max_{0\leqslant t\leqslant T}\|f(t)\|+(2+T)\varepsilon\Big]=\zeta, \\ \|\varphi-\widetilde{\varphi}+(\widetilde{L}-L)\widetilde{y}(0)\| &\leqslant \varepsilon+\varepsilon\|\widetilde{y}(0)\| \leqslant \zeta, \\ \|\psi-\widetilde{\psi}+(\widetilde{R}-R)\widetilde{y}(T)\| &\leqslant \varepsilon+\varepsilon\|\widetilde{y}(T)\| \leqslant \zeta. \end{aligned}$$

By the properties of the solution to the problem (1), (2), we obtain the estimate

$$\|y(t) - \widetilde{y}(t)\| \leqslant K(2+T)\zeta,$$

with the right-hand side being small for small ε. This estimate means that the solution $y(t)$ to the boundary-value problem (1), (2) depends continuously on the coefficients A, L, and R and on the right-hand sides φ, ψ, and $f(t)$. The length T of the segment on which we seek a solution and the bound K for the norms of the Green matrices enter into this estimate.

§13. The Lopatinskii condition

The boundary-value problem on a half-line. The unique solvability condition. The Lopatinskii estimate. An example. The invariant form of the boundary conditions. The Lopatinskii constants. The continuation of the discussion of the example.

In this section, we study one more problem which often must be solved. It is required to determine a bounded solution $y(t)$ to the vector equation

$$\frac{d}{dt}y(t) = Ay(t) + f(t), \quad t \geqslant 0, \tag{1}$$

where $f(t)$ is a continuous bounded vector-valued function ($\|f(t)\| \leqslant F$), with the boundary condition

$$My(0) = \varphi \tag{2}$$

at $t = 0$. The matrix M is rectangular and the number of components of the vector φ coincides with the number of rows of the matrix M. The rows of the matrix M are assumed to be linearly independent. Otherwise, some equalities are linear combinations of remaining equalities in the boundary condition and can be eliminated as unnecessary conditions. Thereby, we can decrease the number of the boundary conditions, i.e., decrease the number of rows of the matrix M. We also assume that the matrix A has no purely imaginary eigenvalues:

$$\begin{gathered} \operatorname{Re}\tau_1 \leqslant -\sigma, \quad \operatorname{Re}\tau_2 \leqslant -\sigma, \quad \dots, \quad \operatorname{Re}\tau_k \leqslant -\sigma, \\ \operatorname{Re}\tau_{k+1} \geqslant \sigma, \quad \dots, \quad \operatorname{Re}\tau_N \geqslant \sigma \quad (\sigma > 0). \end{gathered} \tag{3}$$

Generally speaking, the problem (1), (2) is not well posed for arbitrary boundary condition. To guarantee the existence and uniqueness of a solution, we must impose an additional condition on the matrix M.

THE LOPATINSKII CONDITION. The boundary-value problem for the homogeneous vector equation

$$\frac{d}{dt}v(t) = Av(t) \tag{4}$$

with the boundary condition

$$Mv(0) = \varphi \tag{5}$$

has a unique bounded solution $v(t)$, $0 \leqslant t < \infty$ ($\|\nu(t)\| < \infty$) for any vector φ such that the number of its components coincides with the number of rows of the matrix M.

DEFINITION 1. The constant

$$\mathcal{L} = \|M\| \max_{\|\varphi\|=1} \|v(0)\| \tag{6}$$

is called the external Lopatinskii constant.

Any solution to the problem (4), (5) satisfies the estimate

$$\|v(0)\| \leqslant \frac{\mathcal{L}}{\|M\|} \|\varphi\|. \tag{7}$$

The external Lopatinskii constant $\mathcal{L}$ is the minimal possible constant in the estimate (7).

We will show that under the Lopatinskii condition the number of rows of the matrix M coincides with the number of those eigenvalues of the matrix A that have the negative real parts. We will also give a criterion for fulfillment of the Lopatinskii condition and prove some estimates.

THEOREM 1. *Let the Lopatinskii condition be fulfilled. For any continuous bounded function $f(t)$ ($\|f(t)\| \leqslant F$) defined for $t \geqslant 0$ there exists a unique bounded vector-valued function $y(t)$ which is a solution to the problem* (1), (2).

PROOF. The uniqueness directly follows from the Lopatinskii condition.

To prove the existence, we look for $y(t)$ as the sum $y(t) = u(t) + v(t)$, where $u(t)$, $0 < t < \infty$, is a bounded solution to the equation (1). For example, we can take (cf. §9)

$$u(t) = \int_0^\infty G_0(t - s, A) f(s)\, ds.$$

Then $v(t)$ is a bounded solution to the homogeneous equation (4) and satisfies the boundary condition

$$Mv(0) = \varphi - Mu(0) \tag{8}$$

at $t = 0$. The existence of $v(t)$ follows from the Lopatinskii condition. □

Denote by $\mathcal{M}$ the set of bounded solutions to the homogeneous equation (4) for $t \geqslant 0$, i.e.,

$$\mathcal{M} = \left\{ v(t) \;\middle|\; \frac{dv(t)}{dt} = Av(t),\ \|v(t)\| \leqslant m, \quad t \geqslant 0 \right\}.$$

In this definition the constant m depends on a solution $v(t)$. It is obvious that $\mathcal{M}$ is a linear space.

PROPOSITION 1. *The dimension k of $\mathcal{M}$ is equal to the number of those eigenvalues of the matrix A that have negative real parts.*

PROOF. There exists a unitary matrix U ($U^*U = I$) such that A can be written in the form

$$A = U^* \begin{bmatrix} B & D \\ 0 & C \end{bmatrix} U = U^* \widehat{A} U,$$

where the eigenvalues $\tau_1, \tau_2, \ldots, \tau_k$ of the $k \times k$-matrix B lie in the left half-plane and the eigenvalues $\tau_{k+1}, \tau_{k+2}, \ldots, \tau_N$ of the $(N-k) \times (N-k)$-matrix C lie in the right half-plane. The matrix D has k rows and $N - k$ columns. This assertion follows from the well-known Schur lemma about the possibility of the reduction of any square matrix to the triangular form by conjugating it with a unitary matrix. It is obvious that

$$\|A\| = \|\widehat{A}\|, \quad \|B\| \leqslant \|A\|, \quad \|C\| \leqslant \|A\|, \quad \|D\| \leqslant \|A\|.$$

If $\frac{d}{dt}v(t) = Av(t)$, then, passing to $\widehat{v}(t) = Uv(t)$ ($v(t) = U^*\widehat{v}(t)$, $\|\widehat{v}(t)\| = \|v(t)\|$), we establish that $\widehat{v}(t)$ satisfies the equation

$$\frac{d\widehat{v}(t)}{dt} = \begin{bmatrix} B & D \\ 0 & C \end{bmatrix} \widehat{v}(t).$$

It is obvious that a bounded solution to this equation has the form

$$\widehat{v}(t) = \begin{pmatrix} v_1(t) \\ 0 \end{pmatrix},$$

where $v_1(t)$ has k components and the zero vector has $N-k$ components. Moreover,

$$\frac{dv_1(t)}{dt} = Bv_1(t), \quad \|\widehat{v}(t)\| = \|v_1(t)\|,$$
$$v_1(t) = e^{tB}v_1(0), \quad \|v_1(0)\| = \|\widehat{v}(0)\| = \|v(0)\|.$$

Thus, a bounded solution $v(t)$, $t \geqslant 0$, to the equation $\frac{dv(t)}{dt} = Av(t)$ has the form

$$v(t) = U^* \begin{pmatrix} e^{tB}v_1(0) \\ 0 \end{pmatrix},$$

where $v_1(0)$ is a k-dimensional vector. Consequently, the set $\mathcal{M}$ is a k-dimensional space. □

Now we estimate a bounded solution $v(t)$, $t \geqslant 0$, to the equation (4) as follows:

$$\|v(t)\| = \|e^{tB}v_1(0)\| \leqslant \|e^{tB}\| \, \|v_1(0)\| = \|e^{tB}\| \|v(0)\|$$
$$\leqslant \Omega(k) \left(\frac{\|B\|}{\sigma}\right)^{k-1} e^{-\frac{\sigma}{2}t} \|v(0)\| \leqslant \Omega(k) \left(\frac{\|A\|}{\sigma}\right)^{k-1} e^{-\frac{\sigma}{2}t} \|v(0)\|.$$

Using the estimate (7), we obtain

$$\|v(t)\| \leqslant \Omega(k) \left(\frac{\|A\|}{\sigma}\right)^{k-1} \mathcal{L} \frac{\|\varphi\|}{\|M\|} e^{-\frac{\sigma}{2}t}, \tag{9}$$

where $\Omega(k)$ is a constant depending only on k.

Now we discuss the Lopatinskii condition. Let the Lopatinskii condition be fulfilled. We have shown that the number of boundary conditions (the number of rows of the matrix M) is equal to the number of those eigenvalues of the matrix A that have negative real parts. Let $V(t)$ be an $N \times k$-matrix whose columns form a basis for the space $\mathcal{M}$. We note that the columns of $V(t)$ form a basis for the invariant subspace of the matrix A corresponding to the eigenvalues with negative real parts and

$$V(t) = e^{tA}V(0). \tag{10}$$

Any bounded solution $v(t)$, $t \geqslant 0$, to the equation (4) has the form

$$v(t) = V(t)h, \tag{11}$$

where h is a k-dimensional vector.

The boundary condition (5) holds if the vector h satisfies the condition $MV(0)h = \varphi$. The Lopatinskii condition guarantees the unique solvability of the system with respect to h. But the unique solvability of the system with respect to h for any φ is possible only if $MV(0)$ is a square nonsingular matrix. The matrix $MV(0)$ is a square matrix if M has k rows.

DEFINITION 2. The determinant $\Delta = \det[MV(0)]$ of the matrix $MV(0)$ is called the Lopatinskii determinant.

Thus, we have established the following criterion.

CRITERION 1. *The Lopatinskii condition holds if and only if the number of rows of the matrix M is equal to k and the Lopatinskii determinant $\Delta = \det[MV(0)]$ does not vanish.*

We indicate one more criterion.

CRITERION 2. *The Lopatinskii condition holds if and only if the number of rows of the matrix M is equal to k and the problem*

$$\frac{dv(t)}{dt} = Av(t), \quad t > 0, \qquad Mv(0) = 0 \tag{12}$$

has only zero bounded solution.

Indeed, let the number of rows of the matrix M be equal to k and let the problem (12) have only zero bounded solution. We show that the Lopatinskii determinant does not vanish. Indeed, in the opposite case, the system $MV(0)h = 0$ has a nonzero solution h and the vector-valued function $v(t) = V(t)h$ is a nonzero solution to the homogeneous equation (4) with the boundary condition $Mv(0) = 0$. Thus, the Lopatinskii determinant is not equal to zero. Consequently, the Lopatinskii condition holds.

The value of the Lopatinskii determinant depends on the choice of a basis for the space $\mathcal{M}$. However, the vanishing of the Lopatinskii determinant is independent of the choice of a basis. Indeed, if $W(t)$ is an $N \times k$-matrix whose columns also form a basis for the space $\mathcal{M}$, then $W(t) = V(t)T$, where T is a nonsingular $k \times k$-matrix ($\det T \neq 0$). Then $\det MW(0) = \det MV(0) \det T$. Sometimes it is convenient, instead of the Lopatinskii determinant, to consider the quantity

$$\Delta_0 = \left| \frac{\det[MV(0)]}{\sqrt{\det[V^*(0)V(0)]}} \right|. \tag{13}$$

It is easy to verify that Δ_0 is independent of the choice of $V(t)$.

EXAMPLE 1. We illustrate how to compute the external Lopatinskii constant. We seek a bounded solution to the system

$$\frac{d}{dt}\begin{pmatrix} v_1 \\ v_2 \\ v_3 \end{pmatrix} = \begin{bmatrix} 0 & 1 & 0 \\ 0 & 0 & 1 \\ 1 & 0 & 0 \end{bmatrix} \begin{pmatrix} v_1 \\ v_2 \\ v_3 \end{pmatrix} \tag{14}$$

which is bounded for $t \geqslant 0$ and satisfies the boundary condition

$$\begin{aligned} v_1(0) - v_2(0) + v_3(0) &= \varphi_1, \\ v_1(0) - v_3(0) &= \varphi_2, \end{aligned}$$

where the matrix

$$A = \begin{bmatrix} 0 & 1 & 0 \\ 0 & 0 & 1 \\ 1 & 0 & 0 \end{bmatrix}$$

has the following eigenvalues:

$$\tau_1 = 1, \quad \tau_2 = -\frac{1}{2} - i\frac{\sqrt{3}}{2}, \quad \tau_2 = -\frac{1}{2} + i\frac{\sqrt{3}}{2}.$$

With the eigenvalue τ_2 the eigenvector

$$\begin{pmatrix} -1 - i\sqrt{3} \\ -1 + i\sqrt{3} \\ 2 \end{pmatrix}$$

is associated. Consequently, the system (14) has the solution

$$v(t) = \exp\left(-\frac{1}{2}t - i\frac{\sqrt{3}}{2}t\right)\begin{pmatrix} -1 - i\sqrt{3} \\ -1 + i\sqrt{3} \\ 2 \end{pmatrix}.$$

The real and imaginary parts of this solution

$$v^{[1]}(t) = e^{-\frac{1}{2}t}\left\{\cos\frac{\sqrt{3}}{2}t\begin{pmatrix} -1 \\ -1 \\ 2 \end{pmatrix} + \sin\frac{\sqrt{3}}{2}t\begin{pmatrix} -\sqrt{3} \\ \sqrt{3} \\ 0 \end{pmatrix}\right\},$$

$$v^{[2]}(t) = e^{-\frac{1}{2}t}\left\{-\sin\frac{\sqrt{3}}{2}t\begin{pmatrix} -1 \\ -1 \\ 2 \end{pmatrix} + \cos\frac{\sqrt{3}}{2}t\begin{pmatrix} -\sqrt{3} \\ \sqrt{3} \\ 0 \end{pmatrix}\right\}$$

are also solutions to the system. They are linearly independent and are bounded for $t \geqslant 0$. Consequently, they form a basis for the space $\mathcal{M}$: $V(t) = (v^{[1]}(t), v^{[2]}(t))$. Moreover,

$$V(0) = \begin{bmatrix} -1 & -\sqrt{3} \\ -1 & \sqrt{3} \\ 2 & 0 \end{bmatrix},$$

$$MV(0) = \begin{bmatrix} 1 & -1 & 1 \\ 1 & 0 & -1 \end{bmatrix}\begin{bmatrix} -1 & -\sqrt{3} \\ -1 & \sqrt{3} \\ 2 & 0 \end{bmatrix} = \begin{bmatrix} 2 & -2\sqrt{3} \\ -3 & -\sqrt{3} \end{bmatrix},$$

$$\det MV(0) = -8\sqrt{3} \neq 0, \quad v(0) = V(0)h.$$

The vector h can be found from the system

$$MV(0)h = \varphi$$

or, in detail,

$$\begin{aligned} 2h_1 - 2\sqrt{3}h_2 &= \varphi_1, \\ -3h_1 - \sqrt{3}h_2 &= \varphi_2. \end{aligned}$$

Solving this system, we find

$$h_1 = \frac{1}{8}\varphi_1 - \frac{1}{4}\varphi_2, \quad \sqrt{3}h_2 = -\frac{3}{8}\varphi_1 - \frac{1}{4}\varphi_2.$$

We determine the vector $v(0) = V(0)h$ as follows:

$$v(0) = \begin{pmatrix} v_1(0) \\ v_2(0) \\ v_3(0) \end{pmatrix} = \begin{bmatrix} -1 & -\sqrt{3} \\ -1 & \sqrt{3} \\ 2 & 0 \end{bmatrix} \begin{pmatrix} h_1 \\ h_2 \end{pmatrix} = \begin{pmatrix} \frac{1}{4}\varphi_1 + \frac{1}{2}\varphi_2 \\ -\frac{1}{2}\varphi_2 \\ \frac{1}{4}\varphi_1 - \frac{1}{2}\varphi_2 \end{pmatrix}.$$

Thus,

$$\begin{aligned} v_1(0) &= \frac{1}{4}\varphi_1 + \frac{1}{2}\varphi_2, \\ v_2(0) &= -\frac{1}{2}\varphi_1, \\ v_3(0) &= \frac{1}{4}\varphi_1 - \frac{1}{2}\varphi_2. \end{aligned}$$

In addition,

$$\begin{aligned} \|v(0)\| &= \sqrt{[v_1(0)]^2 + [v_2(0)]^2 + [v_3(0)]^2} \\ &= \sqrt{\left(\frac{1}{4}\varphi_1 + \frac{1}{2}\varphi_2\right)^2 + \left(-\frac{1}{2}\varphi_1\right)^2 + \left(\frac{1}{4}\varphi_1 - \frac{1}{2}\varphi_2\right)^2} = \sqrt{\frac{3}{8}\varphi_1^2 + \frac{1}{2}\varphi_2^2}. \end{aligned}$$

It is clear that

$$\max_{\|\varphi\|=1} \|v(0)\| = \frac{1}{\sqrt{2}}.$$

To find the norm of the matrix

$$M = \begin{bmatrix} 1 & -1 & 1 \\ 1 & 0 & -1 \end{bmatrix}$$

occurring in the condition

$$M \begin{pmatrix} v_1(0) \\ v_2(0) \\ v_3(0) \end{pmatrix} = \begin{pmatrix} \varphi_1 \\ \varphi_2 \end{pmatrix},$$

it is necessary to find the greatest root of the characteristic equation

$$\det[MM^* - \lambda I] = \begin{vmatrix} 3-\lambda & 0 \\ 0 & 2-\lambda \end{vmatrix} = (\lambda - 3)(\lambda - 2) = 0$$

of the matrix MM^* and take the square root:

$$\|M\| = \sqrt{\lambda_{\max}(MM^*)} = \sqrt{3}.$$

In the considered example, the external Lopatinskii constant (cf. (6), (7)) is

$$\mathcal{L} = \|M\| \max_{\|\varphi\|=1} \|v(0)\| = \sqrt{\frac{3}{2}}.$$

If, instead of the boundary condition

$$\begin{aligned} v_1(0) - v_2(0) + v_3(0) &= \varphi_1, \\ v_1(0) - v_3(0) &= \varphi_2, \end{aligned}$$

we take the boundary conditions

$$\begin{aligned} v_1(0) + v_2(0) + v_3(0) &= \varphi_1, \\ v_1(0) - v_3(0) &= \varphi_2, \end{aligned} \tag{15}$$

then the Lopatinskii condition does not hold. In this case, the matrix in the boundary condition has the form

$$M = \begin{bmatrix} 1 & 1 & 1 \\ 1 & 0 & -1 \end{bmatrix},$$

$$MV(0) = \begin{bmatrix} 1 & 1 & 1 \\ 1 & 0 & -1 \end{bmatrix} \begin{bmatrix} -1 & -\sqrt{3} \\ -1 & \sqrt{3} \\ 2 & 0 \end{bmatrix} = \begin{bmatrix} 0 & 0 \\ -3 & -\sqrt{3} \end{bmatrix}.$$

The Lopatinskii determinant vanishes: $\det MV(0) = 0$. The system $MV(0)h = 0$ has the nonzero solution

$$h = \begin{pmatrix} 1 \\ -\sqrt{3} \end{pmatrix}$$

which yields the bounded nonzero solution $v(t) = V(t)h$, $t \geqslant 0$, to the homogeneous problem (14), (15). On the other hand, the problem (14), (15) is solvable only if $\varphi_1 = 0$.

Computing the external Lopatinskii constant $\mathcal{L}$, we must keep in mind that it depends on the choice of the matrix M in the boundary condition $Mv(0) = \varphi$. The choice of M is not unique. Thus, in Example 1, instead of the boundary conditions

$$\begin{aligned} v_1(0) - v_2(0) + v_3(0) &= \varphi_1, \\ v_1(0) - v_3(0) &= \varphi_2 \end{aligned}$$

we could use the equivalent equalities

$$\begin{aligned} &2v_3(0) - v_2(0) = \varphi_1 - \varphi_2 = \varphi_1', \\ &v_1(0) - v_3(0) = \varphi_2 = \varphi_2'. \end{aligned}$$

The matrix

$$M = \begin{bmatrix} 1 & -1 & 1 \\ 1 & 0 & -1 \end{bmatrix} \qquad (\|M\| = \sqrt{3})$$

and the right-hand side φ_1, φ_2 in the boundary condition are replaced by

$$M = \begin{bmatrix} 0 & -1 & 2 \\ 1 & 0 & -1 \end{bmatrix} \qquad (\|M\| = \sqrt{6}),$$

$$\varphi_1' = \varphi_1 - \varphi_2, \quad \varphi_2' = \varphi_2.$$

Hence the external Lopatinskii constant must change. If it was $\mathcal{L} = \sqrt{3}/2$ in the previous computation, then $\mathcal{L} = \frac{1}{2}\sqrt{3(5+\sqrt{13})}$ in the new form of the boundary condition. (Verify this assertion by direct computation.)

In general, leaving the meaning unchanged, we can replace the condition $Mv(0) = \varphi$ by any condition of the form $TMv(0) = T\varphi$, where T is an arbitrary matrix such that its columns are linearly independent and the number of columns coincides with the number of the rows of the matrix M. Sometimes it is convenient to transform the boundary condition by setting $T = M^*(MM^*)^{-1}$. Then the boundary condition takes the form

$$\Pi v(0) = M^*(MM^*)^{-1}Mv(0) = \widetilde{\varphi} = M^*(MM^*)^{-1}\varphi,$$

where $\Pi = M^*(MM^*)^{-1}M$ is a square symmetric ($\Pi^* = \Pi$) $N \times N$-matrix satisfying the relation

$$\Pi^2 = M^*(MM^*)^{-1}MM^*(MM^*)^{-1}M = M^*(MM^*)^{-1}M = \Pi.$$

Such matrices ($\Pi^* = \Pi$, $\Pi^2 = \Pi$) are called the orthogonal projections.

The set of vectors y such that $\Pi y = y$ form the linear subspace Y of the N-dimensional space X ($Y \subset X$). Any vector x can be represented as the sum $x = y + z$, where $y = \Pi x$ and $z = x - \Pi x$. In addition, $\Pi y = \Pi^2 x = \Pi x = y$, i.e., the vector $y = \Pi x$ goes into itself under the transformation Π. In this case, we say that the operator Π projects the whole space onto a subspace. Vectors of this subspace are projected into themselves.

The mapping $x \to z = x - \Pi x = (I - \Pi)x$ is also a projection, but onto another subspace. It is easy to see that

$$(I - \Pi)^2 = I - 2\Pi + \Pi^2 = I - 2\Pi + \Pi = I - \Pi.$$

We show that the projections Π and $(I - \Pi)$ project onto two orthogonal subspaces. To prove this assertion, we use the symmetry property of Π ($\Pi^* = \Pi$). Let the vector u go to Πu under the projection Π and let the vector v go to $v - \Pi v$ under the projection $I - \Pi$. We must prove that $(\Pi u, v - \Pi v) = 0$. The following calculation contains the proof:

$$(\Pi u, v - \Pi v) = (u, \Pi^* v - \Pi^* \Pi v) = (u, \Pi v - \Pi^2 v) = (u, \Pi v - \Pi v) = 0.$$

The boundary conditions

$$Mv(0) = \varphi, \quad \Pi v(0) = M^*(MM^*)^{-1}\varphi = \widetilde{\varphi}$$

are equivalent. Indeed, the first condition implies the second one. To prove that the first condition follows from the second condition, we note that

$$\begin{aligned} Mv(0) - \varphi &= MM^*(MM^*)^{-1}(Mv(0) - \varphi) \\ &= M\big[M^*(MM^*)^{-1}Mv(0) - M^*(MM^*)^{-1}\varphi\big] = M\big[\Pi v(0) - \widetilde{\varphi}\big] = 0, \end{aligned}$$

where we use the linear independence of rows of the matrix M, which guarantees the nonsingularity of the square matrix MM^*.

We note that the set of vectors z satisfying the homogeneous condition $Mz = 0$ remains unchanged under the replacement of M by $T_0 M$, where T_0 is any nonsingular square matrix. Starting with the matrix of the transformed boundary condition $T_0 M$, we construct the corresponding projection Π_{T_0} by the formula

$$\begin{aligned} \Pi_{T_0} &= [T_0 M]^*\big([T_0 M][T_0 M]^*\big)^{-1} T_0 M = M^* T_0^* (T_0 M M^* T_0^*)^{-1} T_0 M \\ &= M^* T_0^* (T_0^*)^{-1} (MM^*)^{-1} T_0^{-1} T_0 M = M^*(MM^*)^{-1}M = \Pi. \end{aligned}$$

The equality $\Pi_{T_0} = \Pi$ shows that the matrix defining the projection is independent of the form of the matrix M. The invariance of Π explains why the boundary condition $\Pi v(0) = \widetilde{\varphi}$ is more convenient than $Mv(0) = \varphi$. However, the vector $\widetilde{\varphi}$ is not arbitrary. Indeed, since $\Pi^2 = \Pi$, we have $\Pi \widetilde{\varphi} = \Pi^2 v(0) = \Pi v(0) = \widetilde{\varphi}$.

We also note that

$$M\widetilde{\varphi} = MT\varphi = M[M^*(MM^*)^{-1}]\varphi = \varphi.$$

To be able to choose the right-hand side of the boundary condition arbitrarily, it is convenient to write the boundary condition in the form $\Pi v(0) = \Pi\psi$, where the notation ψ is used instead of $\widetilde{\varphi}$ so that $\Pi\psi = M^*(MM^*)^{-1}\varphi$. It follows from the equality $\Pi^2 = \Pi$ that the symmetric matrix Π has only eigenvalues 1 and 0. Therefore, $\|\Pi\| = 1$ (if $M \neq 0$). If the Lopatinskii condition holds, the rank of Π must be equal to the number of rows of the matrix M, i.e., to the number k of those eigenvalues of the matrix A that lie in the left half-plane. Therefore, k eigenvalues of Π are equal to 1 and $N-k$ eigenvalues of Π are equal to 0. Hence, $\mathrm{tr}\Pi = k$.

Let $v(t)$, $t > 0$, be a bounded solution to the homogeneous system $\dfrac{dv(t)}{dt} = Av(t)$ such that $\Pi v(0) = \Pi\psi$.

Definition 3. The constant

$$L = \max_{\psi\,(\|\Pi\psi\|\neq 0)} \frac{\|v(0)\|}{\|\Pi\psi\|} \tag{16}$$

is called the internal Lopatinskii constant.

In the estimate

$$\|v(0)\| \leqslant L\|\Pi\psi\|, \tag{17}$$

the internal Lopatinskii constant is the best possible constant.

It is obvious that the internal Lopatinskii constant L is independent of the choice of the matrix M.

Let us show that

$$L \leqslant \mathcal{L}. \tag{18}$$

Indeed, let ψ_0 be such that

$$\frac{\|v(0)\|}{\|\Pi\psi_0\|} = L.$$

Then

$$L = \frac{\|v(0)\|}{\|\Pi\psi_0\|} \leqslant \frac{\|M\|\,\|v(0)\|}{\|M\Pi\psi_0\|} = \frac{\|M\|\,\|v(0)\|}{\|\varphi_0\|} \leqslant \mathcal{L},$$

i.e., L is the minimal value among all possible values of the external Lopatinskii constant.

It is easy to see that $\mathcal{L} = L$ if the rows of the matrix M in the bounded condition are orthonormal, i.e., $MM^* = I_k$.

Example 2. As in Example 1, we consider the system $(t > 0)$

$$\frac{d}{dt}\begin{pmatrix} v_1 \\ v_2 \\ v_3 \end{pmatrix} = \begin{bmatrix} 0 & 1 & 0 \\ 0 & 0 & 1 \\ 1 & 0 & 0 \end{bmatrix}\begin{pmatrix} v_1 \\ v_2 \\ v_3 \end{pmatrix}$$

and the boundary condition of the form

$$Mv(0) = \begin{pmatrix} \varphi_1 \\ \varphi_2 \end{pmatrix},$$

where

$$M = \begin{bmatrix} 1 & -1 & 1 \\ 1 & 0 & -1 \end{bmatrix}.$$

The projection Π corresponding to this boundary condition is

$$\begin{aligned}
\Pi &= M^*(MM^*)^{-1}M \\
&= \begin{bmatrix} 1 & 1 \\ -1 & 0 \\ 1 & -1 \end{bmatrix} \left[\begin{bmatrix} 1 & -1 & 1 \\ 1 & 0 & -1 \end{bmatrix} \begin{bmatrix} 1 & 1 \\ -1 & 0 \\ 1 & -1 \end{bmatrix} \right]^{-1} \begin{bmatrix} 1 & -1 & 1 \\ 1 & 0 & -1 \end{bmatrix} \\
&= \begin{bmatrix} 1 & 1 \\ -1 & 0 \\ 1 & -1 \end{bmatrix} \begin{bmatrix} 3 & 0 \\ 0 & 2 \end{bmatrix}^{-1} \begin{bmatrix} 1 & -1 & 1 \\ 1 & 0 & -1 \end{bmatrix} = \begin{bmatrix} 5/6 & -1/3 & -1/6 \\ -1/3 & 1/3 & -1/3 \\ -1/6 & -1/3 & 5/6 \end{bmatrix},
\end{aligned}$$

and

$$\Pi v(0) = \Pi \begin{pmatrix} \psi_1 \\ \psi_2 \\ \psi_3 \end{pmatrix},$$

where

$$\Pi\psi = M^*(MM^*)^{-1} \begin{pmatrix} \varphi_1 \\ \varphi_2 \end{pmatrix} = \begin{pmatrix} \frac{1}{3}\varphi_1 + \frac{1}{2}\varphi_2 \\ -\frac{1}{3}\varphi_1 \\ \frac{1}{3}\varphi_1 - \frac{1}{2}\varphi_2 \end{pmatrix}$$

$$\|\Pi\psi\| = \left[\left(\frac{1}{3}\varphi_1 + \frac{1}{2}\varphi_2\right)^2 + \left(-\frac{1}{3}\varphi_1\right)^2 + \left(\frac{1}{3}\varphi_1 - \frac{1}{2}\varphi_2\right)^2 \right]^{1/2} = \sqrt{\frac{1}{3}\varphi_1^2 + \frac{1}{2}\varphi_2^2}.$$

As we know, for bounded solutions $v(t)$, $t > 0$, the values $v_1(0)$, $v_2(0)$, $v_3(0)$ are found by the formulas

$$\begin{aligned}
v_1(0) &= \frac{1}{4}\varphi_1 + \frac{1}{2}\varphi_2, \\
v_2(0) &= -\frac{1}{2}\varphi_1, \\
v_3(0) &= \frac{1}{4}\varphi_1 - \frac{1}{2}\varphi_2
\end{aligned}$$

so that $\|v(0)\| = \sqrt{\frac{3}{8}\varphi_1^2 + \frac{1}{2}\varphi_2^2}$. In addition,

$$\frac{\|v(0)\|}{\|\Pi\psi\|} = \frac{\sqrt{\frac{3}{8}\varphi_1^2 + \frac{1}{2}\varphi_2^2}}{\sqrt{\frac{1}{3}\varphi_1^2 + \frac{1}{2}\varphi_2^2}} \leqslant \frac{3}{\sqrt{8}}.$$

Thus, in the considered example, the inequality $\|v(0)\| \leqslant L\|\Pi\psi\|$ holds with the internal Lopatinskii constant $L = 3/\sqrt{8}$.

§14. The Green matrices for boundary-value problems on a half-line

The representation of a solution to a boundary-value problem by special matrices. Conditions defining the matrices. Estimates.

We construct a solution to the boundary-value problem

$$\frac{dy}{dt} = Ay + f(t), \quad t > 0, \tag{1}$$

$$My(0) = \varphi, \tag{2}$$

$$\|y(t)\| \leqslant \text{const} \tag{3}$$

under the assumption that the problem is well posed, i.e., the matrix of coefficients A has no purely imaginary eigenvalues and the Lopatinskii condition holds. We will use the Green matrix $G_0(t, A)$ constructed in §7 and represent a solution in terms of special Green matrices of the considered boundary-value problem.

We recall that the matrix A can be written in the form

$$A = T^{-1} \begin{bmatrix} B & 0 \\ 0 & C \end{bmatrix} T,$$

where T is a suitable nonsingular matrix, the eigenvalues of the matrix B lie in the left half-plane and the eigenvalues of the matrix C lie in the right half-plane. For the Green matrix $G_0(t) = G_0(t, A)$ we have the following representation:

$$G_0(t) = \begin{cases} T^{-1} \begin{bmatrix} e^{tB} & 0 \\ 0 & 0 \end{bmatrix} T & \text{for } t > 0, \\ -T^{-1} \begin{bmatrix} 0 & 0 \\ 0 & e^{tC} \end{bmatrix} T & \text{for } t < 0. \end{cases} \tag{4}$$

In particular, from (4) we obtain the equalities

$$\begin{gathered} G_0(+0) = T^{-1} \begin{bmatrix} I_k & 0 \\ 0 & 0 \end{bmatrix} T, \\ G_0(-0) = -T^{-1} \begin{bmatrix} 0 & 0 \\ 0 & I_{N-k} \end{bmatrix} T, \\ G_0^2(+0) = G_0(+0), \quad -G_0^2(-0) = G_0(-0), \\ G_0(+0)G_0(-0) = G_0(-0)G_0(+0) = 0, \\ G_0(+0)G_0(t) = G_0(t)G_0(+0) = G_0(t) \ \text{ for } t > 0, \\ G_0(-0)G_0(t) = G_0(t)G_0(-0) = G_0(t) \ \text{ for } t < 0; \end{gathered}$$

moreover, any bounded solution $v(t)$, $t > 0$, to the homogeneous equation

$$\frac{d}{dt}v(t) = Av(t) \tag{5}$$

can be represented in the form

$$v(t) = G_0(t)v(0), \tag{6}$$

where $G_0(-0)v(0) = 0$. Any nonzero vector $v(0)$ such that $G_0(-0)v(0) = 0$ can be taken for the initial value of the solution $v(t) = G_0(t)v(0)$ to the equation

$$\frac{d}{dt}v(t) - Av(t) = \left[\frac{d}{dt} - A\right] G_0(t)v(0) = 0. \tag{7}$$

Indeed, by the equality $G_0(+0) - G_0(-0) = I$, we have

$$\begin{aligned} v(0) = v(+0) &= [G_0(+0) - G_0(-0)]v(+0) \\ &= G_0(+0)v(0) - G_0(-0)v(0) = G_0(+0)v(0) \\ &= \lim_{t\to+0} G_0(t)v(0) = \lim_{t\to+0} v(t). \end{aligned}$$

In the k-dimensional space of solutions to the equation $G_0(-0)v(0) = 0$, it is convenient to take $z_1, z_2, \ldots, z_k$ as the orthonormal basis and form the matrix Z with columns z_j. It is obvious that

$$Z^*Z = I_k, \quad G_0(-0)Z = 0, \quad G_0(+0)Z = Z$$

and the vector $v(0)$, being a linear combination of the columns of the matrix Z, can be represented as

$$v(0) = Z\alpha. \tag{8}$$

Since the basis formed by the columns of the matrix Z is orthonormal, we have $\|v(0)\| = \|\alpha\|$. If we want to define $v(0)$ from the boundary condition $Mv(0) = \varphi$, then the vector α must be determined from the equation

$$Mv(0) = MZ\alpha = \varphi.$$

The Lopatinskii condition guarantees the unique solvability of this equation, i.e., the nonsingularity of the $k \times k$-matrix MZ. We have

$$\alpha = [MZ]^{-1}\varphi.$$

It is obvious that the external Lopatinskii constant $\mathcal{L}$ (cf. (6) in §13) in the Lopatinskii estimate

$$\|v(0)\| = \|\alpha\| \leqslant \frac{\mathcal{L}}{\|M\|}\|\varphi\| \tag{9}$$

is calculated as follows:

$$\mathcal{L} = \|[MZ]^{-1}\|\,\|M\|.$$

We recall that the internal Lopatinskii constant L (cf. (16) in §13), being the smallest constant among all constants $\mathcal{L}$ in (9), can be obtained by the replacement of the matrix M by the matrix $\widetilde{M}$ such that the rows of $\widetilde{M}$ are orthonormal and are linear combinations of the rows of M. For $\widetilde{M}$ we can take the matrix

$$\widetilde{M} = (MM^*)^{-1/2}M. \tag{10}$$

In addition,

$$\begin{aligned} \|\widetilde{M}\| = 1, \quad [\widetilde{M}Z]^{-1} &= [(MM^*)^{-1/2}MZ]^{-1} = [MZ]^{-1}[MM^*]^{1/2}, \\ L &= \|[\widetilde{M}Z]^{-1}\|\,\|\widetilde{M}\| = \|[MZ]^{-1}[MM^*]^{1/2}\|. \end{aligned}$$

Thus,

$$v(0) = Z\alpha = Z[MZ]^{-1}\varphi = Y\varphi,$$

where the matrix $Y = Z[MZ]^{-1}$ satisfies the equations $MY = I_k$ and $G(-0)Y = 0$. It is easy to see that Y is uniquely found from these equations; moreover, $G(+0)Y = Y$. From the representation $v(0) = Y\varphi$ it follows that

$$\mathcal{L} = \max_{\varphi\neq 0} \frac{\|M\|\,\|v(0)\|}{\|\varphi\|} = \max_{\varphi\neq 0} \frac{\|M\|\,\|Y\varphi\|}{\|\varphi\|} = \|M\|\,\|Y\|,$$

i.e., the external Lopatinskii constant $\mathcal{L}$ can be computed from Y by the formula

$$\mathcal{L} = \|M\| \, \|Y\|. \tag{11}$$

Consider the bounded partial solution (cf. §7)

$$w(t) = \int_0^\infty G_0(t-s) f(s)\, ds$$

to the nonhomogeneous equation

$$\frac{d}{dt} w(t) = Aw(t) + f(t). \tag{12}$$

We have

$$w(0) = \int_0^\infty G_0(-s) f(s)\, ds. \tag{13}$$

A bounded solution $y(t)$, $t > 0$, to the problem (1), (2) can be found as the sum

$$y(t) = w(t) + v(t) = \int_0^\infty G_0(t-s) f(s)\, ds + G_0(t) Z\alpha. \tag{14}$$

For $y(t)$ we have

$$y(0) = \int_0^\infty G_0(-s) f(s)\, ds + Z\alpha,$$

$$My(0) = M \int_0^\infty G_0(-s) f(s)\, ds + MZ\alpha.$$

To satisfy the boundary condition (2), we must find α from the equation

$$MZ\alpha = \varphi - M \int_0^\infty G_0(-s) f(s)\, ds,$$

i.e., we must set

$$\alpha = (MZ)^{-1}\varphi - (MZ)^{-1} M \int_0^\infty G_0(-s) f(s)\, ds,$$

$$v(t) = G_0(t) Z (MZ)^{-1}\varphi - G_0(t) Z (MZ)^{-1} M \int_0^\infty G_0(-s) f(s)\, ds$$

$$= G_0(t) Z (MZ)^{-1}\varphi - \int_0^\infty G_0(t) Z (MZ)^{-1} M G_0(-s) f(s)\, ds.$$

As a result, we obtain

$$y(t) = \int_0^\infty G_0(t-s)f(s)\,ds - \int_0^\infty G_0(t)Z(MZ)^{-1}MG_0(-s)f(s)\,ds$$
$$+ G_0(t)Z(MZ)^{-1}\varphi.$$

Using the equality $Z(MZ)^{-1} = Y$ and introducing the notation

$$X = Z(MZ)^{-1}M = YM,$$

for a solution to the equations

$$MX = M, \quad G_0(-0)X = 0, \tag{15}$$

we rewrite the representation of $y(t)$ as follows:

$$y(t) = \int_0^\infty G_0(t-s)f(s)\,ds - \int_0^\infty G_0(t)XG_0(-s)f(s)\,ds + G_0(t)Y\varphi.$$

We introduce the notation

$$\begin{aligned} G_1(t,s) &= -G_0(t)XG_0(-s), \\ G(t,s) &= G_0(t-s) + G_1(t,s), \\ G_2(t) &= G_0(t)Y. \end{aligned} \tag{16}$$

The matrices $G(t,s)$ and $G_2(t)$ are called the Green matrices. They allow us to write a solution $y(t)$ to the boundary-value problem (1), (2) on the half-line $t > 0$ in terms of the Green matrices as follows:

$$y(t) = \int_0^\infty G(t,s)f(s)\,ds + G_2(t)\varphi. \tag{17}$$

EXERCISE 1. Verify that for all $t > 0$, $s > 0$ the matrices $G(t,s)$ and $G_2(t)$ are bounded solutions to the following boundary-value problems:

1^0. $MG(0,s) = 0$, $\dfrac{dG(t,s)}{dt} = AG(t,s)$ for $t \neq s$, $G(s+0,s) - G(s-0,s) = I$,

2^0. $MG_2(0) = I_k$, $\dfrac{d}{dt}\,G_2(t) = AG_2(t)$.

Each of these problems has a unique solution if the matrix A has no purely imaginary characteristic values and the Lopatinskii condition holds.

The Lopatinskii constants allow us to estimate the matrices X, Y in the representations (16). First, we already know that $\|Y\| = \frac{\mathcal{L}}{\|M\|}$. Second, $X = YM$ implies the estimate

$$\|X\| \leqslant \|Y\|\,\|M\| = \frac{\mathcal{L}}{\|M\|}\|M\| = \mathcal{L}.$$

This estimate can be made more precise. Indeed, from the system (15) defining the matrix X, we see that X is unchanged if the matrix M is replaced by any matrix of the form RM, where R is an arbitrary nonsingular $k \times k$-matrix. In particular, by the corresponding choice of the matrix R, one can attain that the rows of $\widetilde{M} = RM$ are orthonormal. For $\widetilde{M}$ we have

$$\|\widetilde{M}\| = 1, \quad \|\widetilde{Y}\| = \|Z(\widetilde{M}Z)^{-1}\| = L,$$

where L is the least possible constant $\mathcal{L}$. Hence

$$\|X\| = \|\widetilde{Y}\widetilde{M}\| = \|\widetilde{Y}\| = L.$$

(The equality $\|\widetilde{Y}\widetilde{M}\| = \|\widetilde{Y}\|$ holds since the matrix $\widetilde{M}$ is formed by orthonormal rows.) Thus,

$$\|\widetilde{Y}\| = \frac{\mathcal{L}}{\|M\|}, \quad \|X\| = L.$$

Therefore, for $t > 0$, $s > 0$ we have the estimates

$$\begin{aligned} \|G_2(t)\| &= \|G_0(t)Y\| \leqslant \frac{\mathcal{L}}{\|M\|}\|G_0(t)\|, \\ \|G(t,s)\| &= \|G_0(t-s) - G_0(t)XG_0(-s)\| \\ &\leqslant \|G_0(t-s)\| + L\|G_0(t)\|\,\|G_0(-s)\|. \end{aligned}$$

These estimates for the Green matrices will be used in subsequent chapters.

CHAPTER 2

Quadratic Lyapunov Functions

§1. Sufficient conditions for the global existence of a solution to a vector differential equation

The statement of a local existence theorem and an example. The Lyapunov criterion for global existence of solutions. The proof of the criterion and examples. The domain of attraction and an estimate.

In Chapter 1, we studied only linear equations and systems of equations with constant coefficients. Here we continue this study. However, sometimes we have to consider these simplest differential equations as models for more complicated nonlinear equations. Therefore, we must mention, even if briefly, some aspects of the theory of general differential equations.

We begin with the formulation of a local existence theorem and uniqueness theorem for the Cauchy problem

$$\frac{dy}{dt} = f(t, y), \quad y(t_0) = y_0.$$

Here and in the sequel, $y(t)$ and $f(t, y)$ are N-dimensional vector-valued functions with values in the Euclidean (or unitary) space. In this section, we consider only Euclidean spaces. For unitary spaces the arguments are quite similar, but the norm

$$\|y\| = \sqrt{y_1^2 + y_2^2 + \cdots + y_N^2}$$

in the Euclidean space must be replaced by the norm

$$\|y\| = \sqrt{y_1\overline{y}_1 + y_2\overline{y}_2 + \cdots + y_N\overline{y}_N}$$

in the unitary space. We use one or another norm depending on whether we consider real-valued or complex-valued functions.

Now we formulate the assumptions on the function $f(t, y)$.

1. The vector-valued function $f(t, y)$ is defined and continuous for

$$|t - t_0| \leqslant T, \quad \|y - y_0\| \leqslant \sqrt{\sum_{i=1}^{N}(y_i - y_{i0})^2} \leqslant R.$$

2. The following inequality holds:

$$\|f(t, y)\| \leqslant \sqrt{\sum_{i=1}^{N}(f_i(t, y_1, y_2, \ldots, y_N))^2} \leqslant F.$$

3. All derivatives in the matrix $\dfrac{\partial f(t,y)}{\partial y}$ exist, are continuous, and satisfy the inequality

$$\left\| \frac{\partial f(t,y)}{\partial y} \right\| \leqslant L.$$

Conditions 1–3 are assumed to hold in the cylinder $|y - t_0| \leqslant T$, $\|y - y_0\| \leqslant R$ indicated in Condition 1.

THEOREM 1 (on uniqueness and existence). *Let $T_0 = \min\{T, R/F\}$. Then for $|t - t_0| \leqslant T_0$ there exists a continuous and continuously differentiable vector-valued function $y(t)$ such that*

1) $\dfrac{dy}{dt} = f(t,y)$,

2) $y(t_0) = y_0$,

3) $\|y(t) - y(t_0)\| \leqslant F|t - t_0| \leqslant FT_0 \leqslant R$.

The vector-valued function $y(t)$ is uniquely defined by Conditions 1)–3).

The proof of the theorem can be found in any textbook on the theory of ordinary differential equations (cf., for example, [**21**]).

As an example we consider the Cauchy problem

$$\frac{dy}{dt} = 1 + t^2 + y^2, \quad y(0) = 0,$$

where $y(t)$ is a one-dimensional vector-valued function (a scalar function). The right-hand side $f(t,y) = 1 + t^2 + y^2$ is defined and continuous for any real t and y. However, to apply the existence theorem, we need to choose some T and R and assume that t and y vary only in the rectangle $|t| \leqslant T$, $|y| \leqslant R$ (we have $t_0 = 0$, $y_0 = 0$, $\|y\| = \sqrt{y^2} = |y|$). The values T and R can be chosen arbitrarily, e.g., $T = 2$ and $R = 5$. Assuming that $|t| \leqslant 2$ and $|y| \leqslant 5$, we find

$$\|f(t,y)\| = |f(t,y)| = 1 + t^2 + y^2 \leqslant 1 + 2^2 + 5^2 = 30.$$

We can set $F = 30$. The number T_0 in the bound $|t - t_0| \leqslant T_0$ for the existence interval satisfies the condition

$$T_0 = \min\{T, R/F\} = \min\{2, 5/30\} = 1/6.$$

In this example, the vector-valued function $f(t,y)$ is defined for all t, y. However, the existence theorem allows us to claim that a solution $y = y(t)$ exists only on the interval $|t| < 1/6$. The condition 3 on derivatives of $f(t,y)$ is also valid:

$$\left| \frac{\partial f}{\partial y} \right| = |2y| \leqslant |2| \cdot |5| = 10 = L.$$

It may appear that we have obtained the bound for the length of the existence interval because the bounds for t and y were too tight ($|y| \leqslant 5$, $|t| \leqslant 2$); indeed, in our example $f(t,y) = 1 + t^2 + y^2$ is defined for all t and y.

We extend the domain of $f(t,y)$ by assuming $|y| \leqslant 10$ and $|t| \leqslant 7$. Then

$$\|f(t,y)\| = |1 + t^2 + y^2| \leqslant 150 = F$$

and the bound T_0 for the length of the existence interval is

$$T_0 = \min\{T, R/F\} = \min\{7, 10/150\} = 1/15$$

which is less than the previous value of T_0.

In fact, we are able to show that a solution to the Cauchy problem under consideration cannot be defined on the entire axis. Indeed, the equation

$$\frac{dy}{dt} = 1 + t^2 + y^2$$

for $t \geqslant 0$ implies the inequality

$$\frac{dy}{dt} \geqslant 1 + y^2$$

which can be rewritten in the form

$$\frac{d}{dt} \arctan y(t) = \frac{1}{1+y^2} \frac{dy}{dt} \geqslant 1.$$

Choose t^* such that $0 < t^* < \pi/2$. Assuming that a continuous and continuously differentiable solution $y(t)$ is defined for $0 \leqslant t \leqslant t^*$ ($y(0) = 0$) and integrating the last inequality, we obtain

$$\arctan y(t^*) = \arctan(t^*) - \arctan y(0) = \int_0^{t^*} \frac{d}{dt} \arctan y(t) dt \geqslant \int_0^{t^*} 1\, dt = t^*.$$

Therefore, $y(t^*) \geqslant \tan t^*$. Hence the existence interval cannot be larger than $[0, \pi/2]$ because the function $y(t)$ is unbounded on $[0, \pi/2]$.

The example of the Cauchy problem $\dfrac{dy}{dt} = 1 + t^2 + y^2$, $y(0) = 0$ is very instructive. It shows that even if the right-hand side of the equation is a regular function for all values of the arguments, we cannot assert that the solution is defined for all t. Recall that the existence theorem (cf. Chapter 1, §1) for a linear system with constant coefficients guarantees the existence of a solution for all t.

In applications, it is important to establish, in addition to Conditions 1–3, some sufficient conditions on $f(t, y)$ and the initial conditions which guarantee the existence of a solution $y(t)$ for all positive t ($0 \leqslant t \leqslant \infty$). We consider only equations with the right-hand sides independent of t, $f(t, y) = f(y)$. Systems of such equations

$$\begin{aligned}
\frac{dy_1}{dt} &= f_1(y_1, y_2, \dots, y_N), \\
\frac{dy_2}{dt} &= f_2(y_1, y_2, \dots, y_N), \\
&\cdots\cdots\cdots\cdots\cdots\cdots \\
\frac{dy_N}{dt} &= f_N(y_1, y_2, \dots, y_N)
\end{aligned}$$

or

$$\frac{dy}{dt} = f(y)$$

are said to be autonomous.

We consider the Lyapunov criterion for existence of solutions to autonomous systems on an infinite t-interval. The formulation of the criterion uses some auxiliary functions $\mathcal{H}(y) = \mathcal{H}(y_1, y_2, \dots, y_N)$ satisfying the following conditions.

L1. The function $\mathcal{H}(y_1, y_2, \dots, y_N)$ is defined in the ball

$$\|y\| = \sqrt{y_1^2 + y_2^2 + \cdots + y_N^2} \leqslant R,$$

is continuous inside the ball and on its boundary, and has continuous partial derivatives $\dfrac{\partial \mathcal{H}}{\partial y_j}$ with respect to $y_1, y_2, \dots, y_N$.

L2. The following relations hold:

$$\begin{aligned} \mathcal{H}(y_1, y_2, \dots, y_N) &\geqslant 0 \text{ for } \|y\| \leqslant R, \\ \mathcal{H}(0, 0, \dots, 0) &= 0, \\ \mathcal{H}(y_1, y_2, \dots, y_N) &> 0 \text{ for } 0 < \|y\| \leqslant R. \end{aligned}$$

L3. The continuous function

$$\mathcal{G}(y_1, y_2, \dots, y_N) = -\sum_{i=1}^{N} f_i(y_1, y_2, \dots, y_N) \frac{\partial \mathcal{H}(y_1, y_2, \dots, y_N)}{\partial y}$$

satisfies the inequality $\mathcal{G}(y_1, y_2, \dots, y_N) \geqslant 0$ for $\|y\| \leqslant R$.

Sometimes instead of Condition L3 we need a stronger condition

L3$'$. The continuous function $\mathcal{G}(y_1, y_2, \dots, y_N)$ defined in Condition L3, in addition to the inequality $\mathcal{G}(y_1, y_2, \dots, y_N) \geqslant 0$ for $\|y\| \leqslant R$, satisfies the inequality $\mathcal{G}(y_1, y_2, \dots, y_N) > 0$ for $0 < \|y\| \leqslant R$.

THEOREM 2 (on global existence). *Let there exist a function $\mathcal{H}(y)$ satisfying Conditions* L1–L3. *It is possible to choose ρ_0, $0 < \rho_0 < R$, such that if*

$$\|y_0\| = \sqrt{y_{10}^2 + y_{20}^2 + \cdots + y_{N0}^2} \leqslant \rho_0,$$

then there exists a continuous and continuously differentiable vector-valued function $y(t)$ defined for all $t \geqslant 0$ and satisfying the inequality $\|y(t)\| \leqslant R$ and the conditions

$$\frac{dy}{dt} = f(y), \quad y(0) = y_0. \tag{1}$$

In addition, if Condition L3$'$ *is satisfied, then* $\lim_{t\to\infty} \|y(t)\| = 0$.

In this theorem, we also suppose that $f(y)$ satisfies Conditions 1–3, which guarantees the validity of Theorem 1.

First, we explain the meaning of the function $\mathcal{G}(y_1, y_2, \dots, y_N)$ appearing in Conditions L3 and L3$'$. Suppose that a solution $y(t)$ to the vector equation $\dfrac{dy}{dt} = f(y)$ is defined on some t-interval. We introduce the function $\varphi(t) = \mathcal{H}(y(t))$ of t and compute its derivative:

$$\begin{aligned} \varphi'(t) &= \frac{d}{dt}\mathcal{H}(y_1(t), y_2(t), \dots, y_N(t)) = \sum_{i=1}^{N} \frac{\partial \mathcal{H}}{\partial y_i}\frac{dy_i}{dt} \\ &= \sum_{i=1}^{N} \frac{\partial \mathcal{H}}{\partial y_i} f_i(y_1(t), y_2(t), \dots, y_N(t)) = -\mathcal{G}(y_1(t), y_2(t), \dots, y_N(t)). \end{aligned} \tag{2}$$

The function $-\mathcal{G}(y_1, y_2, \dots, y_N)$ is called the derivative of $\mathcal{H}(y_1, y_2, \dots, y_N)$ with respect to t along the solution $y(t)$.

Thus, if $\mathcal{G}(y_1, y_2, \dots, y_N) \geqslant 0$, then $\varphi(t) = \mathcal{H}(y_1(t), y_2(t), \dots, y_N(t))$ does not increase as t increases. We will show how $\mathcal{H}(y_1(t), y_2(t), \dots, y_N(t))$ can be used to estimate the norm of y. Since $\mathcal{H}$ is not increasing along the solution $y(t)$, the function $y(t)$ is bounded in t. We recall that for the equation $y' = 1 + t^2 + y^2$ the unbounded growth of $y(t)$ on a finite time interval is the reason why the solution cannot be defined for all t.

PROOF OF THEOREM 2. We introduce the following functions:

$$m(\rho) = \min_{\|y\| \geqslant \rho} \mathcal{H}(y), \qquad M(\rho) = \max_{\|y\| \leqslant \rho} \mathcal{H}(y).$$

It is easy to verify that the functions $m(\rho)$ and $M(\rho)$ possess the following properties:

(a) $m(\rho)$ and $M(\rho)$ are continuous,
(b) if $\rho_1 \leqslant \rho_2$, then $m(\rho_1) \leqslant m(\rho_2)$ and $M(\rho_1) \leqslant M(\rho_2)$,
(c) $m(\rho) > 0$ and $M(\rho) > 0$ if $\rho > 0$, $m(0) = 0$, and $M(0) = 0$,
(d) $m(\rho) \leqslant M(\rho)$,
(e) if $\|y\| \leqslant \delta$, then $\mathcal{H}(y) \leqslant M(\delta)$,
(f) if $\mathcal{H}(y) < m(\varepsilon)$, then $\|y\| < \varepsilon$.

We recommend the reader to write down proofs of these obvious properties up to smallest details.

We choose ρ_0 so that $M(\rho_0) < m(R/2)$. The possibility of such a choice follows from the properties (a)–(d) of the functions $m(\rho)$ and $M(\rho)$. It is obvious that $\rho_0 \leqslant R/2$. We construct a solution to the Cauchy problem (1) on some segment $0 \leqslant t \leqslant T$. If $T_0 = R/(2F)$, then, by Theorem 1, for $\|t\| \leqslant T_0$ a solution exists and is estimated as follows:

$$\|y(t) - y_0\| \leqslant F|t - t_0| \leqslant FT_0 = F\frac{R}{2F} = \frac{R}{2},$$
$$\|y(t)\| \leqslant \|y_0\| + \|y(t) - y_0\| \leqslant \frac{R}{2} + \frac{R}{2} = R.$$

We recall that F is the bound for $\|f(t,y)\| = \|f(y)\|$ in Condition 2 which provides Theorem 1. The function $f(y)$ is defined on the set $\|y\| \leqslant R$. Since $\|y_0\| \leqslant R/2$, the function $f(y)$ is defined for $\|y - y_0\| \leqslant R/2$. By the above assumptions, it is obvious that $\|f(y)\| \leqslant F$ for $\|y - y_0\| \leqslant R/2$. Since the system is autonomous ($f(t,y) = f(y)$ is independent of t), for $\|y(t_0)\| \leqslant R/2$, a solution $y(t)$ to the equation $y' = f(y)$ such that $y(t) = y(t_0)$ at $t = t_0$ can be defined on the segment $t_0 \leqslant t \leqslant t_0 + T_0$, $T_0 = R/(2F)$, for any t_0. In this reference to Theorem 1, the condition $t_0 = 0$ stated in this theorem is not essential.

Thus, taking y_0 such that $\|y_0\| \leqslant \rho_0$ we can construct a solution $y(t)$ to the Cauchy problem (1) on the segment $0 \leqslant t \leqslant T_0$. In view of L3 and (2)

$$\frac{d}{dt}\mathcal{H}(y) = -\mathcal{G}(y) \leqslant 0, \tag{3}$$

i.e., the function $\mathcal{H}$ does not increase along the solution. Therefore, $\mathcal{H}(y(T_0)) \leqslant \mathcal{H}(y_0)$. As we know, $\mathcal{H}(y_0) \leqslant M(\rho_0) < m(R/2)$. The inequality $\mathcal{H}(y(T_0)) < m(R/2)$ implies (cf. Property (f) of the functions $M(\rho)$ and $m(\rho)$) that $\|y(T_0)\| < R/2$. We already mentioned that this inequality allows us to construct a solution to the Cauchy problem for the equation

$$\frac{d}{dt}y = f(y)$$

on the segment $T_0 \leqslant t \leqslant T_0 + T_0$, and at $t = T_0$ the value $y(T_0)$ coincides with the value of the solution constructed on the segment $[0, T_0]$. This solution does not leave the ball $\|y(t)\| \leqslant R$ in which the right-hand side is defined.

The solutions constructed on the intervals $[0, T_0]$ and $[T_0, 2T_0]$ take the same value at $t = T_0$. Therefore, it is possible to construct the solution defined on the segment $[0, 2T_0]$. Indeed, since the solutions coincide at $t = T_0$, from the equation $y' = f(y)$ it follows that the derivatives of these solutions coincide at $t = T_0$. Hence we can assume that the solution is continuously differentiable on $[0, 2T_0]$. In view of (3),

$$\mathcal{H}(y(2T_0)) \leqslant \mathcal{H}(y(0)) = \mathcal{H}(y_0) \leqslant M(\rho_0).$$

Consequently,

$$\mathcal{H}(y(2T_0)) \leqslant M(\rho_0) < m(R/2),$$
$$\|y(2T_0)\| < R/2.$$

Assume that we have constructed $y(t)$ on the interval $0 \leqslant t \leqslant kT_0$ and proved the estimate $\|y(kT_0)\| \leqslant R/2$. This allows us to apply Theorem 1 in order to extend the solution to the segment $kT_0 \leqslant t \leqslant (k+1)T_0$. Thereby, we obtain the solution $y(t)$ on the segment $0 \leqslant t \leqslant (k+1)T_0$. Since $\mathcal{H}(y(t))$ is not increasing along this solution, we have

$$\mathcal{H}(y((k+1)T_0)) \leqslant \mathcal{H}(y_0) \leqslant M(\rho_0) < m(R/2),$$
$$\|y((k+1)T_0)\| < R/2.$$

Proceeding by induction and using the last inequality, we conclude that k can be replaced by $k+1$ and, consequently, the vector-valued function $y(t)$ can be defined for all $t > 0$; moreover, $\|y(t)\| < R/2$ for all $t > 0$.

Now we show that if $\|y\| > 0$, then the solution $y(t)$ tends to zero if $\mathcal{G}(y) > 0$ for $\|y\| > 0$. Assume the contrary. There are two possible cases.

Case 1. $\inf \|y(t)\| = 0$ for $0 \leqslant t < \infty$.

Case 2. $\|y(t)\| \geqslant \delta > 0$ for $0 \leqslant t < \infty$.

First, consider Case 1. For any $\varepsilon > 0$, take $\delta > 0$ such that $M(\delta) < m(\varepsilon)$ and find $T^* > 0$ such that $\|y(t^*)\| \leqslant \delta$. By Property (e) of the function $M(\rho)$, the inequality $\mathcal{H}(y(t^*)) \leqslant M(\delta)$ holds. By Condition L3, $\mathcal{H}(y(t))$ is not increasing along the solution, i.e.,

$$\frac{d\mathcal{H}}{dt} = -\mathcal{G} \leqslant 0.$$

Therefore, $\mathcal{H}(y(t)) \leqslant M(\delta) < m(\varepsilon)$ for $t \geqslant t^*$. By Property (f) of the function $m(\rho)$ we conclude that $\|y(t)\| \leqslant \varepsilon$ for $t \geqslant t^*$. Hence, in Case 1, for any $\varepsilon > 0$ there is $t = t^*$ such that $\|y(t)\| < \varepsilon$ if $t \geqslant t^*$. In other words, the limit exists and is equal to zero:

$$\lim_{t \to \infty} \|y(t)\| = 0.$$

In Case 2, for any $t \geqslant 0$ there exists $\delta > 0$ such that $\|y(t)\| \geqslant \delta > 0$. In addition,

$$\mathcal{H}(y(t)) \geqslant m(\delta) > 0.$$

Condition L3′ is equivalent to the inequality $\mu(\rho) > 0$ for $\rho > 0$, where

$$\mu(\rho) = \min_{\|y\| \geqslant \rho} \mathcal{G}(y).$$

It is obvious that $\mathcal{G}(y) \geqslant \mu(\delta) > 0$ along the obtained solution. By the definition of $\mathcal{G}(y)$, along the solution $y(t)$ we have (cf. (2))

$$\frac{d\mathcal{H}(y(t))}{dt} = -\mathcal{G}(y(t)).$$

Integrating the last relation, we find

$$\mathcal{H}(y(t)) = \mathcal{H}(y(0)) - \int_0^t \mathcal{G}(y(\tau))\, d\tau \leqslant \mathcal{H}(y(0)) - t\mu(\delta).$$

Since $\mathcal{H}(y(0)) \geqslant m(\delta) > 0$, the inequality $m(\delta) \leqslant \mathcal{H}(y(0)) - t\mu(\delta)$ must be valid for all $t \geqslant 0$, which is impossible if $m(\delta) > 0$ and $\mu(\delta) > 0$.

Thus, Case 2 contradicts Condition L3′. Hence only Case 1 can be realized. But we have shown that $y(t)$ tends to zero as $t \to \infty$. □

DEFINITION 1. A domain of the initial data $y(0)$ is called the domain of attraction of center $y = 0$ if the domain contains $y = 0$ and any solution $y(t)$ with the initial data $y(0)$ in the domain of attraction tends asymptotically to $y = 0$ as $t \to \infty$.

We have shown that for ρ_0 such that $M(\rho_0) < m(R/2)$ the ball $\|y(0)\| \leqslant \rho_0$ lies in the domain of attraction.

DEFINITION 2. A function $\mathcal{H}(y)$ satisfying Conditions L1, L2, L3 or L1, L2, L3′ is called the Lyapunov function for the system $y' = f(y)$.

DEFINITION 3. A Lyapunov function $\mathcal{H}(y)$ of the form $\mathcal{H}(y) = (Hy, y)$, where H is a positive definite matrix, is called the quadratic Lyapunov function.

The quadratic Lyapunov functions will play an important role hereinafter. It is easy to see that for a quadratic Lyapunov function $\mathcal{H}(y) = (Hy, y)$ we have

$$m(\rho) = \lambda_{\min}(H)\rho^2, \quad M(\rho) = \lambda_{\max}(H)\rho^2,$$

where $\lambda_{\min}(H)$ is the minimal eigenvalue and $\lambda_{\max}(H)$ is the maximal eigenvalue of the matrix H. Therefore, if a quadratic function $\mathcal{H}(y) = (Hy, y)$ is a Lyapunov function for $\|y\| \leqslant R$ and

$$\rho_0 < \sqrt{\frac{\lambda_{\min}(H)}{\lambda_{\max}(H)}}\,\frac{R}{2},$$

then the solution $y(t)$ to the system $y' = f(y)$ satisfies the inequality $\|y(t)\| < R/2$ $(t > 0)$ for $\|y(0)\| \leqslant \rho_0$. If, in addition, Condition L3′ holds, then the ball $\|y(0)\| \leqslant \rho_0$ lies in the domain of attraction of center $y = 0$.

EXAMPLE 1. For the system

$$\frac{dy_1}{dt} = -y_2 - y_1^3,$$
$$\frac{dy_2}{dt} = y_1 - y_2^3$$

Conditions 1–3 in Theorem 1 are valid for any $R > 0$. We set $\mathcal{H}(y) = y_1^2 + y_2^2$ and define

$$\mathcal{G}(y) = -(-y_2 - y_1^3)\frac{\partial \mathcal{H}}{\partial y_1} - (y_1 - y_2^3)\frac{\partial \mathcal{H}}{\partial y_2} = 2(y_1^4 + y_2^4).$$

It is obvious that Conditions L1, L2, L3′ hold for any $R > 0$. A solution to this system exists on the half-line $0 \leqslant t < \infty$ for any initial values $y_1(0) = y_{10}$, $y_2(0) = y_{20}$.

EXERCISE 1. Verify that for ρ_0 any positive number can be taken and any solution to the considered system converges to zero as $t \to \infty$.

EXAMPLE 2. Consider the linear system

$$\frac{dy_1}{dy} = -y_2,$$
$$\frac{dy_2}{dy} = y_1.$$

By setting $\mathcal{H}(y_1, y_2) = y_1^2 + y_2^2$, we define

$$\mathcal{G}(y_1, y_2) = -(-y_2)\frac{\partial \mathcal{H}}{\partial y_1} - y_1\frac{\partial \mathcal{H}}{\partial y_2} = -(-y_2)2y_1 - y_1 2y_2 = 0.$$

Conditions 1–3 (for any $R > 0$) and Conditions L1–L3 hold, but Condition L3′ fails. Theorem 2 is true (this fact is known from the theory of systems of linear equations with constant coefficients; cf. Chapter 1, §1), but we cannot assert that every solution tends to zero as $t \to \infty$. Indeed, a general solution to this system has the form

$$y_1(t) = a\cos(t - \gamma),$$
$$y_2(t) = a\sin(t - \gamma).$$

None of these solutions tends to zero if $a \neq 0$.

§2. The Lyapunov stability

The notions of stability and asymptotic stability. The stability or instability of the zero solution to a linear system, depending on the location of eigenvalues.

We proceed to the study of the notion of the stability of solutions of ordinary differential equations. This notion is very important and has numerous applications. We consider only autonomous systems

$$\frac{dy}{dt} = f(y) \tag{1}$$

or, in the detailed notation,

$$\frac{dy_1}{dt} = f_1(y_1, y_2, \dots, y_N),$$
$$\frac{dy_2}{dt} = f_2(y_1, y_2, \dots, y_N),$$
$$\dots\dots\dots\dots\dots\dots\dots\dots$$
$$\frac{dy_N}{dt} = f_N(y_1, y_2, \dots, y_N).$$

We assume that the right-hand side $f(y)$ vanishes at a point $y = a$ ($y_1 = a_1, y_2 = a_2, \dots, y_N = a_N$), i.e., $f(a) = 0$. It is obvious that the system has the solution

$$\begin{aligned} y_1(t) &= a_1, \\ y_2(t) &= a_2, \\ &\cdots\cdots\cdots \\ y_N(t) &= a_N \end{aligned}$$

whose components are independent of t.

DEFINITION 1. The point $y = a$ is called an equilibrium point for the system (1).

In our study of stability theory, we consider only the stability of equilibrium points for autonomous systems. Without loss of generality, we can assume that the equilibrium point is located at the origin, i.e., $a = 0$. Hereinafter, this assumption is assumed to hold.

Thus, we study an autonomous system of the form (1) such that the right-hand side $f(y)$ vanishes at $y = 0$ ($f(0) = 0$).

Sometimes it is convenient to emphasize the dependence of a solution on the initial condition. Therefore, we also use the notation $y(t, a)$ for a solution to the Cauchy problem for the system (1) with the initial condition $y(0) = a$.

We give two definitions.

DEFINITION 2. A solution $y(t, 0) \equiv 0$ to the system (1) is said to be Lyapunov stable if for every $\varepsilon > 0$ there is $\delta > 0$ such that for any initial vector a with $\|a\| < \delta$ a solution $y(t, a)$ exists and satisfies the estimate $\|y(t, a)\| < \varepsilon$ for all $t > 0$.

DEFINITION 3. A solution $y(t, 0) \equiv 0$ to the system (1) is said to be asymptotically stable if it is Lyapunov stable and there is $\delta_0 > 0$ such that for any initial vector a, $\|a\| < \delta_0$, the solution $y(t, a)$ tends to zero ($\|y(t, a)\| \to 0$) as $t \to \infty$; otherwise, the solution $y(t, 0) \equiv 0$ is said to be Lyapunov unstable.

We begin to study the stability by considering linear systems of equations written, as usual, in the form

$$\frac{dy}{dt} = Ay \tag{2}$$

with a constant coefficients matrix A. A solution to such a system has the form $y(t) = e^{tA} y(0)$. The estimate

$$\|e^{tA}\| \leqslant \gamma(N) \left(\frac{\|A\|}{\sigma} \right)^{N-1} e^{-t\sigma/2}, \quad t \geqslant 0,$$

proved in Chapter 1, §3, and the formula $y(t) = e^{tA} y(0)$ yield the inequality

$$\|y(t)\| \leqslant \gamma(N) \left(\frac{\|A\|}{\sigma} \right) e^{-t\sigma/2} \|y(0)\|$$

provided that all eigenvalues of A lie strictly in the left half-plane ($\operatorname{Re} \tau_j(A) \leqslant -\sigma$, $\sigma > 0$). By this inequality, the zero solution to the system $y' = Ay$ is asymptotically stable. Indeed, for $\varepsilon > 0$ we define

$$\delta = \frac{\varepsilon}{\gamma(N)(\|A\|/\sigma)^{N-1}}$$

and conclude that for any $y(0)$, $\|y(0)\| < \delta$, a solution to the Cauchy problem for (2) ($y(0)$ is fixed) exists for all $t \geqslant 0$. Indeed, by the existence theorem for a linear system of equations with constant coefficients, a solution exists for any initial data. The existence and uniqueness of the solution is guaranteed on the entire half-line $0 \leqslant t < \infty$. For positive t the solution $y(t)$ satisfies the relation

$$\|y(t)\| \leqslant \|e^{tA}\| \, \|y(0)\| < \gamma(N)\left(\frac{\|A\|}{\sigma}\right)^{N-1} e^{-t\sigma/2}\delta$$
$$= \gamma(N)\left(\frac{\|A\|}{\sigma}\right)^{N-1} e^{-t\sigma/2} \frac{\varepsilon}{\gamma(N)\left(\dfrac{\|A\|}{\sigma}\right)^{N-1}} = e^{-t\sigma/2}\varepsilon \leqslant \varepsilon.$$

Hence the Lyapunov stability holds. Furthermore, the inequality

$$\|y(t)\| \leqslant \gamma(N)\left(\frac{\|A\|}{\sigma}\right)^{N-1} e^{-t\sigma/2}\|y(0)\|$$

guarantees that any solution $y(t)$ tends to zero as $t \to \infty$. Consequently, the zero solution is asymptotically stable. Let the matrix A have at least one eigenvalue $\tau_0 = \sigma_0 + i\omega_0$ with positive real part $\sigma_0 > 0$ and let z_0 be the eigenvector corresponding to the eigenvalue τ_0:

$$Az_0 = (\sigma_0 + i\,\omega_0)z_0.$$

It is obvious that the vector-valued function

$$y(t) = \frac{\delta}{2\,\|z_0\|}\, e^{(\sigma_0 + i\omega_0)t} z_0, \quad \sigma > 0,$$

is a solution to the system (2) and $\|y(0)\| = \delta/2 < \delta$. If t increases, then the norm

$$\|y(t)\| = \frac{\delta}{2\,\|z_0\|}\, |e^{(\sigma_0 + i\omega_0)t}| \, \|z_0\| = \frac{\delta}{2} e^{\sigma_0 t}$$

increases unboundedly for any positive $\delta > 0$. Hence the Lyapunov stability does not hold in this case.

Let the matrix A have at least one purely imaginary eigenvalue $\tau_0 = i\omega_0$ corresponding to the eigenvector z_0. In this case, for any $\delta > 0$, the function

$$y(t) = \frac{\delta}{2\,\|z_0\|}\, e^{i\omega_0 t} z_0$$

is a solution, satisfies the condition $\|y(0)\| = \delta/2 < \delta$, and does not tend to zero as t increases:

$$\|y(t)\| = \frac{\delta}{2\,\|z_0\|} |e^{i\omega_0 t}| \|z_0\| = \frac{\delta}{2} = \text{const}.$$

Thus, if the matrix A has at least one purely imaginary eigenvalue, then the zero solution to the system (2) is not asymptotically stable. If the matrix A has at least one purely imaginary eigenvalue $\tau_0 = i\omega_0$ and the Jordan block corresponding to τ_0 is of order greater than one, then the zero solution to the system (2) is necessarily Lyapunov unstable. To prove this assertion, we first note that there exists a nonzero eigenvector z_0 and the adjoined vector z_1 such that

$$Az_0 = i\,\omega_0 z_0, \quad Az_1 = i\,\omega_0 z_1 + z_0.$$

It is easy to verify that the vector-valued function

$$y(t) = \frac{\delta}{2\|z_1\|} e^{i\omega_0 t} z_1 + \frac{\delta}{2\|z_1\|} t e^{i\omega_0 t} z_0$$

is a solution to the Cauchy problem

$$y' = Ay, \quad y(0) = \frac{\delta}{2\|z_1\|} z_1 \quad (\|y(0)\| = \delta/2 < \delta).$$

Indeed, differentiating, we see that

$$\begin{aligned} y'(t) &= \frac{\delta}{2\|z_1\|} i\,\omega_0 e^{i\omega_0 t} z_1 + \frac{\delta}{2\|z_1\|} [i\,\omega_0 e^{i\omega_0 t} t + e^{i\omega_0 t}] z_0 \\ &= \frac{\delta}{2\|z_1\|} e^{i\omega_0 t} [i\,\omega_0 z_1 + z_0] + \frac{\delta}{2\|z_1\|} e^{i\omega_0 t} i\,\omega_0 t z_0. \end{aligned}$$

On the other hand,

$$\begin{aligned} Ay(t) &= A\left[\frac{\delta}{2\|z_1\|} e^{i\omega_0 t} z_1 + \frac{\delta}{2\|z_1\|} e^{i\omega_0 t} t z\right] = \frac{\delta}{2\|z_1\|} e^{i\omega_0 t} A z_1 + \frac{\delta}{2\|z_1\|} e^{i\omega_0 t} t A z_0 \\ &= \frac{\delta}{2\|z_1\|} e^{i\omega_0 t} [i\,\omega_0 z_1 + z_0] + \frac{\delta}{2\|z_1\|} e^{i\omega_0 t} t i\,\omega_0 z_0. \end{aligned}$$

Comparing the expressions for y' and Ay, we see that they are equal. From the formula

$$y(t) = \frac{\delta e^{i\omega_0 t}}{2\|z_1\|} [z_1 + t z_0]$$

it follows that for $t > 0$ we have

$$\|y(t)\| \geqslant \frac{\delta}{2\|z_1\|} (t\|z_0\| - \|z_1\|)$$

($\|z_0\| \neq 0$, $\|z_1\| \neq 0$). Hence the norm of the solution $y(t)$ unboundedly increases as $t \to \infty$.

The existence of solutions $y(t)$ such that $\|y(0)\| < \delta$ for any $\delta > 0$ and $\|y(t)\| \to \infty$ as $t \to \infty$ means that the zero solution to the system (2) is Lyapunov unstable.

We show that if every eigenvalue of A either has negative real part or is purely imaginary and the Jordan blocks corresponding to the purely imaginary eigenvalues are of order one, then the zero solution to the system (2) is Lyapunov stable.

Under the above assumptions the matrix A admits the representation

$$A = W^{-1} \begin{bmatrix} \mathcal{B}_1 & 0 \\ 0 & \mathcal{B}_0 \end{bmatrix} W,$$

where W is a nonsingular matrix and $\mathcal{B}_1$, $\mathcal{B}_2$ are matrices of order N_1, N_2 ($N_1+N_2 = N$) such that all eigenvalues $\tau_1, \tau_2, \dots, \tau_{N_1}$ of $\mathcal{B}_1$ have negative real parts, i.e.,

$$\operatorname{Re}\tau_1 \leqslant -\sigma, \ \operatorname{Re}\tau_2 \leqslant -\sigma, \dots, \operatorname{Re}\tau_N \leqslant -\sigma \quad (\sigma > 0),$$

and $\mathcal{B}_0$ is a diagonal matrix with purely imaginary diagonal elements:

$$\mathcal{B}_0 = \begin{bmatrix} i\,\omega_1 & & & \text{\Large 0} \\ & i\,\omega_2 & & \\ & & \ddots & \\ \text{\Large 0} & & & i\,\omega_{N_0} \end{bmatrix}.$$

This assertion holds because the matrix A can be reduced to the Jordan canonical form.

For the matrix exponential we have the representation

$$e^{tA} = W^{-1}\begin{bmatrix} e^{t\mathcal{B}_1} & 0 \\ 0 & e^{t\mathcal{B}_0} \end{bmatrix} W = W^{-1}\begin{bmatrix} e^{t\mathcal{B}_1} & 0 \\ 0 & 0 \end{bmatrix} W + W^{-1}\begin{bmatrix} 0 & 0 \\ 0 & e^{t\mathcal{B}_0} \end{bmatrix} W$$

which implies that

$$\|e^{tA}\| \leqslant \|W^{-1}\|\,\|W\|\,\|e^{t\mathcal{B}_1}\| + \|W^{-1}\|\,\|W\|\,\|e^{t\mathcal{B}_0}\|$$

for $t > 0$. Since

$$\|e^{t\mathcal{B}_1}\| \leqslant \gamma(N_1)\Big(\frac{\|\mathcal{B}_1\|}{\sigma}\Big)^{N_1-1} e^{-t\sigma/2} \leqslant \gamma(N_1)\Big(\frac{\|\mathcal{B}_1\|}{\sigma}\Big)^{N_1-1},$$

$$e^{t\mathcal{B}_0} = \begin{bmatrix} e^{it\omega_1} & & & \\ & e^{it\omega_2} & & \mathbf{0} \\ & & \ddots & \\ \mathbf{0} & & & e^{it\omega_{N_0}} \end{bmatrix} \qquad \|e^{t\mathcal{B}_0}\| = 1,$$

we have

$$\|e^{tA}\| \leqslant \|W^{-1}\|\,\|W\|\,\gamma(N_1)\Big(\frac{\|\mathcal{B}_1\|}{\sigma}\Big)^{N_1-1} + \|W^{-1}\|\,\|W\| = \widehat{M}.$$

Since the norm of the matrix exponential of a solution to the system (2) is bounded, for $t > 0$ we have $\|y(t)\| \leqslant \widehat{M}\,\|y(0)\|$. It is clear that for any positive ε the condition $\|y(t)\| < \varepsilon$ is satisfied for $t \geqslant 0$ provided that the initial value $y(0)$ satisfies the relation $\|y(0)\| < \delta = \varepsilon/\widehat{M}$. Hence the zero solution to the system (2) is Lyapunov stable.

Our conclusions can be summarized in the table below. The columns I–IV correspond to the following locations of eigenvalues in the complex plane.

I. All eigenvalues $\tau_j(A)$ lie in the left half-plane ($\operatorname{Re}\tau_j < 0$).

II. All eigenvalues $\tau_j(A)$ lie either in the left half-plane or on the imaginary axis; moreover, the Jordan blocks corresponding to the purely imaginary eigenvalues are of order one.

III. All eigenvalues $\tau_j(A)$ lie either in the left half-plane or on the imaginary axis; moreover, there exist Jordan blocks corresponding to the purely imaginary eigenvalues of order greater than one.

IV. There are eigenvalues $\tau_j(A)$ in the right half-plane ($\operatorname{Re}\tau_j > 0$).

<table>
<tr><th>I</th><th>II</th><th>III, IV</th></tr>
<tr><td>The zero solution is asymptotically stable</td><td>The zero solution is not asymptotically stable</td><td rowspan="2">The zero solution is Lyapunov unstable</td></tr>
<tr><td colspan="2">The zero solution is Lyapunov stable</td></tr>
</table>

§3. The matrix Lyapunov equation

The solvability of the matrix Lyapunov equation with two-diagonal matrix of coefficients. The solvability conditions in the general case. Properties of a solution. The integral representation of a solution to the matrix Lyapunov equation. The criterion for asymptotic stability.

We consider the two-diagonal matrix

$$T = \begin{bmatrix} \tau_1 & a_2 & & & \\ & \tau_2 & a_3 & & 0 \\ & & \ddots & \ddots & \\ & & & \tau_{N-1} & a_N \\ 0 & & & & \tau_N \end{bmatrix} = \begin{bmatrix} T_{11} & T_{12} & \dots & T_{1N} \\ T_{21} & T_{22} & \dots & T_{2N} \\ \vdots & \vdots & \dots & \vdots \\ T_{N1} & T_{N2} & \dots & T_{NN} \end{bmatrix}$$

with complex elements

$$T_{kj} = \tau_k \delta_{kj} + a_j \delta_{j-1\,k} \quad (a_1 = 0),$$

where δ_{kj}, as usual, denotes the Kronecker symbol:

$$\delta_{kj} = \begin{cases} 1 & \text{if } k = j, \\ 0 & \text{if } k \neq j. \end{cases}$$

We also consider an arbitrary matrix D with elements D_{jk} and the matrix equation

$$XT + T^* X = -D$$

with an unknown matrix X. The elements X_{kj} of the matrix X are connected by the equalities

$$\sum_{k=1}^{N} (X_{ik} T_{kj} + \overline{T}_{ki} X_{kj}) = -D_{ij}.$$

Using the formulas for T_{kj}, we can rewrite the last equalities in detail:

$$\begin{aligned} (\tau_i + \overline{\tau}_j) X_{ij} + a_j X_{i,j-1} + \overline{a}_i X_{i-1,j} &= -D_{ij} \quad (i, j \geqslant 2), \\ (\tau_1 + \overline{\tau}_j) X_{1j} + a_j X_{1,j-1} &= -D_{1j} \quad (j \geqslant 2), \\ (\tau_i + \overline{\tau}_1) X_{i1} + \overline{a}_i X_{i-1,1} &= -D_{i1} \quad (i \geqslant 2), \\ (\tau_1 + \overline{\tau}_1) X_{11} &= -D_{11}. \end{aligned}$$

We assume that for any i, j the sum $\tau_i + \overline{\tau}_j$ does not vanish: $\tau_i + \overline{\tau}_j \neq 0$. We show that under this assumption the equations for X_{ij} are uniquely solvable.

First, it is clear that X_{11} is uniquely defined:

$$X_{11} = -\frac{D_{11}}{\tau_1 + \overline{\tau}_1}.$$

Second, for $i, j = 1, 2, 3, \dots, N$ we find X_{1j}, X_{i1} by the recurrent formulas

$$\begin{aligned} X_{1j} &= \frac{1}{\tau_1 + \overline{\tau}_j} (-D_{1j} - a_j X_{1,j-1}), \\ X_{i1} &= \frac{1}{\tau_i + \overline{\tau}_1} (-D_{i1} - \overline{a}_i X_{i-1,i}). \end{aligned}$$

Thus, we define the X_{ij} with at least one of the subscripts i, j equal to 1.

We suppose that we already know all X_{ij} such that at least one of the subscripts i, j does not exceed $p-1$ and describe how to define the X_{ij} with one of the subscripts i, j equal to p $(p \geqslant 2)$. First of all, we find

$$X_{pp} = \frac{1}{\tau_p + \overline{\tau}_p}(-D_{pp} - a_p X_{p,p-1} - \overline{a}_p X_{p-1,p}).$$

For $i, j = p+1, p+2, \dots, N$ we recurrently define X_{pj} and X_{ip} by the formulas

$$X_{pj} = \frac{1}{\tau_p + \overline{\tau}_j}(-D_{p,j} - a_j X_{p,j-1} - \overline{a}_p X_{p-1,j}),$$

$$X_{ip} = \frac{1}{\tau_i + \overline{\tau}_p}(-D_{i,p} - a_p X_{i,p-1} - \overline{a}_i X_{i-1,p}).$$

The procedure of defining all X_{ij} such that at least one of the subscripts i, j does not exceed p is completed. Repeating the procedure, for $p = 2, 3, \dots, N$, we consecutively find all X_{ij}, $1 < i, j \leqslant N$. The formulas for X_{ij} show that if all D_{ij} vanish, then all X_{ij} vanish too. From this fact it is easy to establish the uniqueness of X_{ij}. Thus, we have proved that for $\tau_j + \overline{\tau}_j \neq 0$, $i, j = 1, 2, \dots, N$, the matrix equation $XT + T^*X = -D$ is uniquely solvable for any matrix D.

Let

$$A = W^{-1} \begin{bmatrix} \tau_1 & a_2 & & & \\ & \tau_2 & a_3 & & \mathbf{0} \\ & & \ddots & \ddots & \\ & & & \tau_{N-1} & a_N \\ \mathbf{0} & & & & \tau_N \end{bmatrix} W = W^{-1}TW,$$

where $\det W \neq 0$, $\tau_1 + \overline{\tau}_j \neq 0$, $1 \leqslant i, j \leqslant N$, and let C be an arbitrary matrix. We consider the matrix equation

$$HA + A^*H = -C$$

and show that it is uniquely solvable. First we multiply both sides of the equation by nonsingular matrices $(W^{-1})^*$ and W^{-1}. We obtain

$$(W^{-1})^*HAW^{-1} + (W^{-1})^*A^*HW^{-1} = -(W^{-1})^*CW^{-1}$$

and rewrite the result as follows:

$$(W^{-1})^*HW^{-1}WAW^{-1} + (W^{-1})^*A^*W^*(W^{-1})^*HW^{-1} = -(W^{-1})^*CW^{-1}.$$

Denoting $(W^{-1})^*HW^{-1} = X$, $\quad (W^{-1})^*CW^{-1} = D$ and recalling that $WAW^{-1} = T$, we arrive at the familiar equation

$$XT + T^*X = -D$$

which is uniquely solvable with respect to X for any matrix D and, consequently, for any matrix C. A one-to-one correspondence between matrices H and X is given by the formula

$$H = W^*XW, \quad X = (W^{-1})^*HW^{-1}.$$

The possibility of application of these formulas and the fact that they define a one-to-one correspondence follow from the nonsingularity of the matrix W. Finding X and computing H from X, we arrive at a solution to the equation

$$(W^{-1})^*(HA + A^*H)W^{-1} = -(W^{-1})^*CW^{-1}$$

or, which is equivalent, at a solution to the equation $HA + A^*H = -C$. The above arguments prove the uniqueness of the solution H.

Since any matrix can be reduced to the Jordan canonical form, for any matrix A there is a nonsingular matrix W such that

$$A = W^{-1} \begin{bmatrix} \tau_1 & a_2 & & & \\ & \tau_2 & a_3 & & 0 \\ & & \ddots & \ddots & \\ & & & \tau_{N-1} & a_N \\ 0 & & & & \tau_N \end{bmatrix} W;$$

moreover, $\tau_1, \tau_2, \dots, \tau_N$ are eigenvalues of A, i.e., the roots of the characteristic equation $\det[A - \tau I] = 0$. The proved assertion about the solvability of the matrix equation will be formulated as a theorem.

THEOREM 1. *If the eigenvalues $\tau_1, \tau_2, \dots, \tau_N$ of a matrix A are such that $\tau_i + \overline{\tau}_j \neq 0$ for any i, j, then the matrix equation*

$$HA + A^*H = -C \tag{1}$$

is uniquely solvable for any matrix C.

Theorem 1 was proved by Lyapunov.

DEFINITION 1. Equation (1) in Theorem 1 is called the matrix Lyapunov equation.

REMARK 1. It is possible to prove that if matrices A and C are real, then H is also a real matrix. We need only to note that if a system of linear equations with real coefficients and right-hand sides has the same number of equations and unknowns. Then a solution exists and is unique if and only if the determinant of the matrix of coefficients does not vanish. We have established the uniqueness of a solution; therefore, the determinant does not vanish. The solution can be computed by Cramer's rule which yields a real solution in the case of real coefficients. The number of equations is equal to the number of elements of the matrix C and the number of unknowns is equal to the number of elements of the matrix H. It is obvious that these numbers are equal.

We indicate one more important fact. If C is an Hermitian matrix (i.e., $C = C^*$), then H is also an Hermitian matrix ($H = H^*$). To prove this fact, we apply the Hermitian conjugation operation to both sides of equation (1) and obtain the equality $A^*H^* + H^*A = -C^*$. Using the equality $C = C^*$, we find that H^* satisfies the matrix equation $H^*A + A^*H^* = -C$, which is the same as the equation for H. By the uniqueness of a solution to the matrix equation, we obtain the required equality $H^* = H$.

If C and A are real matrices, then the matrix H is real. The matrix H is symmetric in view of the equality $H = H^*$.

From the theorem and remarks made after the proof we obtain the following important assertion.

If all eigenvalues $\tau_1, \tau_2, \dots, \tau_N$ of a matrix A lie strictly in the left half-plane of the complex plane τ ($\operatorname{Re} \tau_j < 0$, $j = 1, 2, \dots, N$), then the matrix Lyapunov equation (1) is uniquely solvable for any Hermitian matrix C and the matrix H is its solution. Indeed, for any i, j $(1 \leqslant i, j \leqslant N)$

$$\operatorname{Re}\tau_i < 0, \quad \operatorname{Re}\tau_j < 0, \quad \operatorname{Re}\overline{\tau}_j = \operatorname{Re}\tau_j < 0, \quad \operatorname{Re}(\tau_i + \overline{\tau}_j) < 0.$$

Consequently, we can apply the above theorem.

We show that if in this case $(\operatorname{Re}\tau_j < 0,\ 1 \leqslant i \leqslant N)$ for the right-hand side we take an arbitrary positive definite Hermitian matrix, then the solution H to the matrix Lyapunov equation (1) is a positive definite matrix.

Rather than proving this assertion by studying the algebraic structure of the matrix Lyapunov equation (1), we proceed indirectly, using properties of the solution to the vector differential equation $y' = Ay$, which can be represented with the help of the exponential as follows:

$$y(t) = e^{tA} y(0).$$

If all eigenvalues τ_j of A lie in the left half-plane $(\operatorname{Re}\tau_j < -\sigma < 0)$, then

$$\|y(t)\| \leqslant \|e^{tA}\|\, \|y(0)\| \leqslant M e^{-t\sigma/2},$$

where M is a constant depending on the quantity $\|A\|/\sigma$ and the order N of the matrix A (cf. Chapter 1, §3).

Let $C = C^*$ be an arbitrary positive definite Hermitian matrix. The Hermitian form in $y(t)$ generated by C satisfies the inequalities

$$(Cy(t), y(t)) \leqslant \|C\|\, \|y(t)\|^2 \leqslant M^2\, \|C\|\, e^{-\sigma t}.$$

This Hermitian form can be rewritten in terms of matrix exponential as follows:

$$(Cy(t), y(t)) = (Ce^{tA}y(0), e^{tA}y(0)) = (e^{tA^*}Ce^{tA}y(0), y(0)).$$

By the estimate

$$(e^{tA^*}Ce^{tA}y(0), y(0)) \leqslant M^2\, \|C\|\, e^{-\sigma t},$$

the integral

$$\int_0^\infty (e^{tA^*}Ce^{tA}y(0), y(0))dt$$

converges and defines the Hermitian form

$$(\widehat{H}y(0), y(0)) = \int_0^\infty (e^{tA^*}Ce^{tA}y(0), y(0))dt.$$

The matrix $\widehat{H}$ of coefficients of this form admits the following integral representation:

$$\widehat{H} = \int_0^\infty e^{tA^*}Ce^{tA}dt.$$

We prove that $\widehat{H}$ is a positive definite matrix. Denote by $\lambda_{\min}(C)$ the minimal eigenvalue of the matrix C. For any y the following inequality holds:

$$(Cy, y) \geqslant \lambda_{\min}(C)\, \|y\|^2.$$

Since C is positive definite, we have $\lambda_{\min}(C) > 0$ and

$$(e^{tA^*}Ce^{tA}y(0), y(0)) = (Ce^{tA}y(0), e^{tA}y(0)) \geqslant \lambda_{\min}(C)\,\|e^{tA}y(0)\|^2.$$

Using the inequality $\|e^{tA}y(0)\| \geqslant e^{-t\|A\|}\|y(0)\|$ (cf. Chapter 1, §3), we see that

$$(e^{tA^*}Ce^{tA}y(0), y(0)) \geqslant \lambda_{\min}(C)\,\|e^{tA}y(0)\|^2 \geqslant \lambda_{\min}(C)e^{-2t\,\|A\|}\,\|y(0)\|^2.$$

Integrating the last inequality over $(0, \infty)$, we obtain

$$\begin{aligned}(\widehat{H}y(0), y(0)) &= \int_0^\infty (e^{tA^*}Ce^{tA}y(0), y(0))dt \\ &\geqslant \lambda_{\min}(C)\,\|y(0)\|^2 \int_0^\infty e^{-2t\,\|A\|}dt = \frac{\lambda_{\min}(C)}{2\,\|A\|}\,\|y(0)\|^2.\end{aligned}$$

The inequality

$$(\widehat{H}y(0), y(0)) \geqslant \frac{\lambda_{\min}(C)}{2\,\|A\|}\,\|y(0)\|^2,$$

which is valid for any $y(0)$, implies the following estimate for the minimal eigenvalue $\lambda_{\min}(\widehat{H})$ of the matrix $\widehat{H}$:

$$\lambda_{\min}(\widehat{H}) \geqslant \frac{\lambda_{\min}(C)}{2\,\|A\|} > 0.$$

Consequently, the matrix $\widehat{H}$ is positive definite.

We consider the expression

$$\begin{aligned}(e^{sA^*}\widehat{H}e^{sA}y(0), y(0)) &= \int_0^\infty (e^{sA^*}e^{tA^*}Ce^{tA}e^{sA}y(0), y(0))dt \\ &= \int_0^\infty (e^{(t+s)A^*}Ce^{(t+s)A}y(0), y(0))dt = \int_s^\infty (e^{tA^*}Ce^{tA}y(0), y(0))dt.\end{aligned}$$

It is easy to see that

$$e^{sA^*}\widehat{H}e^{sA} = \int_s^\infty e^{tA^*}Ce^{tA}dt.$$

We differentiate both sides of the last equality with respect to s,

$$A^*e^{sA^*}\widehat{H}e^{sA} + e^{sA^*}\widehat{H}Ae^{sA} = -e^{sA^*}Ce^{sA},$$

and set $s = 0$. We see that $\widehat{H}$ is a solution to the matrix Lyapunov equation (1). Under the assumption that the eigenvalues of the matrix A have negative real parts, the matrix Lyapunov equation (1) has a unique solution $H = \widehat{H}$. Thus, the solution to the matrix Lyapunov equation (1) admits the representation

$$H = \int_0^\infty e^{tA^*}Ce^{tA}dt$$

which, as was shown above, is a positive definite matrix H.

We formulate once more the above assertions. If all eigenvalues τ_j lie in the left half-plane, then the matrix Lyapunov equation (1) is uniquely solvable for any C. If C is an Hermitian matrix, then the solution H is also an Hermitian matrix. Since C is positive definite, we conclude that H is positive definite; moreover, $\lambda_{\min}(H) \geqslant \lambda_{\min}(C)/(2\|A\|)$.

The converse assertion is also true. If two positive definite Hermitian matrices H and C are connected by the matrix Lyapunov equation (1), then all eigenvalues of A lie in the left half-plane.

Indeed, let λ_0 be an eigenvalue of the matrix A and let z_0 be the corresponding eigenvector, i.e., $Az_0 = \lambda_0 z_0$. By the matrix Lyapunov equation (1) the vector equality

$$HAz_0 + A^*Hz_0 = -Cz_0$$

holds. Taking inner product z_0, we obtain

$$(HAz_0, z_0) + (A^*Hz_0, z_0) = -(Cz_0, z_0)$$

or, which is equivalent,

$$(HAz_0, z_0) + (Hz_0, Az_0) = -(Cz_0, z_0).$$

In the last equality, we replace Az_0 by $\lambda_0 z_0$ and obtain

$$\lambda_0(Hz_0, z_0) + \overline{\lambda}_0(Hz_0, z_0) = -(Cz_0, z_0),$$

i.e.,

$$2\operatorname{Re}\lambda_0(Hz_0, z_0) = -(Cz_0, z_0).$$

Since $(Hz_0, z_0) > 0$ and $(Cz_0, z_0) > 0$, we have

$$\operatorname{Re}\lambda_0 = -\frac{(Cz_0, z_0)}{2(Hz_0, z_0)} < 0.$$

Thus, we have proved the following assertion.

THEOREM 2. *For any positive definite matrix C the matrix Lyapunov equation (1) is uniquely solvable if $\tau_i(A) + \overline{\tau}_j(A) \neq 0$ for any eigenvalues τ_i, τ_j of the matrix A. The solution H is a positive definite matrix if and only if* $\operatorname{Re}\tau_j(A) < 0$ *for all j.*

We will use these results in the study of the Lyapunov theory of stability from the first approximation.

§4. The Lyapunov functions

Properties of the Lyapunov functions. The theorem about asymptotic stability. The Chetaev theorem on instability.

The so-called Lyapunov functions (cf. §1) are widely used in the study of properties of solutions to the autonomous system

$$\frac{dy}{dt} = f(y)$$

which in detailed form is written as the system

$$\frac{dy_1}{dt} = f_1(y_1, y_2, \dots, y_N),$$
$$\frac{dy_2}{dt} = f_2(y_1, y_2, \dots, y_N),$$
$$\dots\dots\dots\dots\dots\dots\dots\dots$$
$$\frac{dy_N}{dt} = f_N(y_1, y_2, \dots, y_N).$$

We recall that by a Lyapunov function $\mathcal{H}(y) = \mathcal{H}(y_1, y_2, \dots, y_N)$ we mean a function satisfying Conditions L1, L2, L3 or L1, L2, L3$'$ (cf. §1).

THEOREM 1 (Lyapunov). *If for the vector equation $y' = f(y)$ with the equilibrium point $y = 0$ ($f(0) = 0$) there exists the Lyapunov function $\mathcal{H}(y)$ satisfying Conditions* L1, L2, *and* L3, *then the solution $y = 0$ is Lyapunov stable. If there exists the Lyapunov function $\mathcal{H}(y)$ satisfying Conditions* L1, L2, *and* L3$'$, *then the solution $y = 0$ is asymptotically stable.*

PROOF. As was shown in §1, if for the equation $y' = f(y)$ there exists a Lyapunov function $\mathcal{H}(y)$ satisfying Conditions L1–L3, then there exists $\rho_0 > 0$ such that for $\|y(0)\| \leqslant \rho_0$ the corresponding solution $y(t)$ exists on the entire half-line $0 \leqslant t < \infty$.

We consider the functions $m(\rho)$ and $M(\rho)$ introduced in §1. We recall that ρ_0 from the above-mentioned existence theorem (cf. §1) was chosen from the condition $M(\rho_0) < m(R/2)$. Let $\varepsilon > 0$. We assume that $\varepsilon \leqslant R/2$. We choose $\delta > 0$ so as to satisfy the inequality $M(\delta) < m(\varepsilon)$. If $\|y(0)\| \leqslant \delta \leqslant \rho_0$, then the corresponding solution $y(t)$ to the equation $y' = f(y)$ exists on the entire half-line $0 \leqslant t < \infty$. In addition,

$$\mathcal{H}(y(0)) \leqslant \max_{\|y\| \leqslant \delta} \mathcal{H}(y) = M(\delta).$$

The function $\varphi(t) = \mathcal{H}(y(t))$ is a nonincreasing function of t because (cf. the proof of this fact in §1)

$$\frac{d\varphi(t)}{dt} = \sum_{j=1}^{N} \frac{\partial \mathcal{H}}{\partial y_j} \frac{dy_j}{dt} = \sum_{j=1}^{N} \frac{\partial \mathcal{H}}{\partial y_j} f_j = -\mathcal{G}(y(t)).$$

By Condition L3, $\mathcal{G}(y) \geqslant 0$ and $\varphi'(t) \leqslant 0$. Therefore, for all $t \geqslant 0$,

$$\mathcal{H}(y(t)) \leqslant \mathcal{H}(y(0)) \leqslant M(\delta) < m(\varepsilon).$$

From the inequality $\mathcal{H}(y(t)) < m(\varepsilon)$ it follows that $\|y(t)\| < \varepsilon$. Thus, we have proved that the zero solution $y_0(t) = 0$ is Lyapunov stable.

If instead of L3 the stronger Condition L3$'$ is satisfied, then any solution $y(t)$ such that $\|y(0)\| < \rho_0$ converges to zero ($\|y(t)\| \to 0$) as $t \to \infty$, which was shown in the global existence theorem (again cf. Theorem 2 in §1). Thus, if Condition L3$'$ is satisfied instead of Condition L3, then the zero solution is asymptotically stable.

As an important example, we consider the linear vector equation $y' = Ay$. The quadratic (or Hermitian) form

$$(Hy, y) = \sum_{i,j=1}^{N} h_{ij} y_i y_j$$

is the Lyapunov function for the equation $\frac{dy}{dt} = Ay$ if the quadratic form (Hy, y) is positive definite and

$$\frac{d}{dt}(Hy,y) = (H\frac{dy}{dt},y)+(Hy,\frac{dy}{dt}) = (HAy,y)+(Hy,Ay) = ((HA+A^*H)y,y) \leqslant 0$$

along solutions to the system, i.e., if the quadratic form (Cy, y) is nonnegative definite, where $C = -(HA + A^*H)$. In this case, the functions $\mathcal{H}(y) = (Hy, y)$ and $\mathcal{G}(y) = (Cy, y)$ satisfy Conditions L1–L3. If the matrix C is positive definite, Condition L3$'$ is also satisfied. In addition, since the positive definite matrices H and C are connected by the matrix equality $HA+A^*H = -C$, all eigenvalues τ_j of A lie strictly in the left half-plane of the complex plane τ ($\operatorname{Re}\tau_j < 0$, $j = 1, 2, \dots, N$) and the zero solution to the vector equation $y' = Ay$ is asymptotically stable. □

Now we analyze instability of the zero solution to an autonomous system. We assume that for the equation $y' = f(y)$ with the equilibrium point $y = 0$ ($f(0) = 0$) there exists a function $\mathcal{K}(y)$ satisfying the following conditions.

H1. The function $\mathcal{K}(y)$ is defined in the ball

$$\|y\| = \sqrt{y_1^2 + y_2^2 + \cdots + y_N^2} \leqslant R,$$

is continuous inside the ball $\|y\| \leqslant R$ and on its boundary and has continuous partial derivatives $\frac{\partial \mathcal{K}(y)}{\partial y_j}$ with respect to $y_1, y_2, \dots, y_N$. In addition, $\mathcal{K}(0) = 0$.

H2. In any neighborhood $\|y\| < \delta$ of the origin, there is $y^{[\delta]} \equiv (y_1^{[\delta]}, y_2^{[\delta]}, \dots, y_N^{[\delta]})$ at which $\mathcal{K}(y^{[\delta]}) \geqslant 0$ and $0 < \|y^{[\delta]}\| < \delta$.

H3. The function

$$\mathcal{F}(y) = \sum_{j=1}^{N} \frac{\partial \mathcal{K}}{\partial y_j} f_j(y_1, y_2, \dots, y_N),$$

is continuous for $0 \leqslant \|y\| \leqslant R$, vanishes at $y = 0$ ($y_1 = 0, y_2 = 0, \dots, y_N = 0$), i.e., $\mathcal{F}(0) = 0$, and is strictly positive ($\mathcal{F}(y) > 0$) at every point y such that $0 < \|y\| \leqslant R$ and $\mathcal{K}(y) > 0$.

We note that $\mathcal{F}(y)$ is the derivative of $\mathcal{K}(y(t))$ with respect to t along solutions $y(t)$ to the equation $y' = f(y)$:

$$\frac{d}{dt}\mathcal{K}(y(t)) = \sum_{j=1}^{N} \frac{\partial \mathcal{K}}{\partial y_j}\frac{dy_j}{dt} = \sum_{j=1}^{N} \frac{\partial \mathcal{K}}{\partial y_j} f_j(y(t)) = \mathcal{F}(y(t)).$$

THEOREM 2 (Chetaev). *Under Conditions* H1–H3, *the zero solution to the vector equation* $y' = f(y)$ *is Lyapunov unstable.*

In the proof of the Chetaev theorem, it is convenient to use Corollaries 1, 2 (below) of Conditions

COROLLARY 1. *If the set* $\mathcal{M}$ *of points* y *such that* $\|y\| \leqslant R$ *and* $\mathcal{K}(y) \geqslant \beta > 0$ *is nonempty, then the lower bound of the values* $\mathcal{F}(y)$ *taken over* $\mathcal{M}$ *is strictly positive:*

$$\inf_{y \in \mathcal{M}} \mathcal{F}(y) = \gamma(\beta) > 0.$$

Before proving the corollary, we note that the continuity of $\mathcal{K}(y)$ implies that the bounded nonempty set $\mathcal{M}$ is closed. Hence there exists a point $y^* \in \mathcal{M}$ at which $\|y\|$ attains the lower bound:

$$\|y^*\| = \inf_{y\in\mathcal{M}} \|y\|, \quad y^* \in \mathcal{M}.$$

PROOF OF COROLLARY 1. It is obvious that $\mathcal{K}(y^*) \geqslant \beta$ and, consequently, $\|y^*\| = \eta_0 > 0$.

For all $y \in \mathcal{M}$ the inequality $\|y\| \geqslant \eta_0 > 0$ holds. (The point y^* cannot coincide with the origin, i.e., $y^* \neq 0$, because $\mathcal{K}(0, 0, \dots, 0) = 0$.)

The nonnegative continuous function $\mathcal{F}(y)$ attains the lower bound over the closed set $\mathcal{M}$ at some point $y = y^{**}$. Since $\mathcal{K}(y^{**}) \geqslant \beta > 0$, we have $\mathcal{F}(y^{**}) = \inf_{y\in\mathcal{M}} \mathcal{F}(y) = \gamma(\beta) > 0$. □

COROLLARY 2. *Let $y(t)$ ($\|y(t)\| \leqslant R$) be a solution to the system $\dfrac{dy}{dt} = f(y)$ defined for $t_1 \leqslant t \leqslant t_2$. We suppose that the functions $\mathcal{K}(y)$ and $\mathcal{F}(y)$ satisfy Conditions* H1–H3 *and $\mathcal{K}(y(t_1)) = \beta > 0$. Then for $t_1 \leqslant t \leqslant t_2$ we have*

$$\mathcal{K}(y(t)) \geqslant \beta + \gamma(\beta)(t - t_1), \quad \gamma(\beta) > 0.$$

PROOF. We begin with the proof of the inequality $\mathcal{K}(y(t)) > 0$ for $t_1 \leqslant t \leqslant t_2$. We assume that the inequality $\mathcal{K}(y(t)) > 0$ fails. Then the set of points t at which $\mathcal{K}(y(t)) = 0$ ($t_1 < t \leqslant t_2$) is nonempty and, by the continuity of $\mathcal{K}(y(t))$, is closed (it may be that the set consists of an infinite number of points, but it can be finite and even consists of a single point). The lower bound

$$t^* = \inf_{\mathcal{K}(y(t))=0} t$$

defines the segment $t_1 \leqslant t < t^*$ of nonzero length on which $\mathcal{K}(y(t)) > 0$; moreover, $\mathcal{K}(y(t^*)) = 0$. By Condition H3, on this segment we have

$$\mathcal{F}(y) > 0, \quad t_1 \leqslant t < t^*,$$
$$\frac{d}{dt}\mathcal{K}(y(t)) = \mathcal{F}(y(t)) > 0, \quad \mathcal{K}(y(t^*)) > \mathcal{K}(y(t_1)) = \beta > 0,$$

which contradicts the equality $\mathcal{K}(y(t^*) = 0$. Hence our assumption that the inequality $\mathcal{K}(y(t)) > 0$ fails is false. Thus, $\mathcal{K}(y(t)) > 0$ and $\mathcal{F}(y(t)) > 0$ on $t_1 \leqslant t \leqslant t_2$. Furthermore, as was shown, these inequalities yield the relation $\mathcal{K}(y(t)) \geqslant \beta > 0$. By Corollary 1, on the entire segment, we have $\mathcal{F}(y(t)) \geqslant \gamma(\beta) > 0$, which leads to the required assertion:

$$\frac{d}{dt}\mathcal{K}(y(t)) = \mathcal{F}(y(t)) \geqslant \gamma(\beta) > 0,$$
$$\mathcal{K}(y(t)) \geqslant \beta + \gamma(\beta)(t - t_1), \quad t_1 \leqslant t \leqslant t_2. \quad \square$$

PROOF OF THEOREM 2. Assume the contrary, i.e., assume that the Lyapunov stability takes place despite the fulfillment of Conditions H1–H3. Given ε $(0 < \varepsilon < R)$, we can find $\delta > 0$ such that any solution $y(t)$ with the initial value in the ball $\|y(0)\| < \delta < R$ can be extended to a function defined for all positive t $(0 \leqslant t < \infty)$ and for all such t the inequality $\|y(t)\| < \varepsilon < R$ holds. Along such a solution the function $\mathcal{K}(y(t)) \leqslant \max_{\|y\| \leqslant R} \mathcal{K}(y)$ is bounded from above. We choose a solution $y(t)$ starting from the initial point $y(0) = y^{[\delta]}$ at $t = 0$ so that $\mathcal{K}(y^{[\delta]}) \geqslant 0$ and $0 < \|y^{[\delta]}\| < \delta$; moreover, $\mathcal{F}(y^{[\delta]}) > 0$. Since $y(t)$ and $\mathcal{F}(y(t))$ are continuous, there exists a segment $0 \leqslant t \leqslant t_1$ on which $\mathcal{F}(y(t)) \geqslant \frac{1}{2}\mathcal{F}(y^{[\delta]}) > 0$. On this segment, we also have

$$\frac{d}{dt}\mathcal{K}(y(t)) = \mathcal{F}(y(t)) \geqslant \frac{1}{2}\mathcal{F}(y^{[\delta]}) > 0,$$

$$\mathcal{K}(y(t)) \geqslant \mathcal{K}(y(0)) + \frac{1}{2}\mathcal{F}(y^{[\delta]})t \geqslant \frac{t}{2}\mathcal{F}(y^{[\delta]}).$$

In particular,

$$\mathcal{K}(y(t_1)) \geqslant \frac{t_1}{2}\mathcal{F}(y^{[\delta]}) \equiv \beta > 0.$$

By Corollary 2, for $t \geqslant t_1$ the inequality

$$\mathcal{K}(y(t)) \geqslant \beta + \gamma(\beta)(t - t_1)$$

holds, which shows that $\mathcal{K}(y(t))$ cannot be bounded along the entire trajectory $y(t)$, $0 \leqslant t < \infty$. As was mentioned, the boundedness of $\mathcal{K}(y(t))$ is a necessary consequence of stability. This contradiction proves the Chetaev theorem. □

The Lyapunov theorem on stability and the Chetaev theorem on instability are of independent interest. They are often used in the analysis of systems of equations. But, in every special case, the construction of $\mathcal{H}(y(t))$ and $\mathcal{K}(y(t))$ requires a certain skill.

However, there is a class of equations for which the construction of $\mathcal{H}(y(t))$ and $\mathcal{K}(y(t))$ is reduced to standard algebraic procedures. For this class, it is possible to give simple and effective criteria for stability and instability. We say that the vector differential equation

$$\frac{dy}{dt} = Ay + \varphi(y)$$

with a constant matrix A is almost linear if the vector-valued function $\varphi(y)$ of the vector-valued variable y is defined for $\|y\| \leqslant Y$ and satisfies the inequality

$$\|\varphi(y)\| \leqslant q\,\|y\|^{1+\omega}, \qquad q > 0, \quad \omega > 0.$$

This class will be studied in §5.

We give two examples that explain how the matrix Lyapunov equation (cf. §3) appears in the theory of linear differential equations. Let $y(t)$ satisfy the equation $y' = Ay$ and let $H = H^*$ be an Hermitian matrix. We find the derivative of the quadratic form $(H(y), y)$ with respect to t:

$$\begin{aligned}\frac{d}{dt}(Hy(t), y(t)) &= (Hy'(t), y(t)) + (Hy(t), y'(t)) = (HAy(t), y(t)) \\ &\quad + (Hy(t), Ay(t)) = ([HA + A^*H]y(t), y(t)) = -(Cy(t), y(t)),\end{aligned}$$

where $C = -(HA + A^*H)$ (it is obvious that $C^* = C$). Thus, we have proved the following lemma.

LEMMA 1. *If the matrices A, $H = H^*$, and $C = C^*$ satisfy the equality $HA + A^*H = -C$ and $y(t)$ is a solution to the equation $y'(t) = Ay(t)$, then*

$$\frac{d}{dt}(Hy(t), y(t)) = -(Cy(t), y(t)).$$

We suppose that all eigenvalues of A lie in the left half-plane ($\operatorname{Re}\lambda_j(A) < 0$, $j = 1, 2, \ldots, N$), $C = C^*$ is an Hermitian matrix, and $y(t)$ is a solution to the Cauchy problem $y'(t) = Ay(t)$, $y(0) = y_0$. We want to compute the integral

$$\int_0^\infty (Cy(t), y(t))dt.$$

It is not necessary to solve the Cauchy problem provided the solution H to the matrix Lyapunov equation $HA + A^*H = -C$ is known (the Lyapunov equation is uniquely solvable because $\operatorname{Re}\lambda_j(A) < 0$). Indeed, by Lemma 1, $\frac{d}{dt}(H(y(t), y(t))) = -(C(y(t), y(t)))$. Integrating the last equality with respect to t from 0 to ∞, we obtain

$$\int_0^\infty (Cy(t), y(t))dt = (Hy_0, y_0).$$

Thus, we have proved the lemma which is formulated below in the case $C = I$.

LEMMA 2. *If $y(t)$ is a solution to the system $y'(t) = Ay(t)$, $\operatorname{Re}\lambda_j(A) < 0$, $j = 1, 2, \ldots, N$, and $HA + A^*H = -I$, then*

$$\int_0^\infty \|y(t)\|^2 dt = (Hy(0), y(0)).$$

Now, for every $N \times N$-matrix A we introduce the characteristic $\varkappa(A)$ which plays an important role in stability theory. Let $y(t)$ be a solution to the equation $y'(t) = Ay(t)$. We define

$$\varkappa(A) = 2\,\|A\| \sup_{T, y(0)} \frac{\int_0^T \|y(t)\|^2 dt}{\|y(0)\|^2} \tag{1}$$

and show that $\varkappa(A)$ is finite if and only if the zero solution to the system $y' = Ay$ is asymptotically stable. Let $\operatorname{Re}\lambda_j(A) < 0$, $j = 1, 2, \ldots, N$, and let H be a solution to the matrix Lyapunov equation $HA + A^*H = -I$. Then

$$\begin{aligned}\varkappa(A) &= 2\,\|A\| \sup_{T, y(0)} \frac{\int_0^T \|y(t)\|^2 dt}{\|y(0)\|^2} = 2\,\|A\| \sup_{y(0)} \frac{\int_0^\infty \|y(t)\|^2 dt}{\|y(0\|^2} \\ &= 2\,\|A\| \sup_{y(0)} \frac{(Hy(0), y(0))}{(y(0), y(0))} = 2\,\|A\|\,\|H\|.\end{aligned}$$

If $\operatorname{Re}\lambda_j(A) \geqslant 0$, $j = 1, 2, \ldots, N$, then $\varkappa(A) = \infty$. Indeed, let y_0 be the eigenvector corresponding to an eigenvalue λ_0, i.e., $Ay_0 = \lambda_0 y_0$. Then $\widetilde{y}(t) = e^{\lambda_0 t} y_0$ satisfies the equation $y' = Ay$; moreover,

$$\|\widetilde{y}(t)\| = |e^{\lambda_0 t}|\,\|y_0\| \geqslant \|y_0\|,$$

$$\sup_T \int_0^T \|\widetilde{y}(t)\|^2 dt = \infty.$$

Consequently, $\varkappa(A) = \infty$.

The proof of the equality $\varkappa(\rho A) = \varkappa(A)$ for any $\rho > 0$ is left to the reader.

We show that $\varkappa(A) \geqslant 1$. Let $\operatorname{Re}\lambda_j(A) < 0$, $j = 1, 2, \ldots, N$, and let H be a solution to the matrix Lyapunov equation $I = -HA - A^*H$. Then

$$1 \leqslant \|HA\| + \|A^*H\| \leqslant 2\|A\|\,\|H\| = \varkappa(A).$$

We recall that for a Hurwitz matrix A, i.e., for a matrix with $\operatorname{Re}\lambda_j(A) < 0$, $j = 1, 2, \ldots, N$, a solution H to the matrix Lyapunov equation $HA + A^*H = -I$ is a positive definite matrix and $\lambda_{\min}(H) \geqslant 1/(2\|A\|)$. On the other hand, for a positive definite matrix the equality $\lambda_{\max}(H) = \|H\|$ is known. Therefore, for the matrix H the ratio of the maximal eigenvalue to the minimal eigenvalue satisfies the estimate

$$\frac{\lambda_{\max}(H)}{\lambda_{\min}(H)} \leqslant \varkappa(A).$$

Let $y(t)$ be a solution to the equation $y'(t) = Ay(t)$. We prove the estimate

$$\|y(t)\| \leqslant \sqrt{\varkappa(A)}\, e^{-t\,\|A\|/\varkappa(A)}\,\|y(0)\| \quad (t > 0).$$

For $\varkappa(A) = \infty$ the estimate is obvious. Let $\varkappa(A) < \infty$ and let H be a solution to the matrix Lyapunov equation $HA + A^*AH = -I$. By Lemma 1,

$$\frac{d}{dt}(Hy(t), y(t)) = -(y(t), y(t)).$$

Using the inequality $(Hy, y) \leqslant \|H\|(y, y)$, we obtain

$$\frac{d}{dt}(Hy(t), y(t)) \leqslant -\frac{(Hy(t), y(t))}{\|H\|}.$$

Multiplying both sides of the last inequality by $e^{t/\|H\|}$, we can write it in the form

$$\frac{d}{dt}[e^{t/\|H\|}(Hy(t), y(t))] \leqslant 0.$$

Therefore,

$$(Hy(t), y(t)) \leqslant e^{-t/\|H\|}(Hy(0), y(0)) \leqslant \|H\|\, e^{-t/\|H\|}(y(0), y(0)).$$

From the inequality $\lambda_{\min}(H) \geqslant 1/(2\|A\|)$ it follows that $(Hy, y) \geqslant \|y\|^2/(2\|A\|)$ for any y. Hence

$$\frac{\|y(t)\|^2}{2\,\|A\|} \leqslant \|H\|\, e^{-t/\|H\|}\,\|y(0)\|^2$$

and, finally,

$$\|y(t)\| \leqslant \sqrt{\varkappa(A)} e^{-t\,\|A\|/\varkappa(A)}\,\|y(0)\|.$$

§5. The theorems on stability from the first approximation

The Lyapunov theorem on stability and instability from the first approximation. The estimate for the domain of attraction. Examples.

As was mentioned above, the vector differential equation

$$\frac{dy}{dt} = Ay + \varphi(y)$$

with a constant matrix A is almost linear if the vector-valued function $\varphi(y)$ of vector variable y is defined for $\|y\| \leqslant Y$ and satisfies the estimate

$$\|\varphi(y)\| \leqslant q\,\|y\|^{1+\omega}, \qquad q > 0, \quad \omega > 0.$$

It is convenient to write this in new notations by setting $q = \|A\|/r$, so that the estimate takes the form

$$\|\varphi(y)\| \leqslant \left(\frac{\|y\|^\omega}{r}\right)\|A\|\,\|y\|.$$

We have $\|\varphi(y)\| \leqslant \|A\|\,\|y\|$ for $\|y\|^\omega < r$ and $\|\varphi(y)\| \ll \|A\|\,\|y\|$ for $\|y\|^\omega \ll r$. Thereby, for $\|y\| \ll r$ the above system almost coincides with the linear system $y' = Ay$. It is natural to call r the radius characterizing the nonlinearity of $\varphi(y)$.

The following two theorems were proved by Lyapunov.

THEOREM 1 (on stability from the first approximation). *If all eigenvalues $\tau_j(A)$ of a matrix A lie strictly in the left half-plane, then the zero solution to an almost linear system is asymptotically stable.*

THEOREM 2 (on instability from the first approximation). *If, among the eigenvalues $\tau_j(A)$ of a matrix A, there is $\tau_{j_0}(A)$ such that the real part of $\tau_{j_0}(A)$ is strictly positive, then the zero solution to an almost linear system is Lyapunov unstable.*

PROOF OF THEOREM 1. We find an Hermitian matrix H as a solution to the Lyapunov equation $HA + A^*H = -I$. As we know, the equation is uniquely solvable and the solution is a positive definite Hermitian matrix H. The Hermitian form $(Hy, y) = \mathcal{H}(y)$ can be regarded as the Lyapunov function for the vector equation $y' = Ay + \varphi(y)$ satisfying Conditions L1, L2, L3$'$. Indeed, it is obvious that Conditions L1, L2 are satisfied. In our case, $\mathcal{H}(y) = (Hy, y)$ is a positive definite quadratic form and the quantity R in Conditions L1, L2 can be arbitrary. Before the verification of Condition L3$'$ we compute $\mathcal{G}(y)$:

$$\begin{aligned}\mathcal{G}(y) &= -\frac{d}{dt}(Hy, y) = -(H[Ay + \varphi(y)], y) - (Hy, [Ay + \varphi(y)]) \\ &= -([HA + A^*H]y, y) - (H\varphi(y), y) - (Hy, \varphi(y)) = (y, y) + \Delta(y),\end{aligned}$$

where $\Delta(y)$ denotes a scalar real function of vector variable y:

$$\Delta(y) = -(H\varphi(y), y) - (Hy, \varphi(y)) = -\overline{(Hy, \varphi(y))} - (Hy, \varphi(y)) = -2\,\mathrm{Re}(Hy, \varphi(y)).$$

It is easy to see that

$$\begin{aligned}|\Delta(y)| &\leqslant 2\,|(Hy, \varphi(y))| \leqslant 2\,\|Hy\|\,\|\varphi(y)\| \\ &\leqslant 2\,\|H\|\,|y\|\left(\frac{\|y\|^\omega}{r}\right)\|A\|\,\|y\| = \varkappa(A)\frac{\|y\|^\omega}{r}\,\|y\|^2,\end{aligned}$$

where $\varkappa(A) = 2\|A\|\,\|H\|$. It is clear that $\Delta(y)$, so also $\mathcal{G}(y)$, are defined for $\|y\| \leqslant Y$.

Thus,

$$\mathcal{G}(y) = \|y\|^2 + \Delta(y) \geqslant \|y\|^2 - \varkappa(A)\frac{\|y\|^\omega}{r}\|y\|^2 = \|y\|^2\left[1 - \varkappa(A)\frac{\|y\|^\omega}{r}\right].$$

Taking

$$R = \min\{Y, (r/(2\varkappa(A)))^{1/\omega}\},$$

for $\|y\| \leqslant R$ we obtain

$$1 - \frac{\varkappa(A)}{r}\|y\|^\omega \geqslant 1 - \frac{\varkappa(A)}{r}\frac{r}{2\varkappa(A)} = \frac{1}{2},$$

$$\mathcal{G}(y) \geqslant \|y\|^2\left[1 - \frac{\varkappa(A)\,\|y\|^\omega}{r}\right] \geqslant \frac{1}{2}\,\|y\|^2.$$

Thus, for $\|y\| \leqslant R$ Condition L3$'$ is satisfied. For the vector equation $\dfrac{dy}{dt} = Ay + \varphi(y)$ we have constructed the quadratic function $\mathcal{H}(y) = (Hy, y)$, which satisfies Conditions L1, L2, L3$'$ for $\|y\| \leqslant R$. The existence of such a function $\mathcal{H}(y)$ implies the asymptotic stability of the zero solution $y = 0$ to the vector equation under consideration. □

REMARK 1. As was shown in §1, if for $\|y\| \leqslant R$ a quadratic function $\mathcal{H}(y) = (Hy, y)$ is a Lyapunov function for the system $y' = f(y)$ and Condition L3$'$ is satisfied, then the ball

$$\|y\| \leqslant \rho_0 = \sqrt{\frac{\lambda_{\min}(H)}{\lambda_{\max}(H)}\frac{R}{2}}$$

lies in the domain of attraction of center $y = 0$.

In §4, we proved the inequality

$$\frac{\lambda_{\min}(H)}{\lambda_{\max}(H)} \leqslant \varkappa(A).$$

Therefore, the ball $\|y\| \leqslant Y_0 = R/(2\sqrt{\varkappa(A)})$ lies in the domain of attraction of center $y = 0$.

We recall that $R = \min\{Y, (r/(2\varkappa(A)))^{1/\omega}\}$. Thus, for Y_0 that bounds the ball in the domain of attraction we obtain the following expression:

$$Y_0 = \min\left\{\frac{Y}{2\sqrt{\varkappa(A)}}, \left(\frac{r}{2}\right)^{1/\omega}\frac{1}{2[\varkappa(A)]^{1/2+1/\omega}}\right\}.$$

REMARK 2. A solution to the Cauchy problem

$$\frac{dy}{dt} = Ay + \varphi(y), \quad y(0) = y_0,$$

where $\|y_0\| \leqslant Y$, satisfies the inequality

$$\frac{d}{dt}(Hy(t), y(t)) \leqslant -\frac{1}{2}\,\|y(t)\|^2.$$

Using this inequality, we can establish the estimate

$$\|y(t)\| \leqslant \sqrt{\varkappa(A)}\,e^{-\frac{t\,\|A\|}{2\varkappa(A)}}\,\|y(0)\|$$

(cf. a similar estimate obtained in §4 in the linear case).

PROOF OF THEOREM 2. We assume that the matrix A has at least one eigenvalue with strictly positive real part. In this case, we show that the zero solution is Lyapunov unstable. To prove this, it suffices to construct a function $\mathcal{K}(y) = \mathcal{K}(y_1, y_2, \dots, y_N)$ satisfying Conditions H1–H3. Then the Chetaev theorem gives the required conclusion.

We note that the matrix $A - \frac{\lambda}{2}I$ has eigenvalues $\tau_j - \lambda/2$, where τ_j are the eigenvalues of A. If there are eigenvalues τ_j with strictly positive real parts, then we can choose a small number $\lambda > 0$ such that there are $\tau_j - \lambda/2$ with positive real parts and there are no sums $(\tau_j - \lambda/2) + (\overline{\tau}_k - \lambda/2) = \tau_j + \overline{\tau}_k - \lambda$ equal to zero. Choosing such λ, we construct a symmetric matrix K ($K = K^*$) as a solution to the Lyapunov equation

$$K\left(A - \frac{\lambda}{2}I\right) + \left(A - \frac{\lambda}{2}I\right)^* K = -I$$

which can be rewritten in the form

$$KA + A^*K = \lambda K - I.$$

This equation is uniquely solvable with respect to K since none of the sums $(\tau_j - \lambda/2) + (\overline{\tau}_k - \lambda/2)$, composed of the eigenvalues $\tau_j - \lambda/2$, $\tau_k - \lambda/2$ of the matrix $A - \frac{\lambda}{2}I$ equals zero (cf. Theorem in §3). The matrix K is symmetric, but it is not positive definite. Indeed, if $K > 0$, then from the properties of the matrix Lyapunov equation (cf. §3) we derive that all eigenvalues of $A - \frac{\lambda}{2}I$ lie strictly in the left half-plane. But at least one of eigenvalues of A has positive real part. Since K is not positive definite, there exists a vector $\widehat{y}$ ($\|\widehat{y}\| > 0$) such that $(K\widehat{y}, \widehat{y}) \leqslant 0$.

We define $\mathcal{K}(y)$ by the formula $\mathcal{K}(y) = -(Ky, y)$ and verify that $\mathcal{K}(y)$ satisfies Conditions H1–H3.

Condition H1 means that the function $\mathcal{K}(y)$ is defined for $\|y\| \leqslant R$, is continuous inside the ball $\|y\| \leqslant R$ and on its boundary, and has continuous partial derivatives $\dfrac{\partial \mathcal{K}}{\partial y_j}$. In addition, $\mathcal{K}(0) = 0$. It is obvious that Condition H1 is satisfied for any $R > 0$.

Condition H2 means that for any $\delta > 0$ there exists a vector $y^{[\delta]}$ such that $0 < \|y^{[\delta]}\| < \delta$ and $\mathcal{K}(y^{[\delta]}) \geqslant 0$. This assertion is also valid in our case. Indeed, it suffices to put

$$y^{[\delta]} = \begin{cases} \frac{\delta}{2\,\|\widehat{y}\|}\widehat{y} & \text{if } \delta \leqslant R, \\ \frac{R}{2\,\|\widehat{y}\|}\widehat{y} & \text{if } R < \delta. \end{cases}$$

Then

$$0 < \|y^{[\delta]}\| \leqslant \delta/2 < \delta,$$

$$(Ky^{[\delta]}, y^{[\delta]}) = \begin{cases} \dfrac{\delta^2}{4\,\|\widehat{y}\|^2}(K\widehat{y}, \widehat{y}) \leqslant 0 & \text{if } \delta \leqslant R, \\ \dfrac{R^2}{4\,\|\widehat{y}\|^2}(K\widehat{y}, \widehat{y}) \leqslant 0 & \text{if } R < \delta, \end{cases}$$

$$\mathcal{K}(y^{[\delta]}) = -(Ky^{[\delta]}, y^{[\delta]}) \geqslant 0.$$

Before proceeding to the verification of Condition H3, we compute

$$\mathcal{F}(y) = \sum_{j=1}^{N} \frac{\partial \mathcal{K}}{\partial y_j} f_j(y).$$

Recall that $\mathcal{F}(y)$ coincides with the derivative $\frac{d}{dt}\mathcal{K}(y(t))$, computed along the trajectories $y(t)$ of the vector equation $\frac{dy}{dt} = f(y)$. We have

$$\mathcal{K}(y) = -(Ky, y), \quad f(y) = Ay + \varphi(y),$$

$$\begin{aligned}\mathcal{F}(y) &= \frac{d}{dt}\mathcal{K}(y(t)) = -\left(K\frac{dy}{dt}, y\right) - \left(Ky, \frac{dy}{dt}\right)\\ &= -(K[Ay + \varphi(y)], y) - (Ky, [Ay + \varphi(y)])\\ &= -([KA + A^*K]y, y) - (K\varphi(y), y) - (Ky, \varphi(y)).\end{aligned}$$

We recall that $f(y)$ and $\varphi(y)$ are defined for $\|y\| \leqslant Y$. Introducing the notation

$$\Delta(y) = (K\varphi(y), y) + (Ky, \varphi(y))$$

and noting that

$$\begin{gathered}|\Delta(y)| \leqslant 2\,\|K\|\,\|\varphi(y)\|\,\|y\| \leqslant 2\,\|K\| q\,\|y\|^{2+\omega},\\ KA + A^*K = -I + \lambda K,\end{gathered}$$

we establish the equality

$$\mathcal{F}(y) = (y, y) - \Delta(y) - \lambda(Ky, y)$$

and the inequality

$$\mathcal{F}(y) \geqslant (1 - 2\,\|K\|\,q\,\|y\|^{\omega})\,\|y\|^2 - \lambda(Ky, y) = (1 - 2\,\|K\|\,q\,\|y\|^{\omega})\,\|y\|^2 + \lambda\mathcal{K}(y).$$

Taking $R = \min\{(1/(5\|K\|q))^{1/\omega}, Y\}$ for $\|y\| \leqslant R$, we obtain

$$\mathcal{F}(y) \geqslant \left(1 - \frac{2}{5}\right)\|y\|^2 + \lambda\mathcal{K}(y) > \frac{1}{2}\,\|y\|^2 + \lambda\mathcal{K}(y).$$

Therefore, for $\|y\| \leqslant R$ and $\mathcal{K}(y) \geqslant 0$ we get the inequality $\mathcal{F}(y) > 0$ which is just Condition H3.

Thus, for an almost linear system we have verified all three Conditions H1–H3 which guarantees the instability in view of the Chetaev theorem. □

In the study of stability of the solution $y = a$ to the autonomous system

$$\frac{dy}{dt} = f(y) \quad (f(a) = 0),$$

the following Lyapunov theorem is often used.

THEOREM 3 (criteria for stability and instability from the first approximation). *Let $f(y)$ have the first and second continuous derivatives in a neighborhood of the point $y = a$ and let A denote the matrix $f_y(a)$ composed of the first order partial derivatives*

$$A = f_y(a) = \left[\begin{array}{cccc} \dfrac{\partial f_1}{\partial y_1} & \dfrac{\partial f_1}{\partial y_2} & \dots & \dfrac{\partial f_1}{\partial y_N} \\ \dfrac{\partial f_2}{\partial y_1} & \dfrac{\partial f_2}{\partial y_2} & \dots & \dfrac{\partial f_2}{\partial y_N} \\ \vdots & \vdots & \dots & \vdots \\ \dfrac{\partial f_N}{\partial y_1} & \dfrac{\partial f_N}{\partial y_2} & \dots & \dfrac{\partial f_N}{\partial y_N} \end{array}\right]\Bigg|_{\substack{y_1=a_1,\\ y_2=a_2,\\ \dots\dots \\ y_N=a_N.}}$$

If all eigenvalues of the matrix A lie strictly in the left half-plane, then the solution $y = a$ is asymptotically stable. If at least one eigenvalue of A has strictly positive real part, then the solution $y = a$ is Lyapunov unstable.

PROOF. Theorem 3 is a simple consequence of the above criteria for stability and instability of the solution $y = 0$ to the almost linear vector equation

$$\frac{dy}{dt} = Ay + \varphi(y), \quad \|\varphi(y)\| \leqslant q\,\|y\|^{1+\omega} \quad (q > 0, \omega > 0).$$

We assume that $\|y\| \leqslant Y$.

Indeed, if in some neighborhood of the point a the function $f(y)$ has the first and second continuous derivatives, which can be assumed to be bounded, and the neighborhood is star-shaped around a (we can assume this without loss of generality because the neighborhood is of the form $\|y - a\| \leqslant Y$), then in this neighborhood we can use the Taylor formula

$$\begin{aligned} f_j(y_1, y_2, \dots, y_N) &= f_j(a_1, a_2, \dots, a_N) + \sum_{k=1}^{N} \frac{\partial f_j}{\partial y_k}\bigg|_{y=a} (y_k - a_k) \\ &\quad + \frac{1}{2}\sum_{k,l} \frac{\partial^2 f_j}{\partial y_k \partial y_l}\bigg|_{y=(1-\rho)a+\rho y} (y_k - a_k)(y_l - a_l) \\ &= 0 + \sum_{k=1}^{N} A_{jk}(y_k - a_k) + \psi_j(y). \end{aligned}$$

We denote by $\Omega_j(y)$ the matrix with elements $\dfrac{1}{2}\dfrac{\partial^2 f_j}{\partial y_k \partial y_l}$. Then

$$|\psi_j(y)| \leqslant \max_{\|y-a\| \leqslant Y} \|\Omega_j(y)\|\,\|y - a\|^2.$$

Consequently, for $\|y\| \leqslant Y$ we have

$$f(y) = A(y - a) + \psi(y),$$

where

$$\psi(y) = \begin{pmatrix} \psi_1(y) \\ \psi_2(y) \\ \vdots \\ \psi_N(y) \end{pmatrix}, \quad \|\psi(y)\| = \sqrt{\sum_{j=1}^{N} |\psi_j(y)|^2} \leqslant q\,\|y - a\|^2.$$

This allows us to rewrite the equation $y' = f(y)$ in the form

$$\frac{d}{dt}(y-a) = A(y-a) + \psi(y),$$

where $\psi(y)$ satisfies the estimate $\|\psi(y)\| \leqslant q\|y-a\|^2$ for $\|y-a\| \leqslant Y$. From this notation of the equation it is obvious that the equation is almost linear ($\omega = 1 > 0$). Consequently, we can use the criteria for stability and instability. They immediately lead to the required assertion. □

EXAMPLE 1. The equilibrium point $y_1 = a_1$, $y_2 = a_2$ of the system

$$\frac{dy_1}{dt} = \sin(y_1 + y_2), \quad \frac{dy_2}{dt} = \cos(y_1 - y_2)$$

can be found from the equalities

$$\sin(a_1 + a_2) = 0, \quad \cos(a_1 - a_2) = 0.$$

There are infinitely many such points:

$$\left.\begin{aligned} a_1 + a_2 &= m\pi \\ a_1 - a_2 &= \left(n + \frac{1}{2}\right)\pi \end{aligned}\right\} \Rightarrow \left\{\begin{aligned} a_1 &= \left(\frac{m+n}{2} + \frac{1}{4}\right)\pi, \\ a_2 &= \left(\frac{m-n}{2} - \frac{1}{4}\right)\pi. \end{aligned}\right.$$

For some integers m and n the matrix A at the point with these m and n has the form

$$\begin{bmatrix} (-1)^{m-1} & (-1)^{m-1} \\ (-1)^{n-1} & -(-1)^{n-1} \end{bmatrix}.$$

The eigenvalues of A are found from the equation

$$\begin{aligned} \det(A - \tau I) &= \begin{vmatrix} (-1)^{m-1} - \tau & (-1)^{m-1} \\ (-1)^{n-1} & -(-1)^{n-1} - \tau \end{vmatrix} \\ &= \tau^2 + \left[(-1)^{n-1} - (-1)^{m-1}\right]\tau - 2(-1)^{m+n} = 0. \end{aligned}$$

For different m and n this equation and its roots are as follows:

Case 1: n is even and m is even. Then $\tau^2 - 2 = 0$, $\tau_1 = \sqrt{2}$, $\tau_2 = -\sqrt{2}$.

Case 2: n is even and m is odd. Then $\tau^2 - 2\tau + 2 = 0$, $\tau_1 = 1 + i$, $\tau_2 = 1 - i$.

Case 3: n is odd and m is even. Then $\tau^2 + 2\tau + 2 = 0$, $\tau_1 = -1 + i$, $\tau_2 = -1 - i$.

Case 4: n is odd and m is odd. Then $\tau^2 - 2 = 0$, $\tau_1 = \sqrt{2}$, $\tau_2 = \sqrt{2}$.

It is clear that the equilibrium point

$$a_1 = \left(\frac{m+n}{2} + \frac{1}{4}\right)\pi, \quad a_2 = \left(\frac{m-n}{2} - \frac{1}{4}\right)\pi$$

is stable only if m is even and n is odd (Case 3).

§6. The Hermite theorem

The generating polynomial for the Hermitian form. Lemmas on properties of auxiliary polynomials and connecting identities. The proof of the theorem and the interpretation from the point of the view of operator theory. Analogy with the Lyapunov theorem.

At the end of §5, we considered the example when the study of stability of the equilibrium points can be reduced to the question of whether all roots of a quadratic equation lie in the left half-plane.

For systems of higher order we must analyze algebraic equations of an arbitrary degree. The analysis is based on a number of deep and important arguments which are considered in detail later. We begin with a beautiful and fundamental theorem due to Hermite.

Let

$$p(x) = x^n + p_1 x^{n-1} + \cdots + p_{n-1}x + p_n = p_0 x^n + p_1 x^{n-1} + \cdots + p_{n-1}x + p_n$$

be a polynomial of degree n with complex coefficients $p_1, p_2, \ldots, p_n$ and let the leading coefficient p_0 be equal to 1 ($p_0 = 1$). We denote $\overline{p(x)} = \overline{p}(x)$. We consider the expression

$$p(x)\overline{p}(y) - \overline{p}(-x)p(-y),$$

where

$$\begin{aligned}
\overline{p}(y) &= \overline{p}_0 y^n + \overline{p}_1 y^{n-1} + \cdots + \overline{p}_{n-1}y + \overline{p}_n, \\
\overline{p}(-x) &= \overline{p}_0(-1)^n x^n + \overline{p}_1(-1)^{n-1}x^{n-1} + \cdots + \overline{p}_{n-1}(-1)x + \overline{p}_n, \\
p(-y) &= p_0(-1)^n y^n + p_1(-1)^{n-1}y^{n-1} + \cdots + p_{n-1}(-1)y + p_n.
\end{aligned}$$

Since

$$\begin{aligned}
x &= \frac{1}{2}(x+y) + \frac{1}{2}(x-y), \\
y &= \frac{1}{2}(x+y) - \frac{1}{2}(x-y),
\end{aligned}$$

it is easy to verify the representation

$$\begin{aligned}
p(x)\overline{p}(y) - \overline{p}(-x)p(-y) &= \sum_{k,l=0}^{n} R_{kl}(x+y)^k(x-y)^l \\
&= \sum_{l=0}^{n} R_{0l}(x-y)^l + (x+y)\sum_{\substack{l=0\\k=1}}^{n} R_{kl}(x-y)^l(x+y)^{k-l}.
\end{aligned}$$

Setting $y = -x$, we find

$$0 = p(x)\overline{p}(-x) - \overline{p}(-x)p(x) = \sum_{l=0}^{n} R_{0l}(2x)^l.$$

It is clear that $R_{00} = R_{01} = R_{02} = \cdots = R_{0n} = 0$. Hence

$$p(x)\overline{p}(y) - \overline{p}(-x)p(-y) = (x+y)\sum_{\substack{l=0\\k=1}}^{n} R_{kl}(x-y)^l(x+y)^{k-l} = (x+y)\mathcal{P}(x,y).$$

We have established that $p(x)\overline{p}(y) - \overline{p}(-x)p(-y)$ is divisible by $x + y$ without remainder. It is easy to see that before the division by $x + y$ this expression is a polynomial in x, y and contains powers of x as well as y of exponent at most n.

Let us compute the coefficient at $x^n y^n$. It is equal to the expression

$$p_0\overline{p}_0 - \overline{p}_0(-1)^n p_0(-1)^n = 0.$$

Therefore, $p(x)\overline{p}(y) - \overline{p}(-x)p(-y)$ is a polynomial whose total degree in x and y does not exceed $2n - 1$. Hence the total degree of $\mathcal{P}(x, y)$ is at most $2n - 2$. For the polynomial

$$p(x)\overline{p}(y) - \overline{p}(-x)p(-y) = \overline{[p(y)\overline{p}(x) - \overline{p}(-y)p(-x)]},$$

which goes to the conjugate polynomial by interchanging y and x, the terms of degree $2n - 1$ have the form

$$\rho x^n y^{n-1} + \overline{\rho} x^{n-1} y^n = \frac{\rho + \overline{\rho}}{2} x^{n-1} y^{n-1}(x + y) + \frac{\rho - \overline{\rho}}{2}(x - y)x^{n-1}y^{n-1}.$$

Since $p(x)\overline{p}(y) - \overline{p}(-x)p(-y)$ is divisible by $x + y$, the collection of terms of the higher degree of this polynomial must be divisible by $x + y$, which is possible only if $\rho = \overline{\rho}$. Therefore, the quotient

$$\mathcal{P}(x, y) = \frac{p(x)\overline{p}(y) - \overline{p}(-x)p(-y)}{x + y}$$

must have only one term of higher degree and it is $\rho x^{n-1} y^{n-1}$.

It is clear that

$$\mathcal{P}(x, y) = \frac{p(x)\overline{p}(y) - \overline{p}(-x)p(-y)}{x + y} = \sum_{r,s=1}^{n} P_{rs} x^{r-1} y^{s-1}. \tag{1}$$

From the equality

$$\overline{\mathcal{P}(y, x)} = \overline{\left[\frac{p(y)\overline{p}(x) - \overline{p}(-y)p(-x)}{x + y}\right]} = \mathcal{P}(x, y),$$

which can be rewritten in the form

$$\sum_{r,s=1}^{n} \overline{P}_{sr} y^{s-1} x^{r-1} = \sum_{r,s=1}^{n} P_{rs} x^{r-1} y^{s-1},$$

it is obvious that $\overline{P}_{sr} = P_{sr}$. It means that the matrix P composed of the coefficients of the polynomial $\mathcal{P}(x, y)$ is a Hermitian matrix:

$$P = P^* = \begin{bmatrix} P_{11} & P_{12} & \cdots & P_{1n} \\ P_{21} & P_{22} & \cdots & P_{2n} \\ \vdots & \vdots & \cdots & \vdots \\ P_{n1} & P_{n2} & \cdots & P_{nn} \end{bmatrix}.$$

Following Hermite, to each polynomial

$$p(x) = x^n + p_1 x^{n-1} + \cdots + p_n$$

we assign the Hermitian matrix P whose elements are coefficients of the generating polynomial

$$\mathcal{P}(x,y)=\sum_{r,s=1}^{n} P_{rs}x^{r-1}y^{s-1}.$$

THEOREM 1 (Hermite). *In order that all roots $\tau_1,\tau_2,\dots,\tau_n$ of the polynomial $p(\tau)$ lie strictly in the left half-plane ($\operatorname{Re}\tau_j<0$), it is necessary and sufficient that the corresponding Hermitian matrix P be positive definite.*

Before the proof of the theorem we prove several lemmas.

LEMMA 1. *The polynomials*

$$\pi_j^{[k]}(x)=\prod_{1\leqslant i\leqslant j-1}(x-\tau_i)\prod_{j+1\leqslant l\leqslant k}(x+\overline{\tau}_l),\quad j=1,2,\dots,k,$$

of degree $k-1$ are linearly independent if $\overline{\tau}_l+\tau_i\neq 0$ for any i, l. In particular, the assertion is true if $\operatorname{Re}\tau_i<0$ for all i.

PROOF. We proceed by induction on k. For $k=1$ there is a single polynomial $\pi_1^{[1]}(x)$ of the form

$$\pi_1^{[1]}(x)=\prod_{1\leqslant i\leqslant 0}(x-\tau_i)\prod_{2\leqslant l\leqslant 1}(x+\overline{\tau}_l)=1.$$

For $k=2$ there are two polynomials

$$\pi_1^{[2]}(x)=\prod_{1\leqslant i\leqslant 0}(x-\tau_i)\prod_{2\leqslant l\leqslant 2}(x+\overline{\tau}_l)=x+\overline{\tau}_2,$$
$$\pi_2^{[2]}(x)=\prod_{1\leqslant i\leqslant 1}(x-\tau_i)\prod_{3\leqslant l\leqslant 2}(x+\overline{\tau}_l)=x-\tau_1$$

which are linearly independent (i.e., not proportional) because

$$\pi_2^{[2]}(\tau_1)=\tau_1-\tau_1=0,\quad \pi_1^{[2]}(\tau_1)=\tau_1+\overline{\tau}_2\neq 0.$$

We suppose that we already prove that the polynomials

$$\pi_1^{[k-1]}(x),\pi_2^{[k-1]}(x),\dots,\pi_{k-1}^{[k-1]}(x)$$

are linearly independent. We note that

$$\pi_j^{[k]}(x)=(x+\overline{\tau}_k)\pi_j^{[k-1]}(x)\quad\text{for } j=1,2,\dots,k-1,$$
$$\pi_k^{[k]}(x)=(x-\tau_{k-1})\pi_{k-1}^{[k-1]}(x).$$

Therefore, $\pi_1^{[k]}(x),\pi_2^{[k]}(x),\dots,\pi_{k-1}^{[k]}(x)$ are linearly independent since they differ from the linearly independent polynomials

$$\pi_1^{[k-1]}(x),\pi_2^{[k-1]}(x),\dots,\pi_{k-1}^{[k-1]}(x)$$

by only the common factor $x+\overline{\tau}_k$. It is clear that $\pi_j^{[k]}(-\overline{\tau}_k)=0$, $1\leqslant j\leqslant k-1$. On the other hand, at $x=-\overline{\tau}_k$ the polynomial

$$\pi_k^{[k]}(x)=(x-\tau_{k-1})(x-\tau_{k-2})\dots(x-\tau_1)$$

does not vanish:

$$\pi_k^{[k]}(x) = (-1)^k(\overline{\tau}_k + \tau_{k-1})(\overline{\tau}_k + \tau_{k-2})\dots(\overline{\tau}_k + \tau_1) \neq 0,$$

whereas the polynomials $\pi_1^{[k]}(x), \pi_2^{[k]}(x), \dots, \pi_{k-1}^{[k]}(x)$, vanish. Therefore, $\pi_k^{[k]}(x)$ cannot be represented as their linear combination.

Thus, we have shown that the linear independence of $\pi_1^{[k-1]}(x), \dots, \pi_{k-1}^{[k-1]}(x)$ implies the linear independence of $\pi_1^{[k]}(x), \dots, \pi_{k-1}^{[k]}(x), \pi_k^{[k]}(x)$ and, thereby, we have proved the induction passage from $k-1$ to k. □

We note that $\pi_1^{[k]}(x), \dots, \pi_k^{[k]}(x)$ are k linearly independent polynomials of degree $k-1$ in x. Consequently, they form a basis for the space of such polynomials.

We also introduce the polynomial $p^{[k]}$ of degree k by the formula

$$p^{[k]}(x) = (x - \tau_1)(x - \tau_2)\dots(x - \tau_k)$$

and note that

$$\begin{aligned}
\pi_k^{[k]}(x) &= p^{[k-1]}(x), \\
p^{[k]}(x) &= (x - \tau_k)\pi_k^{[k]}(x) = (x - \tau_k)p^{[k-1]}(x), \\
\pi_j^{[k]}(x) &= (x + \overline{\tau}_k)\pi_j^{[k-1]}(x), \quad j = 1, 2, \dots, k-1.
\end{aligned}$$

LEMMA 2. *The following equality holds:*

$$\frac{p^{[k]}(x)\overline{p}^{[k]}(y) - \overline{p}^{[k]}(-x)p^{[k]}(-y)}{x+y} = -\sum_{j=1}^{k}(\tau_j + \overline{\tau}_j)\pi_j^{[k]}(x)\overline{\pi}_j^{[k]}(y).$$

It is not necessary to assume that all τ_j lie in the left half-plane.

PROOF. We proceed by induction on k. For $k = 1$ we have

$$\begin{aligned}
\frac{p^{[1]}(x)\overline{p}^{[1]}(y) - \overline{p}^{[1]}(-x)p^{[1]}(-y)}{x+y} &= \frac{(x - \tau_1)(y - \overline{\tau}_1) - (-x - \overline{\tau}_1)(-y - \tau_1)}{x+y} \\
&= -\frac{(\tau_1 + \overline{\tau}_1)(x+y)}{x+y} = -(\tau_1 + \overline{\tau}_1)1 \cdot 1 = -(\tau_1 + \overline{\tau}_1)\pi_1^{[1]}(x)\overline{\pi}_1^{[1]}(y).
\end{aligned}$$

We suppose that the representation

$$\frac{p^{[k-1]}(x)\overline{p}^{[k-1]}(y) - \overline{p}^{[k-1]}(-x)p^{[k-1]}(-y)}{x+y} = -\sum_{j=1}^{k-1}(\tau_j + \overline{\tau}_j)\pi_j^{[k-1]}(x)\overline{\pi}_j^{[k-1]}(y)$$

is proved. We show that it remains valid after the replacement of $k-1$ by k. Indeed,

$$\begin{aligned}
&\frac{p^{[k]}(x)\overline{p}^{[k]}(y)-\overline{p}^{[k]}(-x)p^{[k]}(-y)}{x+y}\\
&=\frac{(x-\tau_k)p^{[k-1]}(x)(y-\overline{\tau}_k)\overline{p}^{[k-1]}(y)-(-x-\overline{\tau}_k)\overline{p}^{[k-1]}(-x)(-y-\tau_k)p^{[k-1]}(-y)}{x+y}\\
&=\frac{(x-\tau_k)(y-\overline{\tau}_k)-(x+\overline{\tau}_k)(y+\tau_k)}{x+y}p^{[k-1]}(x)\overline{p}^{[k-1]}(y)\\
&\quad+(x+\overline{\tau}_k)(y+\tau_k)\frac{p^{[k-1]}(x)\overline{p}^{[k-1]}(y)-\overline{p}^{[k-1]}(-x)p^{[k-1]}(-y)}{x+y}\\
&=-(\tau_k+\overline{\tau}_k)p^{[k-1]}(x)\overline{p}^{[k-1]}(y)-(x+\overline{\tau}_k)(y+\tau_k)\sum_{j=1}^{k-1}(\tau_j+\overline{\tau}_j)\pi_j^{[k-1]}\overline{\pi}_j^{[k-1]}(y)\\
&=-(\tau_k+\overline{\tau}_k)\pi_k^{[k]}(x)\overline{\pi}_k^{[k]}(y)-\sum_{j=1}^{k-1}(\tau_j+\overline{\tau}_j)\pi_j^{[k]}(x)\overline{\pi}_j^{[k]}(y)\\
&=-\sum_{j=1}^{k}(\tau_j+\overline{\tau}_j)\pi_j^{[k]}(x)\overline{\pi}_j^{[k]}(y).
\end{aligned}$$

The justification of the induction and the proof of the lemma are completed. □

We will use Lemmas 1 and 2 only for $k=n$. Therefore, it is convenient to introduce the simplified notation

$$p^{[n]}(x)=p(x)=(x-\tau_1)(x-\tau_2)\dots(x-\tau_N),\qquad \pi_j^{[n]}(x)=\pi_j(x).$$

We recall that $\pi_j^{[n]}(x)$ and, consequently, $\pi_j(x)$ are polynomials of degree $n-1$:

$$\pi_j(x)=\prod_{1\leqslant i\leqslant j-1}(x-\tau_j)\prod_{j+1\leqslant l\leqslant n}(x+\overline{\tau}_l).$$

In the introduced notation, Lemmas 1 and 2 can be reformulated as follows.

LEMMA 1′. *If all roots $\tau_1,\tau_2,\dots,\tau_n$ of the polynomial $p(x)$ lie strictly in the left half-plane,* $\operatorname{Re}\tau_j>0$, *then the polynomials $\pi_1(x),\pi_2(x),\dots,\pi_n(x)$ form a basis for the linear space of all polynomials of degree $n-1$.*

LEMMA 2′. *The following equality holds:*

$$\frac{p(x)\overline{p}(y)-\overline{p}(-x)p(-y)}{x+y}=-\sum_{j=1}^{n}(\tau_j+\overline{\tau}_j)\pi_j(x)\overline{\pi}_j(y).$$

In other words, if

$$\pi_j(x)=\sum_{r=1}^{n}\pi_{jr}x^{r-1},$$

then the matrix π composed of the coefficients of the polynomials π_j is nonsingular:

$$\pi = \begin{bmatrix} \pi_{11} & \pi_{12} & \cdots & \pi_{1n} \\ \pi_{21} & \pi_{22} & \cdots & \pi_{2n} \\ \vdots & \vdots & \vdots & \vdots \\ \pi_{n1} & \pi_{n2} & \cdots & \pi_{nn} \end{bmatrix}, \quad \det \pi \neq 0.$$

We recall the definition of the elements P_{rs} of the matrix P in terms of the generating polynomial

$$\sum_{r,s=1}^{n} P_{rs} x^{r-1} y^{s-1} = \frac{p(x)\overline{p}(y) - \overline{p}(-x)p(-y)}{x+y}$$

and, using the equality from Lemma 2, establish that

$$\begin{aligned} \sum_{r,s=1}^{n} P_{rs} x^{r-1} y^{s-1} &= -\sum_{j=1}^{n} (\tau_j + \overline{\tau}_j) \sum_{r=1}^{n} \pi_{jr} x^{r-1} \sum_{s=1}^{n} \overline{\pi}_{js} y^{s-1} \\ &= \sum_{r,s=1}^{n} \left[-\sum_{j=1}^{n} (\tau_j + \overline{\tau}_j) \pi_{jr} \overline{\pi}_{js} \right] x^{r-1} y^{s-1}. \end{aligned}$$

Consequently,

$$P_{rs} = -\sum_{j=1}^{n} (\tau_j + \overline{\tau}_j) \pi_{jr} \overline{\pi}_{js}.$$

The last equality is valid for $1 \leqslant r \leqslant n$, $1 \leqslant s \leqslant n$ and is equivalent to the following matrix equality:

$$P = \begin{bmatrix} P_{11} & P_{12} & \cdots & P_{1n} \\ P_{21} & P_{22} & \cdots & P_{2n} \\ \vdots & \vdots & \vdots & \vdots \\ P_{n1} & P_{n2} & \cdots & P_{nn} \end{bmatrix} = \begin{bmatrix} \overline{\pi}_{11} & \overline{\pi}_{21} & \cdots & \overline{\pi}_{n1} \\ \overline{\pi}_{12} & \overline{\pi}_{22} & \cdots & \overline{\pi}_{n2} \\ \vdots & \vdots & \vdots & \vdots \\ \overline{\pi}_{1n} & \overline{\pi}_{2n} & \cdots & \overline{\pi}_{nn} \end{bmatrix}$$

$$\times \begin{bmatrix} -(\tau_1 + \overline{\tau}_1) & & & 0 \\ & (-\tau_2 + \overline{\tau}_2) & & \\ & & \ddots & \\ 0 & & & -(\tau_n + \overline{\tau}_n) \end{bmatrix} \begin{bmatrix} \pi_{11} & \pi_{12} & \cdots & \pi_{1n} \\ \pi_{21} & \pi_{22} & \cdots & \pi_{2n} \\ \vdots & \vdots & \vdots & \vdots \\ \pi_{n1} & \pi_{n2} & \cdots & \pi_{nn} \end{bmatrix},$$

$$P = \pi^* \begin{bmatrix} -(\tau_1 + \overline{\tau}_1) & & & 0 \\ & (-\tau_2 + \overline{\tau}_2) & & \\ & & \ddots & \\ 0 & & & -(\tau_n + \overline{\tau}_n) \end{bmatrix} \pi. \tag{2}$$

This equality is always valid for any location of the roots in the complex plane.

PROOF OF THEOREM 1. We use the above representation and Lemma 1 to prove the Hermite theorem. Let $\operatorname{Re} \tau_j < 0$ for all j or, equivalently, $-(\tau_j + \overline{\tau}_j) > 0$. By Lemma 1, the matrix π is nonsingular so that the matrix P is positive definite.

If there is an eigenvalue τ_{j_0} in the right half-plane, then the matrix P cannot be positive definite. Indeed, if π is a nonsingular matrix, then P is nonpositive definite since there are no negative diagonal elements $-(\tau_j + \overline{\tau}_j)$ in the representation (2).

If π is a singular matrix, i.e., $\det \pi = 0$, then $\det P = 0$, which contradicts the fact that P is positive definite. $\square$

We will use Theorem 1 only if the coefficients $p_0, p_1, \dots, p_n$ of the polynomial

$$p(x) = p_0 x^n + p_1 x^{n-1} + \cdots + p_{n-1} x + p_n$$

are real. In this case, the Hermite identity defining the matrix P takes the form

$$p(x)p(y) - p(-x)p(-y) = (x+y) \sum_{r,s=1}^{n} P_{rs} x^{r-1} y^{s-1}.$$

Moreover, the matrix P is a real symmetric matrix.

In the proof of Theorem 1 we assumed that $p_0 = 1$. It is easy to see that this assumption is not essential. Indeed, if we multiply the coefficients of the polynomial $p(x)$ by the same number, then all elements of the matrix P are multiplied by the square of this number and the matrix P is again positive definite. Only the relation $p_0 \neq 0$ is essential.

We note that the Hermite theorem can be thought of as a version of the theorem about the Lyapunov functions adopted to the study of a single equation of higher order. To compare the Hermite theorem with the Lyapunov theorem, we interpret the Hermite identity from the point of view of operator theory by writing this equality in the following equivalent form:

$$p(x)up(y)v - p(-x)up(-y)v = \sum_{r,s=1}^{n} \left[P_{rs} x^r u y^{s-1} v + P_{rs} x^{r-1} u y^s v \right].$$

Let $u(t)$ and $v(t)$ be functions of t having n continuous derivatives

$$\begin{gathered} u(t),\ \frac{d}{dt}u(t),\ \frac{d^2}{dt^2}u(t), \dots, \frac{d^n}{dt^n}u(t), \\ v(t),\ \frac{d}{dt}v(t),\ \frac{d^2}{dt^2}v(t), \dots, \frac{d^n}{dt^n}v(t). \end{gathered}$$

The differentiation operator with respect to t applied to $u(t)$ is denoted by x:

$$xu(t) = \frac{du}{dt},\ x^2 u(t) = \frac{d^2 u}{dt^2}, \dots, x^n u(t) = \frac{d^n u}{dt^n}$$

and the differentiation operator with respect to t applied to $v(t)$ is denoted by y:

$$yv(t) = \frac{dv}{dt},\ y^2 v(t) = \frac{d^2 v}{dt^2}, \dots, y^n v(t) = \frac{d^n v}{dt^n}.$$

In this notation, the Hermite identity becomes the equality

$$\begin{aligned} &\left[p\left(\frac{d}{dt}\right)u\right]\left[p\left(\frac{d}{dt}\right)v\right] - \left[p\left(-\frac{d}{dt}\right)u\right]\left[p\left(-\frac{d}{dt}\right)v\right] \\ &\quad = \sum_{r,s=1}^{n} \left[P_{rs} \frac{d^r u}{dt^r} \frac{d^{s-1} v}{dt^{s-1}} + P_{rs} \frac{d^{r-1} u}{dt^{r-1}} \frac{d^s v}{dt^s} \right] = \frac{d}{dt} \sum_{r,s=1}^{n} \left[P_{rs} \frac{d^{r-1} u}{dt^{r-1}} \frac{d^{s-1} v}{dt^{s-1}} \right], \end{aligned}$$

which is valid for any functions $u(t)$ and $v(t)$. Applying it to the functions $u(t)$ and $v(t)$ which form a solution to the differential equations

$$
\begin{aligned}
p\Big(\frac{d}{dt}\Big)u &= p_0\frac{d^n u}{dt^n} + p_1\frac{d^{n-1}u}{dt^{n-1}} + \cdots + p_{n-1}\frac{du}{dt} + p_n u = 0,\\
p\Big(\frac{d}{dt}\Big)v &= p_0\frac{d^n v}{dt^n} + p_1\frac{d^{n-1}v}{dt^{n-1}} + \cdots + p_{n-1}\frac{dv}{dt} + p_n v = 0,
\end{aligned}
\tag{3}
$$

we can rewrite this equality as follows:

$$
\frac{d}{dt}\left[\sum_{r,s=1}^{n} P_{rs}\frac{d^{r-1}u}{dt^{r-1}}\frac{d^{s-1}v}{dt^{s-1}}\right] + \left[p\Big(-\frac{d}{dt}\Big)u\right]\left[p\Big(-\frac{d}{dt}\Big)v\right] = 0
$$

or

$$
\begin{aligned}
&\frac{d}{dt}\left[\sum_{r,s=1}^{n} P_{rs}\frac{d^{r-1}u}{dt^{r-1}}\frac{d^{s-1}v}{dt^{s-1}}\right]\\
&\qquad + \left[p\Big(\frac{d}{dt}\Big)u - (-1)^n p\Big(-\frac{d}{dt}\Big)u\right]\left[p\Big(\frac{d}{dt}\Big)v - (-1)^n p\Big(-\frac{d}{dt}\Big)v\right] = 0.
\end{aligned}
$$

Assume that $v(t)$ coincides with $u(t)$. Since

$$
p\Big(\frac{d}{dt}\Big)u - (-1)^n p\Big(-\frac{d}{dt}\Big)u = 2\Big(p_1\frac{d^{n-1}u}{dt^{n-1}} + p_3\frac{d^{n-3}u}{dt^{n-3}} + \dots\Big),
$$

we arrive at the equality

$$
\frac{d}{dt}\Big(\sum_{r,s=1}^{n} P_{rs}\frac{d^{r-1}u}{dt^{r-1}}\frac{d^{s-1}v}{dt^{s-1}}\Big) = -4\Big(p_1\frac{d^{n-1}u}{dt^{n-1}} + p_3\frac{d^{n-3}u}{dt^{n-3}} + \dots\Big)^2
\tag{4}
$$

which is valid for any solution $u(t)$ to the first equation in (3). In this identity, the coefficients P_{rs} are defined by means of the generating polynomial $\mathcal{P}(x,y)$. The equality (4) plays the same role for equations (3) as the equality $\frac{d}{dt}(Hy,y) = -(Cy,y)$ in the Lyapunov theorem plays for the vector equation $\frac{d}{dt}y = Ay$. Moreover, the quadratic form in the derivatives

$$
\sum_{r,s=1}^{n} P_{rs}\frac{d^{r-1}u}{dt^{r-1}}\frac{d^{s-1}u}{dt^{s-1}}
$$

is an analog of (Hy,y) and the right-hand side of (4) is similar to $-(Cy,y)$.

We consider the third order equation

$$
\frac{d^3u}{dt^3} + p_1\frac{d^2u}{dt^2} + p_2\frac{du}{dt} + p_3 u = 0
$$

in detail. Introducing the vector

$$
w = \begin{pmatrix} u \\ \dfrac{du}{dt} \\ \dfrac{d^2u}{dt^2} \end{pmatrix} = \begin{pmatrix} w_1 \\ w_2 \\ w_3 \end{pmatrix}
$$

and applying the well-known method, we can reduce the considered equation to the first order system

$$\frac{d}{dt}\begin{pmatrix} w_1 \\ w_2 \\ w_3 \end{pmatrix} = \begin{bmatrix} 0 & 1 & 0 \\ 0 & 0 & 1 \\ -p_3 & -p_2 & -p_1 \end{bmatrix}\begin{pmatrix} w_1 \\ w_2 \\ w_3 \end{pmatrix}$$

or briefly,

$$\frac{dw}{dt} = Aw, \quad A = \begin{bmatrix} 0 & 1 & 0 \\ 0 & 0 & 1 \\ -p_3 & -p_2 & -p_1 \end{bmatrix}.$$

Constructing the matrix P with the help of the generating polynomial $\mathcal{P}(x, y)$, we find that

$$P = \begin{bmatrix} P_{11} & P_{12} & P_{13} \\ P_{21} & P_{22} & P_{23} \\ P_{31} & P_{32} & P_{33} \end{bmatrix} = \begin{bmatrix} 2p_2p_3 & 0 & 2p_3 \\ 0 & 2(p_1p_2 - p_3) & 0 \\ 2p_3 & 0 & 2p_1 \end{bmatrix}.$$

All roots of the equation $p(\tau) = 0$ lie in the left half-plane, which means that the zero solution to the system $\dfrac{dw}{dt} = Aw$ is asymptotically stable if and only if the matrix P is positive definite. Since P is positive definite, we have the following relations:

$$p_2p_3 > 0, \quad \begin{vmatrix} p_2p_3 & 0 \\ 0 & p_1p_2 - p_3 \end{vmatrix} = p_2p_3(p_1p_2 - p_3) > 0,$$

$$\begin{vmatrix} p_2p_3 & 0 & p_3 \\ 0 & p_1p_2 - p_3 & 0 \\ p_3 & 0 & p_1 \end{vmatrix} = (p_1p_2 - p_3)(p_1p_2p_3 - p_3^2) = (p_1p_2 - p_3)^2p_3 > 0.$$

It is clear that for P to be positive definite it is necessary and sufficient to have

$$p_3 > 0, \quad p_2 > 0, \quad p_1 > p_3/p_2.$$

Using Theorem 1, we have deduced the identity

$$\frac{d}{dt}\left[\sum_{r,s=1}^{n} P_{rs}\frac{d^{r-1}u}{dt^{r-1}}\frac{d^{s-1}u}{dt^{s-1}}\right] = -4\left(p_1\frac{d^2u}{dt^2} + p_3u\right)^2$$

which differs only in notation from the equality

$$\frac{d}{dt}(Pw, w) = -4(p_1w_3 + p_3w_1)^2 = -(Cw, w),$$

where

$$C = \begin{bmatrix} 4p_3^2 & 0 & 4p_3p_1 \\ 0 & 0 & 0 \\ 4p_3p_1 & 0 & 4p_1^2 \end{bmatrix}.$$

As is known, this identity is valid on any solution to the equation

$$\frac{d^3u}{dt^3} + p_1\frac{d^2u}{dt^2} + p_2\frac{du}{dt} + p_3u = 0$$

or the equivalent system

$$\frac{d}{dt}w = \frac{d}{dt}\begin{pmatrix} w_1 \\ w_2 \\ w_3 \end{pmatrix} = \begin{bmatrix} 0 & 1 & 0 \\ 0 & 0 & 1 \\ -p_3 & -p_2 & -p_1 \end{bmatrix}\begin{pmatrix} w_1 \\ w_2 \\ w_3 \end{pmatrix} = Aw.$$

Along solutions to the system, we have

$$\begin{aligned}\frac{d}{dt}(Pw, w) &= \left(P\frac{dw}{dt}, w\right) + \left(Pw, \frac{dw}{dt}\right) \\ &= (PAw, w) + (Pw, Aw) = ([PA + A^*P]w, w).\end{aligned}$$

Hence

$$([PA + A^*P]w, w) = -(Cw, w).$$

The last equality must be valid for any w since, by the existence theorem, a solution to the system $\frac{dw}{dt} = Aw$ can be constructed for an arbitrary initial vector at $t = t_0$. Since w is arbitrary, the matrix equality $PA + A^*P = -C$ holds. (Give a strict justification of the last assertion.)

We see that the Hermitian matrix P is a solution to the matrix Lyapunov equation with the right-hand side

$$C = \begin{bmatrix} 4p_3^2 & 0 & 4p_3p_1 \\ 0 & 0 & 0 \\ 4p_3p_1 & 0 & 4p_1^2 \end{bmatrix}.$$

If the matrix C were positive definite, then from the fact that all roots of the equation

$$\det(A - \tau I) = \det\begin{bmatrix} -\tau & 1 & 0 \\ 0 & -\tau & 1 \\ -p_3 & -p_2 & -p_1 - \tau \end{bmatrix} = -(\tau^3 + p_1\tau^2 + p_2\tau + p_3) = 0$$

lie in the left half-plane and the Lyapunov theorem it would follow that P is a positive definite matrix. However, the matrix C is not positive definite because its rank is equal to 1. Therefore, the Lyapunov theorem does not guarantee that the matrix P is positive definite whereas this fact is guaranteed by the Hermite theorem.

On the other hand, the Lyapunov equation $PA + A^*P = -C$ for an arbitrary matrix C is not necessarily solvable if among the roots τ_1, τ_2, τ_3 of the equation $\det(A - \tau I) = 0$, there are τ_i, τ_j such that $\tau_i + \overline{\tau}_j = 0$. Theorem 1 always guarantees the existence of P because of the special choice of C.

§7. The Routh–Hurwitz criterion for stability

The lemma on the decrease of the order of the Hermitian matrix and the recurrence sequence of polynomials. The Routh algorithm and the corresponding Hurwitz table. Examples.

Although the Hermite theorem (Theorem 1 in §6) gives necessary and sufficient conditions for all roots to the equation $p(x) = p_0x^n + p_1x^{n-1} + \cdots + p_n$ to lie strictly in the left half-plane, other criteria are often applied in practice. The most common process uses the Routh algorithm and the Hurwitz inequalities for determinants which are based on this algorithm. Because of this fact we introduce the following definition.

DEFINITION 1. A polynomial $p(x)$ is called a Hurwitz polynomial if its roots lie strictly in the left half-plane.

We prove several lemmas from which the Routh algorithm will be deduced. From the algorithm we obtain the Hurwitz inequalities for determinants. To justify the Routh algorithm we use the Hermite theorem (Theorem 1 in §6). The coefficients $p_1, p_2, \dots, p_n$ of the polynomial $p(x)$ are assumed to be real.

LEMMA 1. *Let*

$$\sum_{r,s=1}^{n} P_{rs}x^{r-1}y^{s-1} = \gamma m_{n-1}(x)m_{n-1}(y) + \sum_{r,s=1}^{n-1} Q_{rs}x^{r-1}y^{s-1},$$

$$m_{n-1}(x) = \mu_n x^{n-1} + \mu_{n-1}x^{n-2} + \cdots + \mu_2 x + \mu_1.$$

The quadratic form $\sum_{r,s=1}^{n} P_{rs}\xi_r\xi_s$ *is positive definite if and only if the following conditions hold:*

(a) $\mu_n \neq 0$,

(b) $\gamma > 0$,

(c) $\sum_{r,s=1}^{n-1} Q_{rs}\xi_r\xi_s$ *is positive definite.*

PROOF. We note that $P_{rs} = \gamma\mu_r\mu_s + Q_{rs}$. In order for this equality to hold for all $1 \leqslant r \leqslant n$ and $1 \leqslant s \leqslant n$, we must take

$$Q_{ns} = Q_{rn} = 0.$$

In particular,

$$P_{nn} = \gamma\mu_n\mu_n = \gamma\mu_n^2 > 0.$$

Hence Conditions (a) and (b) are necessary. The quadratic form $\sum_{r,s=1}^{n-1} Q_{rs}\xi_r\xi_s$ can be represented as the sum of $n-1$ squared independent linear forms with positive, negative, or zero coefficients ω_j:

$$\sum_{r,s=1}^{n-1} Q_{rs}\xi_r\xi_s = \sum_{j=1}^{n-1}\omega_j\left(\sum_{r=1}^{n-1} c_{jr}\xi_r\right)^2.$$

Since

$$\begin{aligned}\sum_{r,s=1}^{n} P_{rs}\xi_r\xi_s &= \sum_{r,s=1}^{n}(\gamma\mu_r\mu_s + Q_{rs})\xi_r\xi_s = \gamma\left(\sum_{r=1}^{n}\mu_r\xi_r\right)^2 + \sum_{r,s=1}^{n} Q_{rs}\xi_r\xi_s \\ &= \gamma\left(\sum_{r=1}^{n}\mu_r\xi_r\right)^2 + \sum_{r,s=1}^{n-1} Q_{rs}\xi_r\xi_s = \gamma\left(\sum_{r=1}^{n}\mu_r\xi_r\right)^2 + \sum_{j=1}^{n-1}\omega_j\left(\sum_{r=1}^{n-1} c_{jr}\xi_r\right)^2\end{aligned}$$

and $\sum_{r,s=1}^{n} P_{rs}\xi_r\xi_s$ is a quadratic form of rank n, the linear forms

$$\mu_n\xi_n + \sum_{r=1}^{n-1}\mu_r\xi_r, \quad \sum_{r=1}^{n-1} c_{jr}\xi_r, \quad j = 1, 2, \dots, n-1,$$

must be linearly independent, which is true if $\mu_n \neq 0$. The necessary condition for $\sum\limits_{r,s=1}^{n} P_{rs}\xi_r\xi_s$ to be positive definite is

$$\gamma > 0, \quad \omega_1 > 0, \omega_2 > 0, \ldots, \omega_{n-1} > 0.$$

Consequently, it is required that $\sum\limits_{r,s=1}^{n-1} Q_{rs}\xi_r\xi_s$ is positive definite. The necessity of the condition (c) is established.

Conversely, Condition (c) implies the representation

$$\sum_{r,s=1}^{n-1} Q_{rs}\xi_r\xi_s = \sum_{j=1}^{n-1} \omega_j \left(\sum_{r=1}^{n-1} c_{jr}\xi_r \right)^2, \quad \omega_j > 0,\ j = 1, 2, \ldots, n-1,$$

with linearly independent forms

$$\sum_{r=1}^{n-1} c_{jr}\xi_r, \quad j = 1, 2, \ldots, n-1.$$

Hence

$$\sum_{r,s=1}^{n} P_{rs}\xi_r\xi_s = \gamma \left(\mu_n\xi_n + \sum_{r=1}^{n} \mu_r\xi_r \right)^2 + \sum_{j=1}^{n-1} \omega_j \left(\sum_{r=1}^{n-1} c_{jr}\xi_r \right)^2.$$

From Condition (a) ($\mu_n \neq 0$) it follows that n linear forms

$$\mu_n\xi_n + \sum_{r=1}^{n-1} \mu_r\xi_r, \quad \sum_{r=1}^{n-1} c_{jr}\xi_r, \quad j = 1, 2, \ldots, n-1,$$

are linearly independent. From Condition (b) ($\gamma > 0$), together with the inequalities $\omega_1 > 0, \ldots, \omega_{n-1} > 0$ following from Condition (c), we conclude that the quadratic form $\sum\limits_{r,s=1}^{n} P_{rs}\xi_r\xi_s$ is positive definite. □

With every polynomial of degree n with real coefficients

$$p_n(x) = p_0x^n + p_1x^{n-1} + \cdots + p_n$$

we associate two polynomials

$$m_n(x) = p_0x^n + p_2x^{n-2} + p_4x^{n-4} + \ldots$$
$$m_{n-1}(x) = p_1x^{n-1} + p_3x^{n-3} + p_5x^{n-5} + \ldots$$

whose sum is $p(x)$:

$$p(x) = m_n(x) + m_{n-1}(x).$$

For each of the polynomials $m_n(x)$ and $m_{n-1}(x)$ only the coefficients at the powers of x of the same evenness differ from zero.

We rewrite the generating polynomial $\mathcal{P}(x,y)$ (cf. §6) in terms of $m_n(x)$ and $m_{n-1}(x)$ as follows:

$$\begin{aligned}\mathcal{P}(x,y) = \sum_{r,s=1}^{n} P_{rs}x^{r-1}y^{s-1} &= \frac{p_n(x)p_n(y) - p_n(-x)p_n(-y)}{x+y}\\ &= \frac{[m_n(x)+m_{n-1}(x)][m_n(y)+m_{n-1}(y)]}{x+y}\\ &\quad - \frac{[m_n(-x)+m_{n-1}(-x)][m_{n-1}(-y)+m_n(-y)]}{x+y}\\ &= 2\frac{m_n(x)m_{n-1}(y)+m_n(y)m_{n-1}(x)}{x+y}.\end{aligned}$$

In the expression $m_n(x)m_{n-1}(y)+m_n(y)m_{n-1}(x)$, the terms of the highest degree $2n-1$ in x and y have the form

$$p_0p_1(x^ny^{n-1}+x^{n-1}y^n) = p_0p_1x^{n-1}y^{n-1}(x+y).$$

It is clear that $P_{nn} = 2p_0p_1$.

If $p_n(x)$ is a Hurwitz polynomial, i.e., $p_0 \neq 0$, and the roots of $p_n(x)$ lie in the left half-plane of the complex plane, then the matrix P formed by the elements P_{rs} is positive definite by the Hermite theorem, which implies $P_{nn} > 0$. Therefore, if $p_n(x)$ is a Hurwitz polynomial, then the leading coefficients of the polynomials $m_n(x)$ and $m_{n-1}(x)$ do not vanish and have the same sign. This allows us to define a strictly positive coefficient $\gamma_n > 0$ so that the polynomial

$$m_{n-2}(x) = m_n(x) - \gamma_n x m_{n-1}(x)$$

having the powers of x of the same evenness as $m_n(x)$ is a polynomial of degree $n-2$. We transform the expression

$$\begin{aligned}&\frac{m_n(x)m_{n-1}(y)+m_n(y)m_{n-1}(x)}{x+y}\\ &\quad= \frac{[\gamma_n x m_{n-1}(x)+m_{n-2}(x)]m_{n-1}(y)+[\gamma_n y m_{n-1}(y)+m_{n-2}(y)]m_{n-1}(x)}{x+y}\\ &\quad= \gamma_n m_{n-1}(x)m_{n-1}(y) + \frac{m_{n-1}(x)m_{n-2}(y)+m_{n-1}(y)m_{n-2}(x)}{x+y}.\end{aligned}$$

By Hermite theorem (cf. Theorem 1 in §6) and Lemma 1, we can conclude that the necessary and sufficient condition for $p_n(x) = m_n(x)+m_{n-1}(x)$ to be a Hurwitz polynomial is that $\gamma_N > 0$ and $p_{n-1}(x) = m_{n-1}(x)+m_{n-2}(x)$ be a Hurwitz polynomial too. We recall that

$$m_{n-2}(x) = m_n(x) - \gamma_n x m_{n-1}(x)$$

and the coefficient γ_n is taken so that the polynomial $m_{n-2}(x)$ is of degree at least $n-2$, i.e., the leading terms of the polynomials $m_n(x)$ and $\gamma_n x m_{n-1}(x)$ coincide.

Starting from the polynomials $m_{n-1}(x)$ and $m_{n-2}(x)$, we choose the coefficient γ_{n-1} so that the polynomial

$$m_{n-3}(x) = m_{n-1}(x) - \gamma_{n-1}x m_{n-2}(x)$$

is of degree less than $n-1$ (moreover, not larger than $n-2$). The above arguments show that $m_{n-1}(x)+m_{n-2}(x)$ is a Hurwitz polynomial if and only if γ_{n-1} is strictly positive. Continuing this recurrence process of defining the polynomials

$$m_n(x), m_{n-1}(x), \ldots, m_1(x), m_0(x),$$

we simultaneously find the constants $\gamma_n, \gamma_{n-1}, \ldots, \gamma_3, \gamma_2$ and establish that

$$p_n(x) = m_n(x) + m_{n-1}(x)$$

is a Hurwitz polynomial if and only if the inequalities

$$\gamma_n > 0, \gamma_{n-1} > 0, \ldots, \gamma_3 > 0, \gamma_2 > 0$$

hold and $p_1(x) = m_1(x) + m_0(x)$ is a Hurwitz polynomial. It is obvious that

$$m_1(x) = \alpha x, \quad m_0(x) = \beta$$

and $p_1(x)$ is a Hurwitz polynomial if and only if α and β both do not vanish and have the same sign. In other words, if γ_1 is taken so as to satisfy the relation

$$m_1(x) - \gamma_1 x m_0 = 0,$$

then $\gamma_1 > 0$.

We have justified the Routh algorithm which allows us to recognize if a given polynomial $p_n(x)$ is a Hurwitz polynomial. The algorithm begins with the representation of the polynomial

$$p_n(x) = p_0 x^n + p_1 x^{n-1} + \cdots + p_{n-1} x + p_n$$

with nonzero leading coefficient $p_0 \neq 0$ as the sum $p_n(x) = m_n(x)+m_{n-1}(x)$, where

$$m_n(x) = p_0 x^n + p_2 x^{n-2} + p_4 x^{n-4} + \ldots,$$
$$m_{n-1}(x) = p_1 x^{n-1} + p_3 x^{n-3} + p_5 x^{n-5} + \ldots,$$

and the verification of the fact that the leading coefficient p_1 of the polynomial $m_{n-1}(x)$ differs from zero; otherwise, $p_n(x)$ is not a Hurwitz polynomial.

After that for $j = n-1, j = n-2, \ldots, 2$ we recurrently find the coefficients γ_{j+1} and the polynomials $m_{j-1}(x)$ by the formulas

$$m_{j-1}(x) = m_{j+1}(x) - \gamma_{j+1} x m_j(x).$$

The coefficient γ_{j+1} is defined so that the polynomial $m_{j-1}(x)$ has degree at most $j-1$. If the leading coefficient of the polynomial m_{j-1} is zero, then the process terminates since in this case we can be sure that $p_n(x)$ is not a Hurwitz polynomial. If we complete the process determining $m_1(x), m_0(x)$ ($m_0(x) = \text{const} \neq 0$), then we can find γ_1 from the condition

$$m_1(x) - \gamma_1 x m_0(x) = 0.$$

If the inequalities

$$\gamma_n > 0, \gamma_{n-1} > 0, \ldots, \gamma_2 > 0, \gamma_1 > 0$$

hold simultaneously, then $p_n(x)$ is a Hurwitz polynomial. However, if at least one of the above inequalities fails, then at least one root of the polynomial $p_n(x)$ lies in the right half-plane. It is obvious that $p_n(x)$ is a Hurwitz polynomial if and only if the leading coefficients of the polynomials $m_j(x)$ do not vanish and have the same sign. In particular, if $p_0 > 0$, then all these coefficients must be positive.

In practical application of the Routh algorithm, it is not necessary to write out the polynomials $m_j(x)$ in detail. It suffices to form the table of their coefficients. It is convenient to form the table following the rule described below.

First we compose a square $(n+1)\times(n+1)$-table whose columns and rows are enumerated by $0, 1, 2, \ldots, n$. In this table, we place the coefficients as follows:

column	0th	1st	2nd	3rd	4th	
0th row	p_0	p_2	p_4	p_6	p_8	$\cdots$
1st	0	p_1	p_3	p_5	p_7	$\cdots$
2nd	0	p_0	p_2	p_4	p_6	$\cdots$
3rd	0	0	p_1	p_3	p_5	$\cdots$
4th	0	0	p_0	p_2	p_4	$\cdots$
5th	0	0	0	p_1	p_3	$\cdots$
6th	0	0	0	p_0	p_2	$\cdots$

Moreover, p_{n+1}, p_{n+2}, and all successive p_j are assumed to vanish. To memorize this matrix we note that on the principal diagonal of the table, the coefficients of the polynomial $p_n(x)$ are placed in order of increasing their indices: $p_0, p_1, \ldots, p_n$. All places not occupied by the coefficients p_i are filled by zeros.

We proceed to transformations of the table. From the 2nd, 4th, 6th, and all even rows we subtract 1st, 3rd, 5th, and all odd rows multiplied by the coefficient p_0/p_1. Then we arrive at the table

0th row	p_0	p_2	p_4	p_6	p_8	$\cdots$
1th	0	p_1	p_3	p_5	p_7	$\cdots$
2nd	0	0	q_0	q_2	q_4	$\cdots$
3rd	0	0	p_1	p_3	p_5	$\cdots$
4th	0	0	0	q_0	q_2	$\cdots$
5th	0	0	0	p_1	p_3	$\cdots$
6th	0	0	0	0	q_0	$\cdots$

Now, we transform this table. From its 3rd, 5th, 7th, and the following odd rows we subtract 2nd, 4th, 6th, and the following even rows multiplied by p_1/q_0. As a result, we obtain the following table:

p_0	p_2	p_4	p_6	p_8	$\cdots$
0	p_1	p_3	p_5	p_7	$\cdots$
0	0	q_0	q_2	q_4	$\cdots$
0	0	0	r_1	r_3	$\cdots$
0	0	0	q_0	q_2	$\cdots$
0	0	0	0	r_1	$\cdots$

As above, we subsequently transform the initial table to the triangle form

p_0	p_2	p_4	p_6	p_8	$\cdots$
0	p_1	p_3	p_5	p_7	$\cdots$
0	0	q_0	q_2	q_4	$\cdots$
0	0	0	r_1	r_3	$\cdots$
0	0	0	0	s_0	$\cdots$

In the jth row of the triangular table, the coefficients of the polynomial $m_{n-j}(x)$ are located:

$$\begin{aligned}
m_n(x) &= p_0x^n + p_2x^{n-2} + p_4x^{n-4} + \dots,\\
m_{n-1}(x) &= p_1x^{n-1} + p_3x^{n-3} + p_5x^{n-5} + \dots,\\
m_{n-2}(x) &= q_0x^{n-2} + q_2x^{n-4} + q_4x^{n-6} + \dots,\\
m_{n-3}(x) &= r_1x^{n-3} + r_3x^{n-5} + r_5x^{n-7} + \dots,\\
m_{n-4}(x) &= s_0x^{n-4} + s_2x^{n-6} + s_4x^{n-8} + \dots,\\
m_{n-5}(x) &= t_1x^{n-5} + t_3x^{n-7} + t_5x^{n-9} + \dots,\\
&\dots\dots\dots\dots\dots\dots\dots\dots\dots\dots
\end{aligned}$$

This is easy to see comparing the described transformation with the rules defining $m_j(x)$.

Thus, in order that an original polynomial $p_n(x)$ be a Hurwitz polynomial, it is necessary that the diagonal terms p_0, p_1, q_0, r_1, s_0, t_1 of the final triangular table have the same sign. If $p_0 > 0$ (e.g., if $p_0 = 1$, which will be assumed hereinafter), then the condition that a matrix is a Hurwitz matrix is equivalent to the inequalities

$$p_0 > 0, \quad p_1 > 0, \quad q_0 > 0, \quad r_1 > 0, \quad s_0 > 0, \quad t_1 > 0, \dots,$$

or the following equivalent inequalities

$$p_0 > 0, \quad p_1p_0 > 0, \quad p_0p_1q_0 > 0, \quad p_0p_1q_0r_1 > 0, \dots.$$

The last collection of inequalities can be interpreted as the condition that the determinants of the principal minors of the final triangular table are positive:

$$p_0 > 0, \quad \begin{vmatrix} p_0 & p_2 \\ 0 & p_1 \end{vmatrix} > 0, \quad \begin{vmatrix} p_0 & p_2 & p_4 \\ 0 & p_1 & p_3 \\ 0 & 0 & q_0 \end{vmatrix} > 0,$$

$$\begin{vmatrix} p_0 & p_2 & p_4 & p_6 \\ 0 & p_1 & p_3 & p_5 \\ 0 & 0 & q_0 & q_2 \\ 0 & 0 & 0 & r_1 \end{vmatrix} > 0, \quad \begin{vmatrix} p_0 & p_2 & p_4 & p_6 & p_8 \\ 0 & p_1 & p_3 & p_5 & p_7 \\ 0 & 0 & q_0 & q_2 & q_4 \\ 0 & 0 & 0 & r_1 & r_3 \\ 0 & 0 & 0 & 0 & s_0 \end{vmatrix} > 0.$$

Recalling the rules which we follow in the transformation of the initial table to the triangular form, we see that at each step the determinants of the principal minors did not change. Therefore,

$$p_0 = p_0, \quad \begin{vmatrix} p_0 & p_2 \\ 0 & p_1 \end{vmatrix} = \begin{vmatrix} p_0 & p_2 \\ 0 & p_1 \end{vmatrix}, \quad \begin{vmatrix} p_0 & p_2 & p_4 \\ 0 & p_1 & p_3 \\ 0 & 0 & q_0 \end{vmatrix} = \begin{vmatrix} p_0 & p_2 & p_4 \\ 0 & p_1 & p_3 \\ 0 & p_0 & p_2 \end{vmatrix},$$

$$\begin{vmatrix} p_0 & p_2 & p_4 & p_6 \\ 0 & p_1 & p_3 & p_5 \\ 0 & 0 & q_0 & q_2 \\ 0 & 0 & 0 & r_1 \end{vmatrix} = \begin{vmatrix} p_0 & p_2 & p_4 & p_6 \\ 0 & p_1 & p_3 & p_5 \\ 0 & p_0 & p_2 & p_4 \\ 0 & 0 & p_1 & p_3 \end{vmatrix},$$

$$\begin{vmatrix} p_0 & p_2 & p_4 & p_6 & p_8 \\ 0 & p_1 & p_3 & p_5 & p_7 \\ 0 & 0 & q_0 & q_2 & q_4 \\ 0 & 0 & 0 & r_1 & r_3 \\ 0 & 0 & 0 & 0 & s_0 \end{vmatrix} = \begin{vmatrix} p_0 & p_2 & p_4 & p_6 & p_8 \\ 0 & p_1 & p_3 & p_5 & p_7 \\ 0 & p_0 & p_2 & p_4 & p_6 \\ 0 & 0 & p_1 & p_3 & p_5 \\ 0 & 0 & p_0 & p_2 & p_4 \end{vmatrix}, \dots.$$

This remark was made by Hurwitz who deduced some inequalities for determinants on the basis of the process due to Routh. These inequalities are usually referred to as the Routh–Hurwitz conditions. They guarantee that all roots of the polynomial $p_n(x)$ lie in the left half-plane:

$$p_0 > 0, \quad \begin{vmatrix} p_0 & p_2 \\ 0 & p_1 \end{vmatrix} > 0, \quad \begin{vmatrix} p_0 & p_2 & p_4 \\ 0 & p_1 & p_3 \\ 0 & p_0 & p_2 \end{vmatrix} > 0,$$

$$\begin{vmatrix} p_0 & p_2 & p_4 & p_6 \\ 0 & p_1 & p_3 & p_5 \\ 0 & p_0 & p_2 & p_4 \\ 0 & 0 & p_1 & p_3 \end{vmatrix} > 0, \quad \begin{vmatrix} p_0 & p_2 & p_4 & p_6 & p_8 \\ 0 & p_1 & p_3 & p_5 & p_7 \\ 0 & p_0 & p_2 & p_4 & p_6 \\ 0 & 0 & p_1 & p_3 & p_5 \\ 0 & 0 & p_0 & p_2 & p_4 \end{vmatrix} > 0, \dots .$$

The number of such inequalities is $n + 1$ for a polynomial $p_n(x)$ of degree n.

We consider several examples.

EXAMPLE 1. A polynomial of degree 2

$$p_2(x) = p_0x^2 + p_1x + p_2$$

is a Hurwitz polynomial if and only if the determinants of the principal minors of the 3×3-matrix

$$\begin{bmatrix} p_0 & p_2 & 0 \\ 0 & p_1 & 0 \\ 0 & p_0 & p_2 \end{bmatrix}$$

are positive:

$$p_0 > 0, \quad \begin{vmatrix} p_0 & p_2 \\ 0 & p_1 \end{vmatrix} = p_0p_1 > 0, \quad \begin{vmatrix} p_0 & p_2 & 0 \\ 0 & p_1 & 0 \\ 0 & p_0 & p_2 \end{vmatrix} = p_0p_1p_2 > 0,$$

which is equivalent to the condition that all coefficients are positive ($p_0 > 0$, $p_1 > 0$, $p_2 > 0$).

EXAMPLE 2. For $p_0 > 0$ a polynomial of degree 3

$$p_3(x) = p_0x^3 + p_1x^2 + p_2x + p_3$$

is a Hurwitz polynomial if and only if the determinants of all principal minors of the 4×4-matrix

$$\begin{bmatrix} p_0 & p_2 & 0 & 0 \\ 0 & p_1 & p_3 & 0 \\ 0 & p_0 & p_2 & 0 \\ 0 & 0 & p_1 & p_3 \end{bmatrix}$$

are positive. It is obvious that this condition is equivalent to the condition that the determinants of all principal minors of the 3×3-matrix

$$\begin{bmatrix} p_1 & p_3 & 0 \\ p_0 & p_2 & 0 \\ 0 & p_1 & p_3 \end{bmatrix}$$

are positive, i.e., it is equivalent to the inequalities

$$p_1 > 0, \quad p_1p_2 - p_0p_3 > 0, \quad p_3(p_1p_2 - p_0p_3) > 0$$

which can be replaced by the following inequalities:

$$p_0 > 0, \quad p_1 > 0, \quad p_3 > 0, \quad p_2 > \frac{p_0 p_3}{p_1}.$$

It turns out that the coefficients p_0, p_1, p_2, p_3 are positive.

EXAMPLE 3. A polynomial of degree 4

$$p_4(x) = p_0 x^4 + p_1 x^3 + p_2 x^2 + p_3 x + p_4$$

is a Hurwitz polynomial (for $p_0 > 0$) if and only if the determinants of all principal minors of the 4×4-matrix

$$\begin{bmatrix} p_1 & p_3 & 0 & 0 \\ p_0 & p_2 & p_4 & 0 \\ 0 & p_1 & p_3 & 0 \\ 0 & p_0 & p_2 & p_4 \end{bmatrix}$$

are positive, which yields the inequalities

$$p_0 > 0, \quad p_1 > 0, \quad p_1 p_2 - p_0 p_3 > 0,$$
$$p_3(p_1 p_2 - p_0 p_3) - p_1^2 p_4 > 0, \quad p_4(p_3(p_1 p_2 - p_0 p_3) - p_1^2 p_4) > 0.$$

Let us note that even without using the theory one can prove that if $p_0 > 0$ and $p_n(x)$ is a Hurwitz polynomial, then $p_1 > 0, p_2 > 0, \ldots, p_{n-1} > 0, p_n > 0$.

CHAPTER 3

Qualitative Properties of Problems and Algorithmic Aspects

§1. The computation of the matrix exponential and solution of the Cauchy problem

A computational procedure based on the use of the Taylor series. A partition of the integration interval. The error estimate. The computation of the matrix exponential. Doubling of the argument. The solution of nonhomogeneous equations. An approximation of the right-hand side by polynomials.

As we know, a solution to the Cauchy problem for a homogeneous system of linear ordinary equations with constant coefficients $\frac{dy}{dt} = Ay$ and the initial condition $y(0) = a$ can be represented as the Taylor series

$$y(t) = a + \frac{t}{1!}Aa + \frac{t^2}{2!}A^2a + \frac{t^3}{3!}A^3a + \dots .$$

To compute partial sums of the series one can use the following procedure.

Setting $a_0 = a$, we find the terms of the Taylor series

$$a_n = \frac{t^n}{n!}A^n a = \frac{t^n}{n!}A^n a_0$$

by the recurrence relations

$$a_k = \frac{t}{k}Aa_{k-1}, \quad k = 1, 2, 3 \dots .$$

The partial sums

$$S_k(t) = a_0 + \frac{t}{1!}Aa_0 + \frac{t^2}{2!}A^2a_0 + \dots + \frac{t^k}{k!}A^k a_0$$

are also found by the recurrence relations

$$S_0 = a_0, \quad S_k = S_{k-1} + a_k.$$

We give a complete set of formulas of the computational procedure. For the input data we can take the vector a and the value t of the argument of the vector-valued function $y(t)$:

$$a_0 = a, \quad S_0 = a_0,$$
$$a_k = \frac{t}{k}Aa_{k-1}, \quad S_k = S_{k-1} + a_k, \quad k = 1, 2, 3 \dots .$$

In practice, only a finite number of steps of the computational procedure is realized: $k = 0, 1, 2, \dots, n$. The choice of n is dictated by the accuracy requirements. We compare the exact solution

$$y(t) = e^{tA}a_0$$

to the Cauchy problem with the approximate solution

$$\widetilde{y}(t) = S_n(t) = a_0 + \frac{t}{1!}Aa_0 + \frac{t^2}{2!}A^2a_0 + \cdots + \frac{t^n}{n!}A^n a_0$$

or

$$\widetilde{y}(t) = \left[I + \frac{t}{1!}A + \frac{t^2}{2!}A^2 + \cdots + \frac{t^n}{n!}A^n\right]a_0.$$

It is clear that

$$\begin{aligned}\|y(t) - \widetilde{y}(t)\| &= \left\|e^{tA}a_0 - \left[I + \frac{t}{1!}A + \frac{t^2}{2!}A^2 + \cdots + \frac{t^n}{n!}A^n\right]a_0\right\| \\ &= \|R_n(t)a_0\| \leqslant \|R_n(t)\|\,\|a_0\|,\end{aligned}$$

where $R_n(t)$ is the remainder of the Taylor series for the matrix exponential e^{tA}:

$$e^{tA} = I + \frac{t}{1!}A + \frac{t^2}{2!}A^2 + \cdots + \frac{t^n}{n!}A^n + R_n(t).$$

It is easy to estimate the norm of $R_n(t)$. Indeed,

$$\begin{aligned}\|R_n(t) &= \left\|\frac{t^{n+1}}{(n+1)!}A^{n+1} + \frac{t^{n+2}}{(n+2)!}A^{n+2} + \frac{t^{n+3}}{(n+3)!}A^{n+3} + \ldots\right\| \\ &= \left\|\frac{t^{n+1}}{(n+1)!}A^{n+1}\left[I + \frac{t}{n+2}A + \frac{t^2}{(n+2)(n+3)}A^2 + \ldots\right]\right\| \\ &\leqslant \frac{|t|^{n+1}}{(n+1)!}\|A\|^{n+1}\left[1 + \frac{|t|}{n+2}\|A\| + \frac{|t|^2}{(n+2)(n+3)}\|A\|^2 + \ldots\right] \\ &\leqslant \frac{|t|^{n+1}}{(n+1)!}\|A\|^{n+1}\left[1 + \frac{|t|}{1!}\|A\| + \frac{|t|^2}{2!}\|A\|^2 + \ldots\right] = \frac{(|t|\,\|A\|)^{n+1}}{(n+1)!}\,e^{|t|\,\|A\|}.\end{aligned}$$

Thus, we have obtained the estimate

$$\|y(t) - \widetilde{y}(t)\| \leqslant \frac{(|t|\,\|A\|)^{n+1}}{(n+1)!}\,e^{|t|\,\|A\|}\,\|a_0\|.$$

We recall (cf. Chapter 1, §2) the estimate from below for $\|y(t)\|$:

$$\|y(t)\| \geqslant e^{-|t|\,\|A\|}\,\|a_0\|.$$

It allows us to establish the inequality

$$\frac{\|y(t) - \widetilde{y}(t)\|}{\|y(t)\|} \leqslant \frac{(|t|\,\|A\|)^{n+1}}{(n+1)!}\,e^{2\,|t|\,\|A\|}.$$

Since for $t \leqslant 1/(2\|A\|)$ we have $e^{2|t|\,\|A\|} \leqslant e < 3$ and $(|t|\,\|A\|)^{n+1} < 1/2^{n+1}$, the last inequality takes the form

$$\frac{\|y(t) - \widetilde{y}(t)\|}{\|y(t)\|} < \frac{3}{(n+1)!\,2^{n+1}}.$$

In particular, for $n = 10$

$$\frac{\|y(t) - \widetilde{y}(t)\|}{\|y(t)\|} < \frac{3}{11!\,2^{11}} \approx 3,7 \cdot 10^{-11}.$$

If $|t|\,\|A\| \leqslant 1/2$, we can indicate k such that the error in the representation of $y(t)$ in terms of $S_k(t)$ is of the same order as the error resulting from the fact that the components of $y(t)$ are "computer numbers" and are represented by a certain number of bits in computers. The above example shows that $k = 10$ provides already a high relative accuracy:

$$\frac{\|y(t) - S_{10}(t)\|}{\|y(t)\|} < 4 \cdot 10^{-11}.$$

But it is not advisable to apply the described method for computation of a solution for large t, e.g., for $|t|\,\|A\| \approx 100$. The necessary number k of summands is larger in this case. Moreover, special attention must be paid to the accumulation of round-off errors and their influence on a final result. On a large interval $0 \leqslant t \leqslant T$, one usually computes $y(t)$ step by step, dividing the entire interval into p subintervals of length $h = T/p$. The number p is chosen so that $h\|A\|$ is not too large. For example, it is convenient to require $h\|A\| < 1/2$. Then, setting $t_m = mh = (mT)/p$, we can use the relation $y(t_{m+1}) = e^{hA}y(t_m)$, which allows us, starting with given A and $y_0 = a$ and using the described procedure, to find consecutively

$$\begin{aligned} &y(t_1) = e^{hA}y_0, \\ &y(t_2) = e^{hA}y(t_1) = (e^{hA})^2 y_0, \\ &\cdots\cdots\cdots\cdots\cdots\cdots\cdots\cdots \\ &y(t_p) = (e^{hA})^p y_0 = e^{TA}y_0. \end{aligned}$$

Sometimes, especially if it is necessary to obtain the table of values of $y(t)$ with very small step h ($h\|A\| \ll 1/2$), only a fixed number n of terms of the Taylor series is considered, e.g., $n = 2$ or even $n = 1$. In these cases, the accuracy is guaranteed by the choice of h only. We estimate the error in such a situation.

Let $y(t_m) = (e^{hA})^m y_0$ be an exact solution to the Cauchy problem at $t = t_m$ and let $\widetilde{y}(t_m)$ denote the approximate solution:

$$\widetilde{y}(t_m) = \left[I + \frac{hA}{1!} + \frac{(hA)^2}{2!} + \cdots + \frac{(hA)^n}{n!}\right]^m y_0, \quad 0 < m < p.$$

To estimate the error, we represent $\widetilde{y}(t_m)$ in the form

$$\widetilde{y}(t_m) = [e^{hA} - R_n(hA)]^m y_0 = [I - e^{-hA}R_n(hA)]^m e^{mhA} y_0 = [I - \frac{1}{m}Q_n(hA)]^m y(t_m),$$

where $Q_n(hA) = me^{-hA}R_n(hA)$. Hence

$$\frac{\|y(t_m) - \widetilde{y}(t_m)\|}{\|y(t_m)\|} \leqslant \left\| I - \left[I - \frac{1}{m}Q_n(hA)\right]^m \right\|.$$

It is obvious that the norm of the matrix polynomial on the right-hand side of this inequality satisfies the estimate

$$\begin{aligned} &\left\| I - \left[I - \frac{1}{m}Q_n\right]^m \right\| \\ &\quad = \left\| \frac{m}{1!}\left(\frac{Q_n}{m}\right) - \frac{m(m-1)}{2!}\left(\frac{Q_n}{m}\right)^2 + \cdots + (-1)^{m+1}\frac{m!}{m!}\left(\frac{Q_n}{m}\right)^m \right\| \\ &\quad \leqslant \frac{\|Q_n\|}{1!} + \frac{\|Q_n\|^2}{2!} + \cdots + \frac{\|Q_n\|^m}{m!} \\ &\quad \leqslant e^{\|Q_n\|} - 1 \leqslant \|Q_n\|\, e^{\|Q_n\|}. \end{aligned}$$

By the above estimate for the remainder of the Taylor series for the matrix exponential, we have

$$\|Q_n(hA)\| \leqslant m\,\|e^{-hA}\|\,\|R_n(hA)\| \leqslant m\frac{(h\,\|A\|)^{n+1}}{(n+1)!}\,e^{2h\,\|A\|}.$$

Since $0 \leqslant t_m = mh \leqslant T$, the relative error in the solution satisfies the estimate

$$\frac{\|y(t_m) - \widetilde{y}(t_m)\|}{\|y(t_m)\|} \leqslant B_n e^{B_n} = \xi,$$

where

$$B_n = Th^n \frac{\|A\|^{n+1}}{(n+1)!} e^{2h\|A\|}.$$

The norm of the absolute error satisfies the estimate

$$\|y(t_m) - \widetilde{y}(t_m)\| \leqslant \xi\, e^{T\|A\|}\,\|y_0\|.$$

For a sufficiently small h the exponential factor e^{B_n} is close to 1. As h decreases, the error tends to zero proportionally to h^n. This allows us to regard the described scheme as the nth order accuracy scheme. In particular, for $n = 2$ (the second order accuracy scheme) we have

$$\|y(t_m) - \widetilde{y}(t_m)\| \leqslant B_2 e^{B_2} e^{T\|A\|}\,\|Q\|,$$

where

$$B_2 = Th^2 \frac{\|A\|^3}{6} e^{2h\|A\|}.$$

The described method can be applied to the numerical solution of the matrix differential equation

$$\frac{d}{dt}Y(t) = AY(t), \quad Y(0) = I,$$

i.e., the computation of the matrix exponential $Y(t) = e^{tA}$. If t is not too large, e.g., $|t|\,\|A\| \leqslant 1/2$, then one can use the Taylor expansion whose terms $A_k(t) = \frac{t^k}{k!}A^k$ and partial sums

$$S_k(t) = I + \frac{t}{1!}A + \frac{t^2}{2!}A^2 + \cdots + \frac{t^k}{k!}A^k = A_0 + A_1 + A_2 + \cdots + A_k$$

are defined by the recurrence relations

$$A_0 = I, \quad A_k = \frac{t}{k}A_{k-1}, \quad S_0 = A_0, \quad S_k = S_{k-1} + A_k.$$

In the matrix case, we can establish the estimate

$$\frac{\|Y(t) - S_k(t)\|}{\|Y(t)\|} \equiv \frac{\|e^{tA} - S_k(t)\|}{\|e^{tA}\|} \leqslant \frac{(|t|\,\|A\|)^{k+1}}{(k+1)!} e^{2|t|\,\|A\|}$$

in the same way as the estimate for the error was derived in the case of a vector equation.

If $T\|A\|$ is not too small, the following elementary method of computation of e^{TA} can be used. First, we find a positive integer n from the inequality $2^{-n}T\|A\| \leqslant 1/2$. For n we take the smallest number satisfying the above inequality. Setting

$\tau = 2^{-n}T$ and using the Taylor series, we find $Y_1 = e^{\tau A}$. The computation is reduced to the successive squaring of the matrices:

$$Y_j = Y_{j-1}^2, \quad j = 2, 3, \dots, n+1.$$

It turns out that $Y_j = e^{2^{j-1}\tau A}$ and $Y_{n+1} = e^{2^n \tau A} = e^{TA}$. Each squaring doubles the argument t of the matrix exponential e^{tA}. Therefore, to increase the argument t by 1024 times we need only ten matrix multiplications ($1024 = 2^{10}$).

We now describe a simple method of numerical solution of the Cauchy problem for the nonhomogeneous equation $y' = Ay + f(t)$ with the initial data $y(0) = a$ on the interval $0 \leqslant t \leqslant T$. We assume that the interval is divided into p subintervals of length $h = T/p$ by the points $t_0 = h$, $t = 2h, \dots, t_p = T = ph$. It suffices to solve the Cauchy problem on each of these subintervals and compose the solution to the original problem on the entire interval of the solutions on the subintervals. For $k = 1, 2, \dots, p$ we find $y(t)$ as a solution to the problem

$$y(t_{k-1}) = \begin{cases} a & \text{if } k = 1, \\ y(t_{k-2} + h) & \text{otherwise} \end{cases}$$

$$y'(t) = Ay(t) + f(t), \quad t_{k-1} \leqslant t \leqslant t_{k-1} + h = t_k.$$

If h is sufficiently small and $f(t)$ is sufficiently smooth, then on the interval $t_{k-1} \leqslant t \leqslant t_k$ the function f can be replaced by the approximate polynomial with required accuracy. For example, if the required accuracy of approximation can be achieved by the use of the polynomial of degree 2, then for $t_{k-1} \leqslant t \leqslant t_k$ we can replace $f(t)$ by $\widehat{f}(t)$, where

$$\begin{aligned} \widehat{f}(t) =& f(t_{k-1}) + \frac{4f(t_k) - f(t_{k+1}) - 3f(t_{k-1})}{2h}(t - t_{k-1}) \\ &+ \frac{f(t_{k+1}) - 2f(t_k) + f(t_{k-1})}{2h^2}(t - t_{k-1})^2 \end{aligned}$$

or

$$\widehat{f}(t) = f(t_{k-1}) + \frac{f(t_k) - f(t_{k-2})}{2h}(t - t_{k-1}) + \frac{f(t_k) - 2f(t_{k-1}) + f(t_{k-2})}{2h^2}(t - t_{k-1})^2.$$

We can use the first formula for $k = 1, 2, \dots, p-1$, and the second one for $k = 2, 3, \dots, p$. The reader can easily verify that $\widehat{f}(t)$ coincide with $f(t)$ at the points $t = t_{k-1}, t_k, t_{k+1}$ if the first formula is used and at the points $t = t_{k-2}, t_{k-1}, t_k$ if the second formula is applied. It is proved in analysis that the error appearing as a result of the replacement of a scalar function $f(t)$ by the interpolating polynomial $\widehat{f}(t)$ admits the estimate

$$|f(t) - \widehat{f}(t)| \leqslant \frac{h^3}{9\sqrt{3}} \max_{0 \leqslant t \leqslant T} |f'''(t)|.$$

The (Euclidean) norm of the error in the interpolation of an N-dimensional vector-valued function $f(t)$ satisfies the estimate

$$\|f(t) - \widehat{f}(t)\| \leqslant \sqrt{N}\frac{h^3}{9\sqrt{3}} \max_{0 \leqslant t \leqslant T} \|f'''(t)\|.$$

This estimate can be taken into account in the choice of the value of h that guarantees the required accuracy of approximation.

Thus, we may always assume that on the interval $t_{k-1} \leqslant t \leqslant t_k$ the interpolating polynomial $\widehat{f}(t)$ is represented by the Taylor formula

$$\widehat{f}(t) = f(t_{k-1}) + \widehat{f}'(t_{k-1})(t - t_{k-1}) + \frac{1}{2}\widehat{f}''(t_{k-1})(t - t_{k-1})^2,$$

where the coefficients $\widehat{f}'(t_{k-1})$, $\widehat{f}''(t_{k-1})$ are found by the formulas

$$\widehat{f}'(t_{k-1}) = \begin{cases} \dfrac{4f(t_k) - f(t_{k+1}) - 3f(t_{k-1})}{2h} & \text{or} \\ \dfrac{f(t_k) - f(t_{k-2})}{2h}, \end{cases}$$

$$\widehat{f}''(t_{k-1}) = \begin{cases} \dfrac{f(t_{k+1}) - 2f(t_k) + f(t_{k-1})}{h^2} & \text{or} \\ \dfrac{f(t_k) - 2f(t_{k-1}) + f(t_{k-2}}{h^2}. \end{cases}$$

The solution to the Cauchy problem on the subinterval $t_{k-1} \leqslant t \leqslant t_k$ for the nonhomogeneous equation $y' = Ay + f(t)$ is replaced by the solution to the Cauchy problem for the equation $y' = Ay + \widehat{f}(t)$ with the same initial data $y(t_{k-1}) = y(t_{k-2} + h)$. In addition,

$$\begin{aligned} y'(t_{k-1}) &= Ay(t_{k-1}) + \widehat{f}(t_{k-1}), \\ y''(t_{k-1}) &= Ay'(t_{k-1}) + \widehat{f}'(t_{k-1}), \\ y'''(t_{k-1}) &= Ay''(t_{k-1}) + \widehat{f}''(t_{k-1}), \\ y^{(m)}(t_{k-1}) &= Ay^{(m-1)}(t_{k-1}) \quad (m \geqslant 4). \end{aligned}$$

Using the notation $a_m = \dfrac{h^m}{m!}y^{(m)}(t_{k-1})$, we arrive at the following recurrent computational procedure of these vectors:

$$\begin{aligned} a_0 &= y(t_{k-2} + h) \equiv y(t_{k-1}), \\ a_1 &= hAa_0 + h\widehat{f}(t_{k-1}), \\ a_2 &= \frac{h}{2}Aa_1 + \frac{h^2}{2}\widehat{f}'(t_{k-1}), \\ a_3 &= \frac{h}{3}Aa_2 + \frac{h^3}{6}\widehat{f}''(t_{k-1}), \\ a_m &= \frac{h}{m}Aa_{m-1} \quad (m \geqslant 4). \end{aligned}$$

The sums $S_0 = a_0$, $S_l = S_{l-1} + a_l$ are partial sums of the Taylor expansion of $y(t_{k-1} + h) \equiv y(t_k)$:

$$y(t_k) = y(t_{k-1}) + hy'(t_{k-1}) + \cdots + \frac{h^m}{m!}y^m(t_{k-1}) + \dots.$$

Therefore, having computed S_l for sufficiently large l, we obtain an approximate value $y(t_k)$ with a certain accuracy.

In this section, we have described computational tools used to derive tables representing solutions to the differential equations $y' = Ay + f$ and to compute the values of the matrix exponential. The effect of computational errors on the final result was discussed very superficially. Certainly, the majority of algorithms used in the numerical integration were not mentioned here. The study of such algorithms and the detailed analysis of their sensitivity to errors inevitable in any realizations of algorithms is the topic of other branches of applied mathematics.

§2. Computational difficulties in the Hurwitz problem and their solutions

An example of a matrix whose eigenvalues change significantly under a very small change of matrix elements. An analysis of the example using the Ostrowski theorem. A parameter characterizing the quality of the asymptotic stability and its properties. Theorem on continuous dependence of the parameter of the quality of stability on the matrix under consideration. The algorithm computing the parameter and the Lyapunov function simultaneously.

In this section, we describe a computational procedure for the study of the stability of the system $y' = Ay$ and discuss the influence of computational errors. As we know, to establish the stability it is necessary to prove that all roots of the characteristic equation

$$\det[\tau I - A] = \tau^N + p_1\tau^{N-1} + \cdots + p_{N-1}\tau + p_N = 0$$

lie in the left complex half-plane, i.e., A is a Hurwitz matrix. This term can be explained by the fact that the coefficients of such a polynomial satisfy the Routh–Hurwitz conditions (cf. Chapter 2, §7).

As an example we consider the 25×25-matrix

$$A = \begin{bmatrix} -1 & 10 & & & & & & & \\ 0 & -1 & 10 & & & & \mathbf{0} & & \\ 0 & 0 & -1 & 10 & & & & & \\ \vdots & \vdots & \vdots & \vdots & \vdots & . & . & & \\ 0 & 0 & \cdot & \cdot & \cdot & \cdot & -1 & 10 \\ \omega & 0 & \cdot & \cdot & \cdot & \cdot & 0 & -1 \end{bmatrix},$$

which depends on the parameter ω. The characteristic equation has the form

$$\det[\tau I - A] = (\tau + 1)^{25} + \omega\, 10^{24}.$$

Let ω run over the small disk $|\omega| \leqslant 10 \cdot 8^{-25} \approx 2,6 \cdot 10^{-22}$. Then the characteristic roots τ run over the disk

$$|1+\tau|^{25} \leqslant \left(\frac{10}{8}\right)^{25} = \left(\frac{5}{4}\right)^{25}, \quad |1+\tau| < 1 + \frac{1}{4}.$$

However, this disk is not small. In particular, for $\omega = -10 \cdot 8^{-25} \approx -2.6 \cdot 10^{-22}$ the characteristic root $\tau = -1 + 5/4 = 1/4$ lies in the right half-plane, whereas for $\omega = 0$ all characteristic roots are equal to -1. Thus, for the matrix A, all characteristic roots have negative real parts if $\omega = 0$, whereas such a property fails if $\omega = -10 \cdot 8^{-25} \approx -2,6 \cdot 10^{-22}$.

For present-day computers the relative error in representation of numbers is of about 10^{-16}. This fact is explained by the way the numbers are represented in computers. Consequently, from the point of view of computers, the matrix A with parameters $\omega = 0$ and $\omega = -10 \cdot 8^{-25}$ and the corresponding systems $y' = Ay$ are assumed to be stable or unstable simultaneously. At first glance, it would seem that the paradox is a consequence of the sharp change of the characteristic roots for the small change of elements of the matrix. Does this fact contradict the theorem on continuous dependence of characteristic roots on elements of the matrix? We give Ostrowski's formulation of this theorem (cf., for example, [**22**]).

THEOREM 1 (Ostrowski). *Let elements b_{ij} of an $n\times n$-matrix B and elements c_{ij} of an $n \times n$-matrix C satisfy the inequalities $|b_{ij}| < 1$ and $|c_{ij}| < 1$. Then for any root λ to the equation*

$$\det[B - \lambda I] = 0$$

there is a root λ' to the equation

$$\det[B + \varepsilon C - \lambda' I] = 0, \quad \varepsilon > 0,$$

such that $|\lambda - \lambda'| \leqslant 2(n+1)^2(n^2\varepsilon)^{1/n}$.

It is convenient to use the following corollary of the Ostrowski theorem.

COROLLARY 1. *Let elements b_{ij} of a matrix B and elements c_{ij} of a matrix C satisfy the inequalities $|b_{ij}| < 10$, $|c_{ij}| < 1$. Then for any root τ to the equation $\det[\tau I - B] = 0$ there exists a root τ' to the equation*

$$\det[\tau' I - B - \varepsilon C] = 0, \quad \varepsilon > 0,$$

such that

$$|\tau' - \tau| \leqslant 20(n+1)^2(n^2\varepsilon)^{1/n}/10.$$

We denote the matrix A with parameter $\omega = 0$ by B, put $|\omega| = \varepsilon$, and denote by C the 25×25-matrix that has the single nonzero element $c_{25,1} = \omega/|\omega|$. Then any eigenvalue $\tau' = \tau(A + \omega + C)$ satisfies the inequality

$$|\tau' - (-1)| \leqslant 20 \cdot (26)^2 \sqrt[25]{\frac{625|\omega|}{10}},$$

which is written in the form

$$|\tau' + 1| \leqslant \frac{20 \cdot (26)^2}{8} \sqrt[25]{625}$$

for $|\omega| = 10 \cdot 8^{-25}$. It is obvious that $\tau' = 1/4$ satisfies this inequality.

Hence the above example does not contradict the theorem on continuous dependence of characteristic roots on elements of the matrix. The point is that the estimate of the module of continuity allows for a significant change of roots for perturbations of the matrix that are too small for computer simulation. This fact leads to the conclusion that the stability criteria based on the location of characteristic roots are not convenient for applications. The question arises whether it is possible to find criteria which will be more suitable for analysis by computers. We describe one such criterion.

We suggest taking the quantity $\varkappa(A)$ as the characteristic of the quality of stability. This characteristic of a matrix A was introduced in Chapter 2, §4. We recall the definition (cf. the formula (1) in Chapter 2, §4)

$$\varkappa(A) = 2\,\|A\| \sup_{T, y(0)} \frac{\int_0^T \|y(t)\|^2 dt}{\|y(0)\|^2}.$$

As was shown, $\varkappa(A)$ is finite if and only if all eigenvalues of A lie strictly in the left half-plane. In this case, the matrix Lyapunov equation $HA + A^*H = -I$

is uniquely solvable, the solution H is symmetric and positive definite, and the estimate $\varkappa(A) = 2\|A\|\,\|H\| \geqslant 1$ holds. As was shown in Chapter 2, any solution to the equation $y' = Ay$ satisfies the estimate

$$\|y(t)\| \leqslant \|y(0)\| \sqrt{\varkappa(A)}\, e^{-\frac{t\|A\|}{\varkappa(A)}} \quad t > 0.$$

Now we study the continuity of the dependence of $\varkappa(A)$ on A. We begin with auxiliary considerations. Let all eigenvalues of an $N \times N$-matrix A lie strictly in the left half-plane, i.e., $\varkappa(A) < \infty$, and let B be an $N \times N$-matrix such that

$$\frac{\|B\|}{\|A\|} < \frac{1}{3\varkappa^2(A)}.$$

We introduce the matrix X as the solution to the integral equation

$$X = H + \int_0^\infty e^{tA^*}[B^*X + XB]e^{tA}dt,$$

where

$$H = \int_0^\infty e^{tA^*}e^{tA}dt.$$

Let us show that a solution to this equation exists, is unique, and can be obtained by successive approximations. We set $X_0 = 0$ and define the sequence of matrices $X_1, X_2, \dots, X_k, \dots$ by the recurrence relations

$$X_k = H + \int_0^\infty e^{tA^*}[B^*X_{k-1} + X_{k-1}B]e^{tA}dt.$$

In addition, $X_1 - X_0 = H$ and

$$X_k - X_{k-1} = \int_0^\infty e^{tA^*}[B^*(X_{k-1} - X_{k-2}) + (X_{k-1} - X_{k-2})B]e^{tA}dt$$

for $k > 2$. We note that for any $N \times N$-matrix C the matrix

$$Z = \int_0^\infty e^{tA^*}Ce^{tA}dt$$

satisfies the estimate

$$\|Z\| \leqslant \|C\| \left\| \int_0^\infty e^{tA^*}e^{tA}dt \right\| = \|C\|\,\|H\| = \|C\| \frac{\varkappa(A)}{2\,\|A\|}.$$

Indeed, this estimate is derived from the following relation which connects the quadratic forms (Zy, y) and (Hy, y):

$$|(Zy,y)| = \left|\left(\left[\int_0^\infty e^{tA^*} C e^{tA} dt\right] y, y\right)\right| = \left|\int_0^\infty (Ce^{tA}y, e^{tA}y)dt\right| \leqslant \|C\| \left|\int_0^\infty (e^{tA}y, e^{tA}y)dt\right|$$

$$= \|C\| \int_0^\infty (e^{tA^*} e^{tA} y, y)dt = \|C\| \, (Hy, y).$$

In view of the above estimate, we conclude that

$$\|X_1\| = \|X_1 - X_0\| = \|H\|,$$

$$\|X_k - X_{k-1}\| \leqslant 2\,\|B\|\,\|X_{k-1} - X_{k-2}\| \frac{\varkappa(A)}{2\,\|A\|}$$

$$= \varkappa(A)\frac{\|B\|}{\|A\|}\,\|X_{k-1} - X_{k-2}\|, \quad k \geqslant 2.$$

Consequently,

$$\|X_k - X_{k-1}\| \leqslant \left[\varkappa(A)\frac{\|B\|}{\|A\|}\right]^{k-1} \|H\|, \quad k \geqslant 1.$$

By assumption,

$$\frac{\|B\|}{\|A\|} < \frac{1}{3\varkappa^2(A)}, \qquad \varkappa(A)\frac{\|B\|}{\|A\|} < \frac{1}{3\varkappa(A)} \leqslant \frac{1}{3}.$$

The sequence of matrices X_k converges to a matrix X; moreover,

$$\|X\| \leqslant \|H\| \sum_{k=1}^{\infty} \left[\varkappa(A)\frac{\|B\|}{\|A\|}\right]^{k-1} = \|H\| \Big/ \left[1 - \frac{\|B\|}{\|A\|}\varkappa(A)\right],$$

$$\|X - H\| = \|X - X_1\| \leqslant \|H\| \sum_{k=2}^{\infty} \left[\varkappa(A)\frac{\|B\|}{\|A\|}\right]^{k-1} = \frac{\varkappa(A)\frac{\|B\|}{\|A\|}}{1 - \varkappa(A)\frac{\|B\|}{\|A\|}}\,\|H\|,$$

$$\|X - X_j\| \leqslant \|H\| \sum_{k=j+1}^{\infty} \left[\varkappa(A)\frac{\|B\|}{\|A\|}\right]^{k-1} = \frac{\left[\varkappa(A)\frac{\|B\|}{\|A\|}\right]^j}{1 - \varkappa(A)\frac{\|B\|}{\|A\|}}\,\|H\|.$$

These inequalities allow us to pass to the limit on both sides of the equality

$$X_j = H + \int_0^\infty e^{tA^*}[B^* X_{j-1} + X_{j-1}B]e^{tA}dt$$

as $j \to \infty$ and establish that the obtained matrix X satisfies the equation

$$X = H + \int_0^\infty e^{tA^*}[B^* X + XB]e^{tA}dt.$$

The existence of a solution to the integral equation is proved.

If the solution to this equation were not unique, then there would have existed a matrix $Y \neq X$ such that

$$Y = H + \int_0^\infty e^{tA^*}[B^*Y + YB]e^{tA}dt$$

and

$$Y - X = \int_0^\infty e^{tA^*}[B^*(Y - X) + (Y - X)B]e^{tA}dt,$$

$$\|Y - X\| > 0, \quad \|Y - X\| \leqslant \frac{\varkappa(A)\,\|B\|}{\|A\|}\,\|Y - X\|.$$

Cancelling the positive factor $\|Y - X\|$ in both sides of the last inequality, we see that this inequality holds only if $1 \leqslant \varkappa(A)\,\|B\|/\|A\|$. However, by assumption,

$$\frac{\varkappa(A)\,\|B\|}{\|A\|} < \frac{1}{3\varkappa(A)} \leqslant \frac{1}{3}.$$

Hence the assumption $Y \neq X$ ($\|Y - X\| > 0$) leads to a contradiction. The uniqueness of a solution to the integral equation is proved.

Using the representation

$$H = \int_0^\infty e^{tA^*}e^{tA}dt,$$

we can rewrite the equation

$$X = H + \int_0^\infty e^{tA^*}[B^*X + XB]e^{tA}dt$$

in the form

$$X = \int_0^\infty e^{tA^*}[I + B^*X + XB]e^{tA}dt.$$

The inequalities

$$\|X\| \leqslant \frac{\|H\|}{1 - \varkappa(A)\frac{\|B\|}{\|A\|}} = \frac{\varkappa(A)\frac{1}{2\,\|A\|}}{1 - \varkappa(A)\frac{\|B\|}{\|A\|}},$$

$$\|B^*X\| \leqslant \|B^*\|\,\|X\| = \|B\|\,\|X\|, \quad \|XB\| \leqslant \|B\|\,\|X\|,$$

$$\|B^*X + XB\| \leqslant 2\,\|B\|\,\|X\| \leqslant \frac{\varkappa(A)\frac{\|B\|}{\|A\|}}{1 - \varkappa(A)\frac{\|B\|}{\|A\|}} \leqslant \frac{1/3}{1 - 1/3} = \frac{1}{2}$$

imply the estimate

$$([I + B^*X + XB]y, y) \geqslant (y, y) - \frac{1}{2}\,(y, y) = \frac{1}{2}\,(y, y),$$

i.e., the symmetric matrix $I + B^*X + XB$ is positive definite. Using the theory of matrix Lyapunov equations (cf. Chapter 2, §3) and the assumption that A is a

Hurwitz matrix, we can assert that the matrix X admitting the integral representation

$$X = \int_0^\infty e^{tA^*}[I + B^*X + XB]e^{tA}dt$$

is positive definite. In the study of the Lyapunov theory (cf. Chapter 2, §3), we established that the matrix

$$K = \int_0^\infty e^{tA^*}Ce^{tA}dt$$

(A is a Hurwitz matrix) satisfies the equation $KA + A^*K = -C$. Setting $C = I + B^*X + XB$ and deriving the equality $K = X$ from the integral equation, we get the matrix equation

$$XA + A^*X = -I - XB - B^*X$$

which can be rewritten in the standard form of the Lyapunov equation:

$$X(A + B^*) + (A + B)X = -I.$$

By the theory of matrix Lyapunov equations, $A + B$ is a Hurwitz matrix because I and X are positive definite. We have

$$\begin{aligned}\varkappa(A+B) &= 2\,\|A+B\|\,\|X\| \leqslant 2\,\|A\|\left(1+\frac{\|B\|}{\|A\|}\right)\frac{\|H\|}{1-\varkappa(A)\frac{\|B\|}{\|A\|}}\\ &= \varkappa(A)\frac{1+\frac{\|B\|}{\|A\|}}{1-\varkappa(A)\frac{\|B\|}{\|A\|}} \leqslant \varkappa(A)\frac{1+\varkappa(A)\frac{\|B\|}{\|A\|}}{1-\varkappa(A)\frac{\|B\|}{\|A\|}} = \varkappa(A) + \frac{2\varkappa^2(A)\frac{\|B\|}{\|A\|}}{1-\varkappa(A)\frac{\|B\|}{\|A\|}}\\ &\leqslant \varkappa(A)\frac{2\varkappa^2(A)\frac{\|B\|}{\|A\|}}{1-1/3} = \varkappa(A) + 3\varkappa^2(A)\frac{\|B\|}{\|A\|},\end{aligned}$$

$$\begin{aligned}\varkappa(A+B) &= 2\,\|A+B\|\,\|X\| \geqslant 2\,\|A\|\left(1-\frac{\|B\|}{\|A\|}\right)\left[\|H\| - \frac{\varkappa(A)\frac{\|B\|}{\|A\|}}{1-\varkappa(A)\frac{\|B\|}{\|A\|}}\,\|H\|\right]\\ &= \varkappa(A)\frac{1-\frac{\|B\|}{\|A\|}}{1-\varkappa(A)\frac{\|B\|}{\|A\|}}\left(1-2\varkappa(A)\frac{\|B\|}{\|A\|}\right)\\ &\geqslant \varkappa(A)\frac{1-\varkappa(A)\frac{\|B\|}{\|A\|}}{1-\varkappa(A)\frac{\|B\|}{\|A\|}}\left(1-2\varkappa(A)\frac{\|B\|}{\|A\|}\right)\\ &= \varkappa(A) - 2\varkappa^2(A)\frac{\|B\|}{\|A\|} \geqslant \varkappa(A) - 3\varkappa^2(A)\frac{\|B\|}{\|A\|},\end{aligned}$$

$$\varkappa(A) - 3\varkappa^2(A)\frac{\|B\|}{\|A\|} \leqslant \varkappa(A+B) \leqslant \varkappa(A) + 3\varkappa^2(A)\frac{\|B\|}{\|A\|}.$$

Thus, we have proved the following theorem.

THEOREM 2 [**1**]. *Let* $\varkappa(A)$ *be finite and*

$$\frac{\|B\|}{\|A\|} < \frac{1}{3\varkappa^2(A)}.$$

Then $\varkappa(A+B)$ *is finite and*

$$|\varkappa(A+B) - \varkappa(A)| < 3\varkappa^2(A)\frac{\|B\|}{\|A\|}.$$

Theorem 2 differs from Theorem 1 by the fact that the size of the matrix A is not mentioned in the formulation. In order to use $\varkappa(A)$ in the analysis of some matrices, it is necessary to know how to compute it. We now present the procedure of computation of $\varkappa(A) = 2\|A\|\,\|H\|$. It suffices to indicate the rule of computing H. It is possible to try to find H by solving the Lyapunov equation $HA+AH = -I$, but this equation can be unsolvable or the solution H can be not positive definite (in these cases, $\varkappa(A) = \infty$). It is convenient to compute H as the improper integral

$$H = \int_0^\infty e^{tC^*} e^{tC}\,dt, \quad C = \frac{A}{\|A\|}$$

which diverges if $\varkappa(A) = \varkappa(C) = \infty$. We introduce the notation

$$H_k = \int_0^{2^k} e^{tC^*} e^{tC}\,dt.$$

Then either $\|H_k\| \to \infty$ if not all eigenvalues of A (consequently, C) lie strictly in the left half-plane (in this case, $\varkappa(A) = \infty$) or $\lim_{k\to\infty} \|H_k\| = H$ and, in this case,

$$\varkappa(A) = 2\,\|C\|\,\|H\| = 2\,\|H\| = 2\lim_{k\to\infty} \|H_k\|$$

since $\|C\| = \left\|\frac{1}{\|A\|}A\right\| = 1$. The matrices H_k can be found by an elegant procedure described below. First, we compute $B_0 = e^C$ in some way, e.g., by using the Taylor series as in §1. To compute

$$H_0 = \int_0^1 e^{tC^*} e^{tC}\,dt$$

we can again use the Taylor series. We denote

$$H_0(s) = \int_0^s e^{tC} e^{tC}\,dt$$

and verify the relations

$$\begin{aligned}
&H_0(0) = 0,\\
&\frac{dH_0(s)}{ds} = e^{sC^*} e^{sC} = M_1(s),\\
&\frac{d^2 H_0(s)}{ds^2} = C^* e^{sC^*} e^{sC} + e^{sC^*} e^{sC} C = C^* M_1(s) + M_1(s) C = M_2(s),\\
&\frac{d^3 H_0(s)}{ds^3} = \frac{dM_2(s)}{ds} = C^* \frac{dM_1(s)}{ds} + \frac{dM_1(s)}{ds} C = C^* M_2(s) + M_2(s) C = M_3(s),\\
&\dots\dots\dots\dots\dots\dots\dots\dots\dots\dots\dots\dots\dots\dots\dots\dots\\
&\frac{d^m H_0(s)}{ds^m} = M_m(s) = C^* M_{m-1}(s) + M_{m-1}(s) C,\\
&\dots\dots\dots\dots\dots\dots\dots\dots\dots\dots\dots\dots\dots\dots\dots\dots
\end{aligned}$$

Let

$$M_j(0) = L_j = \left.\frac{d^j H_0(s)}{ds^j}\right|_{s=0}.$$

By the above arguments, it is obvious that

$$\begin{gathered}
L_1 = I, \quad L_j = C^* L_{j-1} + L_{j-1} C,\\
\|L_j\| \leqslant \|C^*\| \, \|L_{j-1}\| + \|L_{j-1}\| \, \|C\| \leqslant 2 \, \|L_{j-1}\| \leqslant 2^{j-1},\\
H_0 = H_0(1) = L_1 + \frac{1}{2!} L_2 + \frac{1}{3!} L_3 + \dots .
\end{gathered}$$

Denoting $P_0 = I$, $P_k = \frac{1}{k!} L_k$, we can write the computational formulas in the form

$$\begin{gathered}
P_0 = I, \quad R_0 = I,\\
P_{k+1} = \frac{1}{k+1}(C^* P_k + P_k C), \quad R_{k+1} = R_k + P_{k+1};
\end{gathered}$$

moreover,

$$H_0 = \int_0^1 e^{tC^*} e^{tC} \, dt = \lim_{k \to \infty} R_k.$$

The existence of the limit and the estimates for the rate of convergence follow from the inequalities

$$\|L_j\| \leqslant 2 \, \|C\| \, \|L_{j-1}\| = 2 \, \|L_{j-1}\|, \quad \|L_j\| \leqslant 2^{j-1}, \quad \|P_k\| \leqslant \frac{2^{k-1}}{k!}.$$

Having computed B_0 and H_0, we compute $H_1, H_2, \dots, H_j$. The process can be realized by the simple formula

$$H_k = H_{k-1} + B_{k-1}^* H_{k-1} B_{k-1},$$

where $B_k = B_{k-1}^2$. Indeed, from the last formula and the definition of B_0 it follows that

$$B_k = e^{2^k C} = [e^C]^{2^k}.$$

On the other hand,

$$\begin{aligned} H_k &= \int_0^{2^k} e^{tC^*} e^{tC}\,dt = \int_0^{2^{k-1}} e^{tC^*} e^{tC}\,dt + \int_{2^{k-1}}^{2^k} e^{tC^*} e^{tC}\,dt \\ &= \int_0^{2^{k-1}} e^{tC^*} e^{tC}\,dt + \int_0^{2^{k-1}} e^{(t+2^{k-1})C^*} e^{(t+2^{k-1})C}\,dt \\ &= H_{k-1} + e^{2^{k-1}C^*}\left[\int_0^{2^{k-1}} e^{tC^*} e^{tC}\,dt\right] e^{2^{k-1}C} = H_{k-1} + B_{k-1}^* H_{k-1} B_{k-1}. \end{aligned}$$

For concrete processes it is natural that the process described by the system $y' = Ay$ is regarded as unstable if $\varkappa(A)$ is sufficiently large, e.g., $\varkappa(A) > 10^{10}$. Therefore, for a given $\varkappa^*$ (e.g., $\varkappa^* = 10^{10}$) it is natural to try to establish the inequality $\varkappa(A) > \varkappa^*$ or compute the value of $\varkappa(A)$. If $\varkappa(A) < \varkappa^*$ and k_* is the least integer satisfying the inequality

$$2^k \geqslant \varkappa^* \ln[2\sqrt{\varkappa^*}],$$

then

$$\begin{aligned} \|B_{k_*}\| &= \|e^{2^{k_*}C}\| \leqslant \sqrt{\varkappa(C)}e^{-2^{k_*}/\varkappa(C)} = \sqrt{\varkappa(A)}e^{-2^{k_*}/\varkappa(A)} \\ &\leqslant \sqrt{\varkappa^*}e^{-2^{k_*}/\varkappa^*} \leqslant \sqrt{\varkappa^*}e^{-\ln(2\sqrt{\varkappa^*})} = \frac{\sqrt{\varkappa^*}}{2\sqrt{\varkappa^*}} = \frac{1}{2}, \end{aligned}$$

i.e., $\|B_{k_*}\| \leqslant 1/2$. If $\|B_{k_*}\| > 1/2$, then we can guarantee that $\varkappa(A) > \varkappa^*$ and terminate the study with the conclusion that the process is practically unstable. If $\|B_{k_*}\| \leqslant 1/2$, then we continue to compute the matrices

$$H_k = H_{k-1} + B_{k-1}^* H_{k-1} B_{k-1}.$$

For $k > k_* + 1$ we obtain the estimates

$$\begin{gathered} \|B_k\| \leqslant \|B_{k-1}\|^2, \quad \|B_k\| \leqslant \left(\frac{1}{2}\right)^{2^{k-k_*}}, \\ \|H_k - H_{k-1}\| \leqslant \left(\frac{1}{4}\right)^{2^{k-k_*}} \|H_{k-1}\|, \\ \|H_k\| \leqslant \left[1 + \left(\frac{1}{4}\right)^{2^{k-k_*}}\right] \|H_{k-1}\|, \end{gathered}$$

which guarantee the fast convergence in the computation of $H = \lim_{k\to\infty} H_k$. In this case, $\varkappa(A) = \varkappa(C)$ can be computed by the formula $\varkappa(A) = 2\|H\|$. If $\varkappa(A) > \varkappa^*$, then we conclude that the process is practically unstable. Otherwise, we conclude that the process is stable and can indicate $\varkappa(A)$ which characterizes the quality of stability.

§3. Solving of boundary-value problems by the orthogonal sweep method

A naive approach to solving a boundary-value problem and an example which shows that this approach cannot be applied. The orthogonalization of bases formed by partial solutions. Triangular matrices appearing in the orthogonalization. A partial solution to a nonhomogeneous equation which is orthogonal to all solutions to the homogeneous equation at fixed points. The representation of a general solution in terms of basic solutions on different intervals. The use of boundary conditions. The backward sweep method.

We describe a method of numerical solution of the boundary-value problem

$$Lx(0) = \varphi, \quad \frac{dx}{dt} = Ax + f(t), \quad Rx(t) = \psi.$$

The method is based on a representation of a solution in terms of solutions to the Cauchy problems. There are different and formally ideal variants of such a representation. However, in many cases they turn out to be very sensitive to small perturbations and, consequently, cannot be suitable for practical computations by computers. We present ideas of a method which has the reputation of an effective and reliable tool of numerical solution to the problem under consideration.

We begin with the description of the simplest method which, at first glance, is natural and yields a solution to the boundary-value problem. Let L and R consist of $N-k$ and k linearly independent rows respectively. It is clear that $x(0)$ can be represented as

$$x(0) = z_0 + \sum_{j=1}^{k} \alpha_j z_j,$$

where z_0 is a partial solution to the system $Lz_0 = \varphi$ and $z_1, z_2, \dots, z_k$ are linearly independent solutions to the homogeneous equations $Lz_j = 0$. It is convenient to assume that the vectors $z_1, z_2, \dots, z_k$ are orthonormal:

$$(z_j, z_k) = \begin{cases} 1 & \text{for } j = k, \\ 0 & \text{for } j \neq k \end{cases}$$

and the vector z_0 is orthogonal to z_j, $j = 1, 2, \dots, k$:

$$(z_0, z_j) = 0, \quad j = 1, 2, \dots, k.$$

It is not hard to ensure that these conditions be satisfied.

Let the vectors $z_0, z_1, z_2, \dots, z_k$ be the initial conditions for the numerical integration of the following family of the Cauchy problems:

$$\frac{dx_0}{dt} = Ax_0 + f(t), \quad x_0(0) = z_0,$$
$$\frac{dx_j}{dt} = Ax_j, \quad x_j(0) = z_j, \qquad j = 1, 2, \dots k.$$

In Chapter 3, §1, we described procedures that help to obtain detailed tables of the vector-valued functions $x_j(t)$ with a certain accuracy. A linear combination of partial solutions

$$x(t) = x_0(t) + \sum_{j=1}^{k} \alpha_j x_j(t)$$

satisfies (for any $\alpha_1, \alpha_2, \ldots, \alpha_k$) the left boundary condition

$$Lx(0) = Lx_0(0) + \sum_{j=1}^{k} \alpha_j Lx_j(0) = Lz_0 + \sum_{j=1}^{k} \alpha_j Lx_j = \varphi$$

and the differential equations

$$\frac{dx}{dt} = Ax + f(t).$$

Indeed,

$$\begin{aligned}\frac{dx}{dt} &= \frac{dx_0}{dt} + \sum_{j=1}^{k} \alpha_j \frac{dx_j}{dt} = Ax_0 + f + \sum_{j=1}^{k} \alpha_j Ax_j \\ &= A\Big[x_0 + \sum_{j=1}^{k} \alpha_j x_j\Big] + f(t) = Ax(t) + f(t).\end{aligned}$$

Any admissible value $x(0)$ can be represented as such a linear combination and any solution to the equation $\frac{d}{dt}x(t) = Ax + f(t)$ is uniquely determined by $x(0)$. Therefore, any solution satisfying the left boundary condition can be represented as

$$x_0(t) + \sum_{j=1}^{k} \alpha_j x_j(t).$$

If the boundary-value problem is solvable and the solution is unique, then the coefficients $\alpha_1, \alpha_2, \ldots, \alpha_k$ are uniquely found from the condition $Rx(T) = \psi$. The latter is the system of equations

$$\sum_{j=1}^{k} \alpha_j Rx_j(T) = \psi - Rx_0(T)$$

for α_j, $j = 1, 2, \ldots, k$. The system contains exactly k equations since the matrix R has k rows and the vector ψ has k components. The assumption that the system is uniquely solvable is equivalent to the condition that the determinant of the system does not vanish. Determining $\alpha_1, \alpha_2, \ldots, \alpha_k$ from this system, we can, using the tables for $x_0(t), x_1(t), \ldots, x_k(t)$, form a table for the required solution

$$x(t) = x_0(t) + \sum_{j=1}^{k} \alpha_j x_J(t).$$

In some cases, the described procedure leads to a definite answer. However, in many cases, it is impracticable in view of the reasons we will illustrate by an example. Let us find a three-dimensional vector $x(t)$ that is a solution to the boundary-value problem

$$Lx_0 = 3 \equiv \varphi, \quad \frac{dx}{dt} = Ax + f(t), \quad Rx(1) = \begin{pmatrix} 4 \\ 3 \end{pmatrix} \equiv \psi,$$

where

$$A = \begin{bmatrix} 20 & 0 & 0 \\ 0 & 0 & 0 \\ 0 & 0 & -20 \end{bmatrix}, \quad L = (1, 0, -1), \quad R = \begin{bmatrix} 1 & 0 & 0 \\ 1 & -1 & 0 \end{bmatrix}, \quad f(t) = \begin{pmatrix} 1 \\ t \\ 1 \end{pmatrix}.$$

Solving the Cauchy problem

$$\frac{dx_0}{dt} = Ax_0 + \begin{pmatrix} 1 \\ t \\ 1 \end{pmatrix}, \quad x_0(0) = z_0 = \begin{pmatrix} \frac{3}{2} \\ 0 \\ -\frac{3}{2} \end{pmatrix},$$

we find

$$x_0(t) = \begin{pmatrix} \dfrac{31e^{20t} - 1}{20} \\ \dfrac{t^2}{2} \\ \dfrac{-31e^{-20t} - 1}{20} \end{pmatrix}.$$

The partial solutions $x_1(t)$ and $x_2(t)$ to the homogeneous equation $\frac{d}{dt}x_j(t) = Ax_j(t)$ with given $x_1(0)$ and $x_2(0)$ have the form

$$x_1(t) = \begin{pmatrix} \dfrac{1}{2}e^{20t} \\ \dfrac{1}{\sqrt{2}} \\ \dfrac{1}{2}e^{-20t} \end{pmatrix}, \quad x_1(2) = \begin{pmatrix} \dfrac{1}{2}e^{20t} \\ -\dfrac{1}{\sqrt{2}} \\ \dfrac{1}{2}e^{-20t} \end{pmatrix}.$$

Hence we find the solution $x(t) = x_0(t) + \alpha_1 x_1(t) + \alpha_2 x_2(t)$ to the boundary-value problem provided that we find α_1 and α_2 from the right boundary condition $Rx(1) = \psi$. In our case, it has the form

$$(1) \quad \begin{bmatrix} 1 & 0 & 0 \\ 1 & -1 & 0 \end{bmatrix} \left[\begin{pmatrix} \dfrac{31e^{20} - 1}{20} \\ \dfrac{1}{2} \\ \dfrac{-31e^{-20} - 1}{20} \end{pmatrix} + \alpha_1 \begin{pmatrix} \dfrac{1}{2}e^{20} \\ \dfrac{1}{\sqrt{2}} \\ \dfrac{1}{2}e^{-20} \end{pmatrix} + \alpha_2 \begin{pmatrix} \dfrac{1}{2}e^{20} \\ -\dfrac{1}{\sqrt{2}} \\ \dfrac{1}{2}e^{-20} \end{pmatrix} \right] = \begin{pmatrix} 4 \\ 3 \end{pmatrix}.$$

This problem is uniquely solvable (cf. Chapter 1, §11) and the solution $x(t)$ satisfies the estimate

$$\|x(t)\| \leqslant K[\,\|\varphi\| + \|\psi\|\,] + \max_{0 \leqslant t \leqslant 1} \|f(t)\|,$$

where $K = \sqrt{3}$. In the above example, $T = 1$, $\|\varphi\| = |\varphi| = 3$, $\|\psi\|\sqrt{4^2 + 3^2} = 5$, $\|f(t)\| = \sqrt{1 + t^2 + 1} \leqslant \sqrt{3}$, and we can be sure that

$$\|x(t)\| \leqslant \sqrt{3}\,(3 + 5 + \sqrt{3}\,) = 8\sqrt{3} + 3 < 19.$$

The left boundary condition will be satisfied if we choose

$$x(0) = \begin{pmatrix} 3/2 \\ 0 \\ -3/2 \end{pmatrix} + \alpha_1 \begin{pmatrix} 1/2 \\ 1/\sqrt{2} \\ 1/2 \end{pmatrix} + \alpha_2 \begin{pmatrix} 1/2 \\ -1/\sqrt{2} \\ 1/2 \end{pmatrix} \equiv z_0 + \alpha_1 z_1 + \alpha_2 z_2.$$

The vectors

$$z_0 = \begin{pmatrix} 3/2 \\ 0 \\ -3/2 \end{pmatrix}, \quad z_1 = \begin{pmatrix} 1/2 \\ 1/\sqrt{2} \\ 1/2 \end{pmatrix}, \quad z_2 = \begin{pmatrix} 1/2 \\ -1/\sqrt{2} \\ 1/2 \end{pmatrix}$$

are pairwise orthogonal and moreover, z_1 and z_2 are normalized. Moreover, these vectors satisfy the equations

$$Lz_0 \equiv (1,0,-1)\begin{pmatrix} 3/2 \\ 0 \\ -3/2 \end{pmatrix} = 3,$$

$$Lz_1 \equiv (1,0,-1)\begin{pmatrix} 1/2 \\ 1/\sqrt{2} \\ 1/2 \end{pmatrix} = 0,$$

$$Lz_2 \equiv (1,0,-1)\begin{pmatrix} 1/2 \\ -1/\sqrt{2} \\ 1/2 \end{pmatrix} = 0.$$

Hence $Lx(0) = 3$ for any α_1 and α_2. Using the right boundary condition (1), we get the system of two linear equations with respect to α_1 and α_2:

$$\frac{e^{20}}{2}\alpha_1 + \frac{e^{20}}{2}\alpha_2 = 4 - \frac{31e^{20}-1}{20},$$

$$\left(\frac{e^{20}}{2} - \frac{1}{\sqrt{2}}\right)\alpha_1 + \left(\frac{e^{20}}{2} + \frac{1}{\sqrt{2}}\right)\alpha_2 = 3 - \frac{31e^{20}-1}{20} + \frac{1}{2}.$$

We note that e^{20} is a very large number ($e^{20} \approx 0,485 \cdot 10^9$). Consequently, in the first and second equations, the corresponding coefficients are equal with sufficiently high relative accuracy. Therefore, the system is ill conditioned. Replacing the coefficients and the right-hand sides with the corresponding values, we get the system

$$0.2425825977 \cdot 10^9\alpha_1 + 0.2425825977 \cdot 10^9\alpha_2 = -0.7520060488 \cdot 10^9,$$

$$0.2425825970 \cdot 10^9\alpha_1 + 0.2425825984 \cdot 10^9\alpha_2 = -0.7520060493 \cdot 10^9.$$

The coefficients and the right-hand sides of the first and second equations turn out to be indistinguishable within the precision used.

Even if a number of significant digits is sufficiently large, to solve such a system using a computer is not easy and a result can be imprecise due to effects of computational errors. The reason for these difficulties consists in the fact that the original orthonormal basis formed by z_1 and z_2 and used as the initial data for linearly independent solutions $x_1(t)$, $x_2(t)$ to the homogeneous equation $\frac{dx}{dt} = Ax$ cannot provide the orthogonality of the vectors $x_1(t)$, $x_2(t)$ for all t. These vectors increase and the angle between them decreases as t increases. The basis formed by these vectors begins to " flatten". Furthermore, the angles between $x_1(t)$, $x_0(t)$ and between $x_2(t)$, $x_0(t)$ also decreases, where x_0 is a partial solution to the non-homogeneous equation $\frac{dx}{dt} = Ax + f(t)$. As t increases, all three vectors $x_0(t)$, $x_1(t)$, and $x_2(t)$ rotate, lengthen, and become "more collinear" to the eigenvector corresponding to the greatest eigenvalue $a = 20$ of the matrix A (this eigenvector has components (1, 0, 0)). To overcome "flattening", one can orthogonalize the basis several times. Preliminarily, the entire interval $(0, T)$ is divided into several subintervals so that after the multiplication by $\|A\|$ the length of each subinterval is of about 1 (or 2 or 3): $\|A\|T/n \approx 1$–3.

As above, on the left subinterval $0 \leqslant t \leqslant T/n$ we define a partial solution $x_0^{(1)}(t)$ to the nonhomogeneous system $\dfrac{dx}{dt} = Ax + f(t)$ and solutions

$$x_1^{(1)}(t), x_2^{(1)}(t), \dots, x_k^{(1)}(t)$$

to the homogeneous system $\dfrac{dx}{dt} = Ax$ with the initial data $x_0^{(1)}(0) = z_0$ and $x_1^{(1)}(0) = z_1$, $x_2^{(1)}(0) = z_2, \dots, x_k^{(1)}(0) = z_k$.

We assume that the initial values $z_0, z_1, \dots, z_k$ are orthonormal. This assumption is valid in the above example. We define the vectors $z_0^{(1)}, z_1^{(1)}, z_2^{(1)}, \dots, z_k^{(1)}$ so that $x_0(T/n), x_1(T/n), \dots, x_k(T/n)$ obtained by numerical integration are linear combinations of the form

$$\begin{aligned}
x_1^{(1)}\left(\frac{T}{n}\right) &= \omega_{11}^{(1)} z_1^{(1)},\\
x_2^{(1)}\left(\frac{T}{n}\right) &= \omega_{21}^{(1)} z_1^{(1)} + \omega_{22}^{(1)} z_2^{(1)},\\
&\dots\dots\dots\dots\dots\dots\dots\dots\dots\dots\\
x_k^{(1)}\left(\frac{T}{n}\right) &= \omega_{k1}^{(1)} z_1^{(1)} + \omega_{k2}^{(1)} z_2^{(1)} + \dots + \omega_{kk}^{(1)} z_k^{(1)},\\
x_0^{(1)}\left(\frac{T}{n}\right) &= \omega_{01}^{(1)} z_1^{(1)} + \omega_{02}^{(1)} z_2^{(1)} + \dots + \omega_{0k}^{(1)} z_k^{(1)} + z_0^{(1)}.
\end{aligned}$$

The coefficients $\omega_{ij}^{(1)}$ are taken so that all vectors $z_j^{(1)}$ are pairwise orthogonal ($(z_i^{(1)}, z_j^{(1)}) = 0$ if $i \neq j$) and all $z_j^{(1)}$, except $z_0^{(1)}$, are orthonormal ($(z_j^{(1)}, z_1^{(1)}) = 1$, $j = 1, 2, \dots, k$). To choose the coefficients $\omega_{ij}^{(1)}$ one can apply the Gram–Schmidt orthogonalization process which is known from the standard course of linear algebra.

At the first step, we normalize the vector $x_1^{(1)}(T/n)$. To do this, we compute

$$\omega_{11} = \left\| x_1^{(1)}\left(\frac{T}{n}\right) \right\| = \sqrt{\left(x_1^{(1)}\left(\frac{T}{n}\right), x_1^{(1)}\left(\frac{T}{n}\right) \right)}$$

and set

$$z_1^{(1)} = \frac{1}{\omega_{11}} x_1^{(1)}\left(\frac{T}{n}\right).$$

At the second step, we construct the projection p_2 of $x_2^{(1)}(T/n)$ on $z_1^{(1)}$:

$$p_2 = \omega_{21} z_1^{(1)}, \quad \omega_{21} = \left(x_2^{(1)}\left(\frac{T}{n}\right), z_1^{(1)} \right).$$

Since $x_2^{(1)}(T/n) - p_2 = x_2^{(1)}(T/n) - \omega_{21} z_1^{(1)}$ is orthogonal to $z_1^{(1)}$, for the second vector $z_2^{(1)}$ we can take the vector obtained by the normalization of $x_2^{(1)}(T/n) - p_2$:

$$z_2^{(1)} = \frac{1}{\omega_{22}} \left\{ x_1^{(1)}\left(\frac{T}{n}\right) - \omega_{21} z_1^{(1)} \right\},$$

$$\omega_{22} = \sqrt{\left(x_2^{(1)}\left(\frac{T}{n}\right) - p_2, x_2^{(1)}\left(\frac{T}{n}\right) - p_2 \right)} = \sqrt{\left(x_2^{(1)}\left(\frac{T}{n}\right), x_2^{(1)}\left(\frac{T}{n}\right) \right) - \omega_{21}^2}.$$

At the third step, we construct the projection p_3 of the vector $x_3^{(1)}(T/n)$ on the space spanned by the orthonormal vectors $z_1^{(1)}$ and $z_2^{(1)}$. The projection p_3 is composed of the projections of $x_3^{(1)}(T/n)$ on $z_1^{(1)}$ and on $z_2^{(1)}$ (because $z_1^{(1)}$ and $z_2^{(1)}$ are orthogonal):

$$p_3 = \omega_{31} z_1^{(1)} + \omega_{32} z_2^{(1)}, \quad \omega_{31} = \left(x_3^{(1)}\left(\frac{T}{n}\right), z_1^{(1)} \right), \quad \omega_{32} = \left(x_3^{(1)}\left(\frac{T}{n}\right), z_2^{(1)} \right).$$

Since $x_3^{(1)}(T/n) - p_3 = x_3^{(1)}(T/n) - \omega_{31} z_1^{(1)} - \omega_{32} z_2^{(1)}$ is orthogonal to $z_1^{(1)}$ and $z_2^{(1)}$, for the third vector $z_3^{(1)}$ of the basis we can take the vector obtained by the normalization of $x_3^{(1)}(T/n) - p_3$:

$$\begin{gathered} z_3^{(1)} = \frac{1}{\omega_{33}} \left\{ x_3^{(1)}\left(\frac{T}{n}\right) - \omega_{31} z_1^{(1)} - \omega_{32} z_2^{(1)} \right\}, \\ \omega_{33} = \left\| x_3^{(1)}\left(\frac{T}{n}\right) - p_3 \right\| = \sqrt{\left(x_3^{(1)}\left(\frac{T}{n}\right), x_3^{(1)}\left(\frac{T}{n}\right) \right) - \omega_{31}^2 - \omega_{32}^2}. \end{gathered}$$

Continuing the orthogonalization process, we consecutively obtain

$$\begin{aligned} z_4^{(1)} &= \frac{1}{\omega_{44}} \left\{ x_4^{(1)}\left(\frac{T}{n}\right) - \omega_{41} z_1^{(1)} - \omega_{42} z_2^{(1)} - \omega_{43} z_3^{(1)} \right\}, \\ z_5^{(1)} &= \frac{1}{\omega_{55}} \left\{ x_5^{(1)}\left(\frac{T}{n}\right) - \omega_{51} z_1^{(1)} - \omega_{52} z_2^{(1)} - \omega_{53} z_3^{(1)} - \omega_{54} z_4^{(1)} \right\}, \\ &\cdots\cdots\cdots\cdots\cdots\cdots\cdots\cdots\cdots\cdots\cdots\cdots \\ z_k^{(1)} &= \frac{1}{\omega_{kk}} \left\{ x_k^{(1)}\left(\frac{T}{n}\right) - \omega_{k1} z_1^{(1)} - \omega_{k2} z_2^{(1)} - \cdots - \omega_{k,k-1} z_{k-1}^{(1)} \right\}. \end{aligned}$$

We note that all ω_{jj}, which appear in the process as divisors, do not vanish. Indeed, if $\omega_{jj} = 0$ for some j, then

$$x_j^{(1)}\left(\frac{T}{n}\right) = \omega_{j1} z_1^{(1)} + \omega_{j2} z_2^{(1)} + \cdots + \omega_{j\,j-1} z_{j-1}^{(1)}.$$

Hence, recalling the definition of $z_1^{(1)}, z_2^{(1)}, \ldots, z_{j-1}^{(1)}$, we conclude that $x_j^{(1)}(T/n)$ is a linear combination of $x_1^{(1)}(T/n), x_2^{(1)}(T/n), \ldots, x_{j-1}^{(1)}(T/n)$. By the Liouville theorem, $x_j^{(1)}(0)$ is also a linear combination of $x_1^{(1)}(0), x_2^{(1)}(0), \ldots, x_{j-1}^{(1)}(0)$. However, this contradicts the fact that $x_1^{(1)}(0), x_2^{(1)}(0), \ldots, x_{j-1}^{(1)}(0)$ are orthonormal by construction. In fact, ω_{jj} can be very small in practical realization, but if T/n satisfies the condition $T\|A\|/n \approx 1$–3, this cannot happen.

As a result of the orthogonalization process, we obtain the following relations connecting the vectors $x_i^{(1)}(T/n)$ and $z_j^{(1)}$:

$$\begin{aligned}
x_1^{(1)}\left(\frac{T}{n}\right) &= \omega_{11} z_1^{(1)}, \\
x_2^{(1)}\left(\frac{T}{n}\right) &= \omega_{21} z_1^{(1)} + \omega_{22} z_2^{(1)}, \\
x_3^{(1)}\left(\frac{T}{n}\right) &= \omega_{31} z_1^{(1)} + \omega_{32} z_2^{(1)} + \omega_{33} z_3^{(1)}, \\
&\dots\dots\dots\dots\dots\dots \\
x_k^{(1)}\left(\frac{T}{n}\right) &= \omega_{k1} z_1^{(1)} + \omega_{k2} z_2^{(1)} + \dots + \omega_{kk} z_k^{(1)}.
\end{aligned}$$

It is convenient to use the matrix notation for this relation.

Let $\widehat{X}^{(1)}$ and $\widehat{Z}^{(1)}$ be $N \times k$-matrices whose columns are the vectors $x_j^{(1)}(T/n)$ and $z_j^{(1)}$ respectively:

$$\begin{aligned}
\widehat{X}^{(1)} &= \left[x_1^{(1)}\left(\frac{T}{n}\right), x_2^{(1)}\left(\frac{T}{n}\right), \dots, x_k^{(1)}\left(\frac{T}{n}\right)\right], \\
\widehat{Z}^{(1)} &= \left[z_1^{(1)}, z_2^{(1)}, \dots, z_k^{(1)}\right].
\end{aligned}$$

Let $\widehat{\Omega}^{(1)}$ denote the upper triangular $k \times k$-matrix

$$\widehat{\Omega}^{(1)} = \begin{bmatrix} \omega_{11} & \omega_{21} & \omega_{31} & \cdots & \omega_{k1} \\ & \omega_{22} & \omega_{32} & \cdots & \omega_{k2} \\ & & \omega_{33} & \cdots & \omega_{k3} \\ & 0 & & \ddots & \vdots \\ & & & & \omega_{kk} \end{bmatrix}.$$

We see that $\widehat{X}^{(1)} = \widehat{Z}^{(1)} \times \widehat{\Omega}^{(1)}$. The condition that the columns of $\widehat{Z}^{(1)}$ are orthonormal can be written in the matrix notation as follows:

$$[\widehat{Z}^{(1)}]^* \widehat{Z}^{(1)} = I_k.$$

We take a solution $x_0^{(1)}(T/n)$ to the nonhomogeneous equation and compute its projection p_0 on the subspace spanned by the vectors $z_1^{(1)}, z_2, \dots, z_k^{(1)}$:

$$p_0 = \omega_{01} z_1^{(1)} + \omega_{02} z_2^{(1)} + \dots + \omega_{0k} z_k^{(1)},$$

where

$$\begin{aligned}
\omega_{01} &= \left(x_0^{(1)}\left(\frac{T}{n}\right), z_1^{(1)}\right), \\
\omega_{02} &= \left(x_0^{(1)}\left(\frac{T}{n}\right), z_2^{(1)}\right), \\
&\dots\dots\dots\dots\dots\dots \\
\omega_{0k} &= \left(x_0^{(1)}\left(\frac{T}{n}\right), z_k^{(1)}\right).
\end{aligned}$$

To the orthonormal family of the vectors $z_1^{(1)}, z_2^{(1)}, \dots, z_k^{(1)}$ we add the vector $z_0^{(1)} = x_0^{(1)}(T/n) - p_0$, where

$$x_0^{(1)}\left(\frac{T}{n}\right) = \omega_{01} z_1^{(1)} + \omega_{02} z_2^{(1)} + \dots + \omega_{0k} z_k^{(1)} + z_0^{(1)}.$$

Let $X^{(1)}$ and $Z^{(1)}$ be $N \times (k+1)$-matrices with columns $x_1^{(1)}(T/n), \dots, x_k^{(1)}(T/n)$, $x_0^{(1)}(T/n)$ and $z_1^{(1)}, \dots, z_k^{(1)}, z_0^{(1)}$ respectively:

$$X^{(1)} = \left[x_1^{(1)}\left(\frac{T}{n}\right), \dots, x_k^{(1)}\left(\frac{T}{n}\right), x_0^{(1)}\left(\frac{T}{n}\right)\right] = \left[\widehat{X}^{(1)}, x_0^{(1)}\left(\frac{T}{n}\right)\right],$$
$$Z^{(1)} = [z_1^{(1)}, z_2^{(1)}, \dots, z_k^{(1)}, z_0^{(1)}] = [\widehat{Z}^{(1)}, z_0^{(1)}].$$

Then

$$X^{(1)} = Z^{(1)}\Omega^{(1)}, \quad \Omega^{(1)} = \left[\begin{array}{ccc|c} & \widehat{\Omega}^{(1)} & & \omega_{01} \\ & & & \omega_{02} \\ & & & \vdots \\ & & & \omega_{0k} \\ \hline 0 & \cdots & 0 & 1 \end{array}\right];$$

here $\Omega^{(1)}$ is an upper triangular $(k+1) \times (k+1)$-matrix. Moreover,

$$[Z^{(1)}]^* Z^{(1)} = \left[\begin{array}{c|c} I_k & 0 \\ \hline 0 & * \end{array}\right].$$

We construct the solutions $x_0^{(2)}(t), x_1^{(2)}(t), \dots, x_k^{(2)}(t)$ to the following Cauchy problems on the interval $T/n < t \leqslant (2T)/n$:

$$\frac{dx_0^{(2)}}{dt} = Ax_0^{(2)} + f(t), \quad x_0^{(2)}\left(\frac{T}{n}\right) = z_0^{(1)},$$
$$\frac{dx_j^{(2)}}{dt} = Ax_j^{(2)}, \quad x_j^{(2)}\left(\frac{T}{n}\right) = z_j^{(1)}, \quad j = 1, 2, \dots, k,$$

and introduce the $N \times (k+1)$-matrix

$$X^{(2)} = \left[x_1^{(2)}\left(\frac{2T}{n}\right), x_2^{(2)}\left(\frac{2T}{n}\right), \dots, x_k^{(2)}\left(\frac{2T}{n}\right), x_0^{(2)}\left(\frac{2T}{n}\right)\right].$$

We orthogonalize the vectors $x_1^{(2)}(2T/n), x_2^{(2)}(2T/n), \dots, x_k^{(2)}(2T/n), x_0^{(2)}(2T/n)$ in the same way as the vectors $x_1^{(1)}(T/n), x_2^{(1)}(T/n), \dots, x_k^{(1)}(T/n), x_0^{(1)}(T/n)$ and represent the result as the matrix decomposition $X^{(2)} = Z^{(2)}\Omega^{(2)}$, where the columns of the matrix $Z^{(2)}$ are orthogonal and are normalized (except the last column):

$$[Z^{(2)}]^* Z^{(2)} = \left[\begin{array}{c|c} I_k & 0 \\ \hline 0 & * \end{array}\right],$$

and $\Omega^{(2)}$ is an upper triangular matrix

$$\Omega^{(2)} = \left[\begin{array}{cccc|c} & & & & \times \\ & & & & \times \\ & & \widehat{\Omega}^{(2)} & & \vdots \\ & & & & \times \\ \hline 0 & 0 & \cdots & 0 & 1 \end{array}\right].$$

If a solution $x(t)$ to the nonhomogeneous system $\dfrac{dx}{dt} = Ax + f$ satisfying the left boundary condition $Lx(0) = \varphi$ is given by the linear combination

$$x(t) = \alpha_1^{(1)} x_1^{(1)}(t) + \alpha_2^{(1)} x_2^{(1)}(t) + \cdots + \alpha_k^{(1)} x_k^{(1)}(t) + x_0^{(1)}(t)$$

for $0 < t \leqslant T/n$, then

$$x\left(\frac{T}{n}\right) = \alpha_1^{(1)} x_1\left(\frac{T}{n}\right) + \cdots + \alpha_k^{(1)} x_k\left(\frac{T}{n}\right) + x_0\left(\frac{T}{n}\right) = X^{(1)} \begin{pmatrix} \alpha_1^{(1)} \\ \alpha_2^{(1)} \\ \vdots \\ \alpha_k^{(1)} \\ 1 \end{pmatrix} \equiv X^{(1)} \alpha^{(1)}.$$

Let the vector $\alpha^{(1)}$ have components $\alpha_1^{(1)}, \alpha_2^{(1)}, \ldots, \alpha_k^{(1)}, \alpha_0^{(1)}$. The formula $X^{(1)} = Z^{(1)}\Omega^{(1)}$ implies $x(T/n) = Z^{(1)}\alpha^{(2)}$, where $\alpha^{(2)} = \Omega^{(1)}\alpha^{(1)}$, i.e.,

$$x(T/n) = \alpha_1^{(2)} z_1^{(1)} + \alpha_2^{(2)} z_2^{(1)} + \cdots + \alpha_k^{(2)} z_k^{(1)} + z_0^{(1)}.$$

On the interval $T/n < t \leqslant 2T/n$ the partial solutions

$$x_1^{(2)}(t), x_2^{(2)}(t), \ldots, x_k^{(2)}(t), x_0^{(2)}(t)$$

to the homogeneous and nonhomogeneous equations were constructed from the initial data $z_1^{(1)}, z_2^{(1)}, \ldots, z_k^{(1)}, z_0^{(1)}$. Therefore, if we extend by continuity the solution

$$x(t) = \alpha_1^{(2)} x_1^{(2)}(t) + \alpha_2^{(2)} x_2^{(2)}(t) + \cdots + \alpha_k^{(2)} x_k^{(2)}(t) + x_0^{(2)}(t)$$

to the left endpoint $t = T/n$ of the interval, then the extended function coincides with the solution $x(t)$ constructed on $0 < t \leqslant T/n$. On each of the intervals $0 < t \leqslant T/n$ and $T/n < t \leqslant 2T/n$, we choose the basis of solutions so that the vectors of the basis are pairwise orthogonal at the left endpoint of the interval. Since the length T/n of the intervals is not too large, the orthogonality property cannot fail significantly at the remaining points.

On different intervals, the same solution $x(t)$ has representations with different coefficients:

$$\begin{aligned} x(t) &= \alpha_1^{(1)} x_1^{(1)}(t) + \alpha_2^{(1)} x_2^{(1)}(t) + \cdots + \alpha_k^{(1)} x_k^{(1)}(t) + x_0^{(1)}(t), \quad 0 < t \leqslant T/n, \\ x(t) &= \alpha_1^{(2)} x_1^{(2)}(t) + \alpha_2^{(2)} x_2^{(2)}(t) + \cdots + \alpha_k^{(2)} x_k^{(2)}(t) + x_0^{(2)}(t), \quad T/n < t \leqslant 2T/n \end{aligned}$$

since the bases are different on these intervals, but the coefficients are connected by the relation $\alpha^{(2)} = \Omega^{(1)}\alpha^{(1)}$.

We assume that on the interval $(l-2)T/n < t \leqslant (l-2)T/n$ we have constructed the solutions $x_1^{(l-1)}(t), x_2^{(l-1)}(t), \dots, x_k^{(l-1)}(t), x_0^{(l-1)}(t)$ to the equations

$$\frac{dx_j^{(l-1)}}{dt} = Ax_j^{(l-1)}, \quad 1 \leqslant j \leqslant k, \qquad \frac{dx_0^{(l-1)}}{dt} = Ax_0^{(l-1)} + f(t)$$

such that, on this interval, any solution to the nonhomogeneous equation $\frac{dx}{dt} = Ax + f(t)$ is represented as

$$x(t) = \alpha_0^{(l-1)} x_0^{(l-1)} + \sum_{j=1}^{k} \alpha_j^{(l-1)} x_j^{(l-1)}.$$

Consider the matrix $X^{(l-1)}$ with the columns $x_1^{(l-1)}, x_2^{(l-1)}, \dots, x_k^{(l-1)}, x_0^{(l-1)}$:

$$X^{(l-1)} = [x_1^{(l-1)}, x_2^{(l-1)}, \dots, x_k^{(l-1)}, x_0^{(l-1)}]$$

and represent it as the matrix product

$$X^{(l-1)} = Z^{(l-1)} \Omega^{(l-1)},$$

where $Z^{(l-1)}$ is a matrix whose columns are pairwise orthogonal and, except the last column, are normalized:

$$[Z^{(l-1)}]^* Z^{(l-1)} = \left[\begin{array}{ccc|c} & & & 0 \\ & I_k & & \vdots \\ & & & 0 \\ \hline 0 & \cdots & 0 & * \end{array}\right],$$

and $\Omega^{(l-1)}$ is an upper triangular matrix:

$$\Omega^{(l-1)} = \left[\begin{array}{cccc|c} & \widehat{\Omega}^{(l-1)} & & & \times \\ & & & & \times \\ & & & & \vdots \\ & & & & \times \\ \hline 0 & 0 & \cdots & 0 & 1 \end{array}\right].$$

We denote the columns of $Z^{(l-1)}$ by $z_1^{(l-1)}, z_2^{(l-1)}, \dots, z_k^{(l-1)}, z_0^{(l-1)}$. We recall that such a decomposition can be realized, for example, with the help of process.

Sometimes (cf. §5) it is convenient to eliminate the last columns $x_0^{(l-1)}$ and $z_0^{(l-1)}$ from the matrices $X^{(l-1)}$ and $Z^{(l-1)}$. Such (shortened) matrices will be denoted by $\widehat{X}^{(l-1)}$ and $\widehat{Z}^{(l-1)}$. In addition,

$$\widehat{X}^{(l-1)} = \widehat{Z}^{(l-1)} \widehat{\Omega}^{(l-1)}, \quad [\widehat{Z}^{(l-1)}]^* [\widehat{Z}^{(l-1)}] = I_k$$

and $\widehat{\Omega}^{(l-1)}$ is an upper triangular matrix.

On the interval $(t-1)T/n < t \leqslant lT/n$, let $x_1^{(l)}(t), x_2^{(l)}(t), \dots, x_k^{(l)}(t), x_0^{(l)}(t)$ be the solutions to the Cauchy problems

$$\frac{dx_0^{(l)}}{dt} = Ax_0^{(l)} + f(t), \quad x_0^{(l)}\left(\frac{(l-1)T}{n}\right) = z_0^{(l-1)},$$
$$\frac{dx_j^{(l)}}{dt} = Ax_j^{(l)}, \quad x_j^{(l)}\left(\frac{(l-1)T}{n}\right) = z_j^{(l-1)}, \quad 1 \leqslant j \leqslant k.$$

If a solution $x(t)$ to the system $\dfrac{dx}{dt} = Ax + f(t)$ is represented as

$$x(t) = x_0^{(l-1)}(t) + \sum_{j=1}^{k} \alpha_j^{(l-1)} x_j^{(l-1)}(t)$$

on the interval $(l-2)T/n < t \leqslant (l-1)T/n$ and

$$x(t) = x_0^{(l)}(t) + \sum_{j=1}^{k} \alpha_j^{(l)} x_j^{(l)}(t)$$

on the interval $(l-1)T/n < t \leqslant lT/n$, then both representations yield the same value $x((l-1)T/n)$ at $t = (l-1)T/n$, i.e.,

$$x\left(\frac{(l-1)T}{n}\right) = x_0^{(l-1)}\left(\frac{(l-1)T}{n}\right) + \sum_{j=1}^{k} \alpha_j^{(l-1)} x_j^{(l-1)}\left(\frac{(l-1)T}{n}\right) = X^{(l-1)}\alpha^{(l-1)},$$
$$x\left(\frac{(l-1)T}{n}\right) = z_0^{(l)} + \sum_{j=1}^{k} \alpha_j^{(l)} z_j^{(l)} = Z^{(l)}\alpha^{(l)},$$
$$X^{(l-1)}\alpha^{(l-1)} = Z^{(l-1)}\Omega^{(l-1)}\alpha^{(l-1)} = Z^{(l-1)}\alpha^{(l)}.$$

Hence the vectors

$$\alpha^{(l-1)} = \begin{pmatrix} \alpha_1^{(l-1)} \\ \alpha_2^{(l-1)} \\ \vdots \\ \alpha_k^{(l-1)} \\ \alpha_0^{(l-1)} \end{pmatrix} \qquad \alpha^{(l)} = \begin{pmatrix} \alpha_1^{(l)} \\ \alpha_2^{(l)} \\ \vdots \\ \alpha_k^{(l)} \\ \alpha_0^{(l)} \end{pmatrix}$$

are connected by the triangular matrix $\Omega^{(l)}$ as follows: $\alpha^{(l)} = \Omega^{(l-1)}\alpha^{(l-1)}$.

The described procedure allows us to construct basic solutions $x_0^{(l)}, x_j^{(l)}$, $j = 1, 2, \dots, k$ for $l = 1, 2, \dots, n$, i.e., on each of n intervals of length T/n which form the partition of the domain of the required solution. Moreover, we simultaneously find the triangular matrices $\Omega^{(1)}, \Omega^{(2)}, \dots, \Omega^{(n-1)}$ and the matrix

$$X^{(n)} = [x_1^{(n)}(T), x_2^{(n)}(T), \dots, x_k^{(n)}(T), x_0^{(n)}(T)].$$

On the last interval $(n-1)T/n < t \leqslant T$, any solution $x(t)$ to the system $\dfrac{dx}{dt} = Ax + f(t)$ satisfying the boundary condition $Lx(0) = \varphi$ at $t = 0$ admits the representation

$$x(t) = x_0^{(n)}(t) + \sum_{j=1}^{k} \alpha_j^{(n)} x_j^{(n)}(t).$$

Therefore,

$$x(T) = x_0^{(n)}(T) + \sum_{j=1}^{k} \alpha_j^{(n)} x_j^{(n)}(T) = X^{(n)}\alpha^{(n)}.$$

It is convenient to orthogonalize the columns of the matrix $Z^{(n)}$ by setting $X^{(n)} = Z^{(n)}\Omega^{(n)}$ and assuming that the columns of $Z^{(n)}$ are pairwise orthogonal and, except the last column, are normalized.

In the orthogonalization process, we obtain the upper triangular matrix $\Omega^{(n)}$ whose lower right element is equal to 1. (The orthogonalization process is similar to the above orthogonalization of the columns of $X^{(l)}$, $1 \leqslant l \leqslant n-1$.) Moreover,

$$x(T) = X^{(n)}\alpha^{(n)} = Z^{(n)}\Omega^{(n)}\alpha^{(n)} = Z^{(n)}\alpha^{(n+1)}.$$

The boundary-value problem under consideration is uniquely solvable. Hence the coefficient vector $\alpha^{(n+1)}$ can be uniquely found from the boundary condition

$$Rx(T) = RZ^{(n)}\alpha^{(n+1)} = \psi, \quad \alpha^{(n+1)} = [RZ^{(n)}]^{-1}\psi.$$

The unique solvability implies the invertibility of the matrix $[RZ^{(n)}]^{-1}$, i.e., its nonsingularity. (Verify that this matrix is a square $k \times k$-matrix.) The process of defining $X^{(1)}, X^{(2)}, \dots, X^{(n)}, \Omega^{(1)}, \Omega^{(2)}, \dots, \Omega^{(n)}$ is carried out consecutively from left to right and is called the direct move of the sweep method or direct sweep for brevity. The term "sweep" can be explained by the fact that the recurrence formulas connecting basic solutions on successive intervals provide the validity of the left boundary condition $Lx(0) = \varphi$ for the linear combination

$$x(t) = x_0^{(l)}(t) + \sum_{j=1}^{k} \alpha_j^{(l)} x_j^{(l)}(t),$$

i.e., they "sweep" this boundary condition from one interval to the next interval from the right. Each step of such a "sweep" is realized by the orthogonalization of basic solutions. Therefore, the procedure is called the orthogonal sweep method.

To determine the solution $x(t)$ for all t $(0 \leqslant t \leqslant T)$, it is necessary to find consecutively $\alpha^{(n)}, \alpha^{(n-1)}, \dots, \alpha^{(1)}$ from the recurrence relations

$$\alpha^{(j)} = [\Omega^{(j)}]^{-1}\alpha^{(j+1)}, \quad j = n, \dots, 1,$$

and, on the interval $(l-1)T/n < t \leqslant lT/n$, find $x(t)$ as the linear combination

$$x(t) = x_0^{(l)}(t) + \sum_{j=1}^{k} \alpha_j^{(l)} x_j^{(l)}(t)$$

of the basic solutions $x_0^{(l)}(t), x_1^{(l)}(t), \dots, x_l^{(l)}$, which have been computed by the direct sweep. The computation of $\alpha_j^{(l)}$ is carried out from right to left and is called the "backward sweep."

Recently a modification of the orthogonal sweep method was suggested in [**17**], where the Gram–Schmidt orthogonalization process is replaced by the decomposition of a matrix into the product of two matrices, one of which has orthonormal columns and another is an upper triangular matrix. Such a procedure is called the QR-decomposition. It was developed in detail in numerical methods of linear algebra and is based on orthogonal reflections (the Householder transformations). It is stable with respect to round-off errors appearing in computer realizations of the method. In this version, the stability with respect to the computational errors can be established under the assumption that the boundary-value problem has the Green matrices $G(t,s)$, $G_L(t)$, and $G_R(t)$ bounded by a not-too-large constant K (cf. Chapter 1, §§10,11). Although the proof is not too complicated, it is cumbersome and we omit it.

To conclude the section, we note that the orthogonal sweep method can be used without essential changes in order to solve boundary-value problems for linear equations with not only constant but variable coefficients as well.

We emphasize that for equations with constant coefficients and the zero right-hand side $f(t) = 0$ there are versions of the orthogonal sweep method which are less laborious [**19, 20**]. In these versions it suffices to carry out the QR-decomposition only $\log_2 n$ times (instead of n times which is required by the above procedure).

§4. The two-way sweep for the computation of the Green matrices

Properties of the Green matrices for boundary-value problems (reminder). The transmission of boundary conditions from the right boundary to the left boundary by two-way sweep. The use of a result for the computation of the Green matrices.

The integral representation of solutions to the boundary-value problems in terms of the Green matrices is not often used in computations because to compute the Green matrices is more laborious than to find a specific solution. However, some questions require the knowledge of the Green matrices $G_R(t)$, $G(t,s)$, and $G_L(t)$ which appear in the representation

$$x(t) = G_L(t)\varphi + \int_0^T G(t,s)f(s)ds + G_R(t)\psi$$

of a solution $x(t)$ to the boundary-value problem

$$\begin{aligned} &\frac{dx}{dt} = Ax + f, \quad 0 \leqslant t \leqslant T, \\ &Lx(0) = \varphi, \quad Rx(T) = \psi. \end{aligned}$$

We present a version of the orthogonal sweep method which is adapted for the computation of $G_L(t)$, $G(t,s)$, and $G_R(t)$. As in §3, we assume that A is an $N \times N$-matrix and the rows of the matrices L and R are linearly independent. The matrix L has k rows and N columns, and the matrix R has $N-k$ rows and N columns. We recall (cf. Chapter 1, §10) that $G_L(t)$, $G(t,s)$, and $G_R(t)$ are matrix-valued solutions to the following boundary-value problems:

$$\begin{gathered} LG_L(0) = I_k,\ \frac{d}{dt}G_L(t) = AG_L(t),\ RG_L(t) = O_{(N-k)\times k}, \\ LG_R(0) = O_{k\times(N-k)},\ \frac{d}{dt}G_R(t) = AG_R(t),\ RG_R(T) = I_{N-k}, \\ \frac{d}{dt}G(t,s) = AG(t,s), \quad t \neq s, \\ G(s+0,0) - G(s-0,0) = I_N, \\ LG(0,s) = O_{k\times N}, \quad RG(0,s) = O_{(N-k)\times N}, \end{gathered}$$

where $0 \leqslant t \leqslant T$, I_m denotes the identity matrix of order m, and $O_{m\times n}$ denotes the zero $m \times n$-matrix.

Below we will omit the subscripts in the notation of the zero matrices.

Let $\widehat{X}(t)$ be an $N \times (N-k)$-matrix whose linearly independent columns are formed by the coordinates of the vectors satisfying the vector equation $\dfrac{dx}{dt} = Ax$ and the homogeneous left boundary condition $Lx(0) = 0$, so that

$$\frac{d}{dt}\widehat{X}(t) = A\widehat{X}(t), \quad L\widehat{X}(0) = O.$$

Any column of the matrices $G(t,s)$ and $G_R(t)$ is a linear combination of columns of $\widehat{X}(t)$. Therefore, the Green matrices $G(t,s)$ and $G_R(t)$ are represented in the form

$$G(t,s) = \widehat{X}(t)M(s), \quad G_R(t) = \widehat{X}(t)M_R, \qquad t \leqslant s.$$

In this representation, the matrices $M(s)$ and M_R have N and $N-k$ columns respectively. The number of rows of the matrices $M(s)$ and M_R is equal to $N-k$. Describing the orthogonal sweep method in §3, we emphasized that on each subinterval $(j-1)T/n \leqslant t \leqslant lT/n$ of the interval $[0,T]$ it is convenient to choose its own matrix $\widehat{X}(t)$ of basic solutions satisfying the left boundary condition. The matrix $\widehat{X}(t)$ acting on the jth subinterval is denoted by $\widehat{X}^{(j)}(t)$. It is obvious that the choice of the matrices $M(s)$ and M_R must depend on the choice of a subinterval. These matrices are supplied with superscripts indicating the number of the corresponding subinterval. On the subinterval $(j-1)T/n \leqslant t \leqslant jT/n$, we use the representation

$$G(t,s) = \widehat{X}^{(j)}(t)M^{(j)}(s), \quad t \leqslant s, \qquad G_R(t) = \widehat{X}^{(j)}M_R^{(j)}.$$

We recall (cf. §3) that the boundary values of $\widehat{X}^{(j)}(t)$ at the left endpoints $t = (j-1)T/n$ of the intervals $(j-1)T/n < t \leqslant jT/n$ were denoted by $Z^{(j)}$. Moreover, to find these values we used the decomposition of the matrix

$$\widehat{X}^{(j-1)} \equiv \widehat{X}^{(j-1)}\left(\frac{(j-1)T}{n}\right) \equiv e^{\frac{T}{n}A}\widehat{Z}^{(j-1)}$$

in the product

$$e^{\frac{T}{n}A}\widehat{Z}^{(j-1)} = \widehat{Z}^{(j)}\widehat{\Omega}^{(j-1)},$$

where the columns of the matrix $\widehat{Z}^{(j)}$ are orthonormal and the matrix $\widehat{\Omega}^{(j)}$ is upper triangular ($[\widehat{Z}^{(j)}]^*\widehat{Z}^{(j)} = I_{N-k}$).

If $(j-1)T/n < s$, then

$$\widehat{X}^{(j)}\left(\frac{(j-1)T}{n}\right) = \widehat{Z}^{(j)}, \quad \widehat{X}^{(j-1)}\left(\frac{(j-1)T}{n}\right) = \widehat{X}^{(j-1)}$$

and the value $G((j-1)T/n, s)$ admits representations connected with the chosen bases on the interval $(j-2)T/n < t \leqslant (j-1)T/n$ and on the interval $(j-1)T/n < t \leqslant jT/n$ as well.

In the same way, for $G_R((j-1)T/n)$ both representations

$$G_R((j-1)T/n) = \widehat{X}^{(j-1)}M_R^{(j-1)} = \widehat{Z}^{(j)}M_R^{(j)}$$

hold. Using the decomposition $\widehat{X}^{(j-1)} = \widehat{Z}^{(j-1)}\widehat{\Omega}^{(j)}$, we get the equalities

$$\widehat{Z}^{(j)}\widehat{\Omega}^{(j)}M^{(j-1)}(s) = \widehat{Z}^{(j)}M^{(j)}(s),$$
$$\widehat{Z}^{(j)}\widehat{\Omega}^{(j)}M_R^{(j-1)} = \widehat{Z}^{(j)}M_R^{(j)}.$$

After multiplying these equalities from the left by $[\widehat{Z}_R^{(j)}]^*$ and using the equality $[\widehat{Z}^{(j)}]^*\widehat{Z}^{(j)} = I$, we get the relations between the consecutive matrices $M^{(j-1)}(s)$, $M^{(j)}(s)$ and $M_R^{(j-1)}$, $M_R^{(j)}$:

$$\widehat{\Omega}^{(j)}M^{(j-1)}(s) = M^{(j)}(s), \quad \widehat{\Omega}^{(j)}M_R^{(j-1)} = M_R^{(j)}.$$

We show how to compute the Green matrix $G_R(t)$, which describes the influence of the right-hand side ψ in the right boundary condition $Rx(t) = \psi$ on the solution $x(t)$. First, the above arguments show that

$$G_R(T) = \widehat{Z}^{(n+1)}M_R^{(n+1)}.$$

Second, we know that $RG_R(T) = I$. Hence

$$[R\widehat{Z}^{(n+1)}]M_R^{(n+1)} = I_k.$$

The matrix $R\widehat{Z}^{(n+1)}$ is a square $k\times k$-matrix, which, obviously, is nonsingular (since the product of R and $M_R^{(n+1)}$ is equal to the nonsingular matrix I_k). Therefore, we can write

$$M_R^{(n+1)} = [R\widehat{Z}^{(n+1)}]^{-1}$$

and use this formula to compute $M_R^{(n+1)}$. We note that

$$\begin{aligned}\|G_R(T)\psi\|^2 &= \|\widehat{Z}^{(n+1)}M_R^{(n+1)}\psi\|^2 = ([M_R^{(n+1)}]^*[\widehat{Z}^{(n+1)}]\widehat{Z}^{(n+1)}M_R^{(n+1)}\psi,\psi)\\ &= (M_R^{(n+1)}\psi, M_R^{(n+1)}\psi) = \|M_R^{(n+1)}\psi\|^2.\end{aligned}$$

Therefore,

$$\|G_R(T)\| = \|M_R^{(n+1)}\| = \|\,[R\widehat{Z}^{(n+1)}]^{-1}\|.$$

On the other hand,

$$\|R\widehat{Z}^{(n+1)}\| \leqslant \|R\|\,\|\widehat{Z}^{(n+1)}\|,$$
$$\|\widehat{Z}^{(n+1)}\xi\|^2 = ([\widehat{Z}^{(n+1)}]^*\widehat{Z}^{(n+1)}\xi,\xi) = (\xi,\xi) = \|\xi\|^2,$$
$$\|\widehat{Z}^{(n+1)}\| = 1.$$

Consequently,

$$\|R\widehat{Z}^{(n+1)}\| \leqslant \|R\|.$$

Finally, we obtain the estimate for the condition number of the matrix $[R\widehat{Z}^{(n+1)}]$ (this matrix should be inverted to compute $M_R^{(n+1)}$):

$$\mu(R\widehat{Z}^{(n+1)}) = \|R\widehat{Z}^{(n+1)}\|\,\|[R\widehat{Z}^{(n+1)}]^{-1}\| \leqslant \|R\|\,\|G_R(T)\| \leqslant K\,\|R\|.$$

In Chapter 1, §§10,11, we assumed that the bound K for the Green matrix is known. We recall that the number K was introduced in Chapter 1, §§10,11 in order to characterize the dependence of the solution to the boundary-value problem on perturbations of coefficients and the right-hand side.

If $R\widehat{Z}^{(n+1)}$ is practically singular and, consequently, is not invertible, or inverting the matrix $R\widehat{Z}^{(n+1)}$ we obtain the matrix $M_R^{(n+1)}$ with very large elements, then the constant K is very large, i.e., the problem is ill conditioned and we cannot hope to obtain the solution with the required accuracy.

Thus, determining $\widehat{Z}^{(1)}, \widehat{Z}^{(2)}, \dots, \widehat{Z}^{(n+1)}$ and $\widehat{\Omega}^{(1)}, \widehat{\Omega}^{(2)}, \dots, \widehat{\Omega}^{(n+1)}$ by the direct sweep, we find

$$M_R^{(n+1)} = [R\widehat{Z}^{(n+1)}]^{-1}, \quad G_R(T) = \widehat{Z}^{(n+1)} M_R^{(n+1)}.$$

The values of the Green matrix $G_R(t)$ at the intermediate points $t = jT/n$ can be found by the recurrence relations

$$M_R^{(j)} = [\widehat{\Omega}^{(j+1)}]^{-1} M_R^{(j+1)}, \quad G_R((j-1)T/n) = \widehat{Z}^{(j)} M_R^{(j)}.$$

We will discuss the computation of $G(t,s)$ later. We now consider the computation of $G_L(t)$ which is similar to the above computation of $G_R(t)$. We replace t by $T - t$. Then the left boundary condition becomes the right one, the right boundary condition becomes the left one, and the equation $\dfrac{dx}{dt} = Ax$ takes the form $\dfrac{dx}{d(T-t)} = -Ax$. At the first step of the computational procedure for $G_L(t)$, we find the basis composed of linearly independent solutions to the homogeneous system of equations $Ry = 0$. The basis vectors $y_1, y_2, \dots, y_{N-k}$ are assumed to be orthonormal. The $N \times (N-k)$-matrix

$$\widehat{Y}^{(n)} = [y_1, y_2, \dots, y_{N-k}]$$

is taken for the initial matrix in the direct sweep which will be carried out from right to left (and not from left to right as in the computation of $G_R(t)$).

We recurrently determine the matrices $\widehat{Y}^{(n-1)}, \widehat{Y}^{(n-2)}, \dots, \widehat{Y}^{(1)}, \widehat{Y}^{(0)}$ with the orthonormal columns $((\widehat{Y}^{(j-1)})^*(\widehat{Y}^{(j-1)}) = I_{N-k})$ by using the decompositions

$$e^{-\frac{T}{n}A}\widehat{Y}^{(j)} = \widehat{Y}^{(j-1)}\Phi^{(j-1)},$$

where $\Phi^{(j-1)}$ is an upper triangular matrix. It is easy to verify that

$$G_L(jT/n) = \widehat{Y}^{(j)} F_L^{(j)},$$

where $F_L^{(j)}$ is a square $(N-k) \times (N-k)$-matrix, and $F_L^{(j-1)}$, $F_L^{(j)}$ are connected by the triangular matrix $\widehat{\Phi}^{(j-1)}$ as follows:

$$F_L^{(j)} = [\widehat{\Phi}^{(j-1)}] F_L^{(j-1)}.$$

In particular, $G_L(0) = \widehat{Y}^{(0)} F_L^{(0)}$. Having computed $\widehat{Y}(0)$, from the boundary condition $LG_L(0) = I_{N-k}$ we get

$$F_L^{(0)} = [L\widehat{Y}^{(0)}]^{-1}, \quad G_L(0) = \widehat{Y}^{(0)} F_L^{(0)}.$$

Using the backward sweep, we find

$$F_L^{(j)} = [\widehat{\Phi}^{(j-1)}]^{-1} F_L^{(j-1)}, \quad G_L(jT/n) = \widehat{Y}^{(j)} F_L^{(j)}.$$

Having completed the description of the computation of $G_R(t)$ and $G_L(t)$, we pass to the Green matrix $G(t,s)$. As was shown, the equation $\dfrac{d}{dt}G(t,s) = AG(t,s)$ for $t < s$ and the boundary condition $LG(0,s) = 0$ imply the representation

$$G((j-1)T/n, s) = \widehat{Z}^{(j)} M^{(j)}(s), \quad (j-1)T/n < s.$$

In the same way, from the equation $\frac{d}{dt}G(t,s) = AG(t,s)$ for $t > s$ and the boundary condition $RG(T,s) = 0$ it follows that for

$$G(jT/n, s) = \widehat{Y}^{(j)}F^{(j)}(s), \quad s < jT/n.$$

Let $(j-1)T/n \leqslant s \leqslant jT/n$ and

$$\begin{aligned} G(s-0,s) &= \exp((s-(j-1)T/n)A)G(jT/n,s) \\ &= \exp((s-(j-1)T/n)A)\widehat{Z}^{(j)}M^{(j)}(s), \\ G(s+0,s) &= \exp((jT/n - s)A)\widehat{Y}^{(j)}F^{(j)}(s). \end{aligned}$$

From the condition $G(s+0,s) - G(s-0,s) = I$ it follows that $M^{(j)}(s)$ and $F^{(j)}(s)$ are connected by the equation

$$\exp((jT/n - s)A)\widehat{Y}^{(j)}F^{(j)}(s) - \exp((s-(j-1)T/n)A)\widehat{Z}^{(j)}M^{(j)}(s) = I_N.$$

This matrix equation consists of N^2 scalar equations with unknown elements of the matrices $F^{(j)}(s)$ and $M^{(j)}(s)$. As was mentioned, the matrix $M^{(j)}(s)$ has N columns and $N-k$ rows. It is easy to verify that the matrix $F^{(j)}(s)$ has k rows and N columns. Thus, the number of unknowns (the sum of numbers of elements of the matrices $M^{(j)}(s)$ and $F^{(j)}(s)$) is equal to N^2 and coincides with the number of equations. If the existence conditions for the Green matrix are valid, then the above matrix equation is uniquely solvable and the matrices $M^{(j)}(s)$, $F^{(j)}(s)$ can be found from this equation.

It is important to note that we must carry out the direct sweep in both directions. We determine the matrices

$$\widehat{Z}^{(0)}, \widehat{Z}^{(1)}, \widehat{Z}^{(2)}, \dots, \widehat{Z}^{(j)}, \quad \widehat{\Omega}^{(1)}, \widehat{\Omega}^{(2)}, \dots, \widehat{\Omega}^{(j)}$$

moving in the direction from left to right (increasing superscripts) whereas the matrices

$$\widehat{Y}^{(n)}, \widehat{Y}^{(n-1)}, \dots, \widehat{Y}^{(j)}, \quad \widehat{\Phi}^{(n-1)}, \dots, \widehat{\Phi}^{(j)}$$

are computed from right to left (decreasing superscripts). This explains why the process is referred to as the "two-way sweep."

After computing $M^{(j)}(s)$ and $F^{(j)}(s)$, we can pass to the backward sweep which consists in the successive computation of

$$M^{(j-1)}(s), \quad F^{(j+1)}(s), \quad M^{(j-2)}(s), \quad F^{(j+2)}(s).$$

We indicate the computational formulas in the backward sweep:

$$\begin{aligned} M^{(j-1)}(s) &= [\widehat{\Omega}^{(j)}]^{-1}M^{(j)}(s), \quad G((j-2)T/n, s) = \widehat{Z}^{(j-1)}M^{(j-1)}(s), \\ F^{(j+1)}(s) &= [\widehat{\Phi}^{(j)}]^{-1}F^{(j)}(s), \quad G((j+1)T/n, s) = \widehat{Y}^{(j+1)}F^{(j+1)}(s). \end{aligned}$$

We have described the procedure of computation of the values of the Green matrices for the boundary-value problem on the finite interval $[0,T]$. In §6, we consider a method of computation of the Green matrix $G(t)$; the latter is used to compute solutions to $\frac{dx}{dt} = Ax + f(t)$ which are bounded on the entire line.

§5. Integral and local estimates for the Green matrix

Properties of the Green matrix (reminder). The dichotomy of the matrix spectrum by the imaginary axis as the existence condition for the Green matrix. A parameter characterizing the quality of dichotomy. The estimate of the Green matrix in terms of the parameter. The proof of the estimate.

This section is auxiliary. We recall properties of the Green matrix $G(t, A)$ which are used for the construction of solutions to the equation $\dot{x} = Ax + f(t)$ that are bounded on the entire line $-\infty < t < \infty$ and establish an estimate for the rate of decrease of $\|G(t, A)\|$ as $t \to \pm\infty$ in terms of an integral characteristic of $G(t, A)$. These inequalities will be used to justify an algorithm for computing $G(t, A)$.

The theory of the Green matrix $G(t, A)$ was developed in Chapter 1, §7. The integral representation of $G(t, A)$ was obtained in Chapter 1, §10. The Green matrix can be constructed only if it has no purely imaginary eigenvalues, i.e., if there is a nonsingular matrix T such that

$$A = T^{-1} \begin{bmatrix} B & 0 \\ 0 & C \end{bmatrix} T$$

and the spectra of the blocks B and C are located in the left and right half-planes respectively. Moreover,

$$G(t, A) = \begin{cases} T^{-1} \begin{bmatrix} e^{tB} & 0 \\ 0 & 0 \end{bmatrix} T & \text{for } t > 0, \\ T^{-1} \begin{bmatrix} 0 & 0 \\ 0 & -e^{tC} \end{bmatrix} T & \text{for } t < 0. \end{cases}$$

If all eigenvalues of A lie in the same (right or left) half-plane, then one of the blocks B and C is absent. It turns out that

$$G(t, A) = \begin{cases} e^{tA} & \text{for } t > 0, \\ 0 & \text{for } t < 0 \end{cases}$$

if $\operatorname{Re} \tau_j(A) < 0$ for all j and

$$G(t, A) = \begin{cases} 0 & \text{for } t > 0, \\ -e^{tA} & \text{for } t < 0 \end{cases}$$

if $\operatorname{Re} \tau_j(A) > 0$ for all j.

The Green matrix $G(t, A)$ can be defined as a matrix function of the real argument t which is continuous together with its derivatives for $t \neq 0$ and satisfies the conditions

$$\frac{d}{dt} G(t, A) = AG(t, A) \quad \text{for } t \neq 0,$$
$$G(+0, A) - G(-0, A) = I,$$
$$\|G(t, A)\| \to 0 \quad \text{as } t \to \pm\infty.$$

These conditions uniquely define $G(t, A)$ provided that A has no purely imaginary eigenvalues.

If h is a vector, then for $t \neq 0$ the vector-valued function

$$x(t) = G(t, A)h$$

satisfies the equation $\frac{dx}{dt} = Ax$ and the conditions $x(+0) - x(-0) = h$, $\|x(t)\| \to 0$ as $t \to \pm\infty$. These three properties uniquely define $x(t)$:

$$x(+0) = G(+0, A)h, \quad x(-0) = G(-0, A)h.$$

As we know, if $\tau_j = \tau_j(A)$ are points of the spectrum of A and $t \neq 0$, then $G(t, A)$ is a polynomial in A:

$$\begin{aligned} G(t, A) = g_1(t)I + g_2(t)[A - \tau_1 I] + g_3(t)[A - \tau_1 I][A - \tau_2 I] + \dots \\ + g_N(t)[A - \tau_1 I][A - \tau_2 I] \dots [A - \tau_{N-1} I]. \end{aligned}$$

Therefore,

$$AG(t, A) = G(t, A)A, \quad G(s, A)G(t, A) = G(t, A)G(s, A).$$

In particular,

$$AG(\pm 0, A) = G(\pm 0, A)A, \quad G(t, A)G(\pm 0, A) = G(\pm 0, A)G(t, A).$$

From the representation

$$G(t, A) = \begin{cases} T^{-1} \begin{bmatrix} e^{tB} & 0 \\ 0 & 0 \end{bmatrix} T & (t > 0), \\ T^{-1} \begin{bmatrix} 0 & 0 \\ 0 & -e^{tC} \end{bmatrix} T & (t < 0) \end{cases}$$

it follows that

$$G(-t, -A) = -G(t, A),$$

$$G(+0, A) = T^{-1} \begin{bmatrix} I & 0 \\ 0 & 0 \end{bmatrix} T, \quad e^{tA} G(+0, A) = G(t, A) \quad (t > 0),$$

$$G(-0, A) = T^{-1} \begin{bmatrix} 0 & 0 \\ 0 & -I \end{bmatrix} T, \quad e^{tA} G(-0, A) = G(t, A) \quad (t < 0)$$

and

$$\begin{gathered} G(+0, A)G(-0, A) = G(-0, A)G(+0, A) = 0, \\ G(t, A)G(+0, A) = \begin{cases} G(t, A) & \text{for } t > 0, \\ 0 & \text{for } t < 0, \end{cases} \\ G(t, A)G(-0, A) = \begin{cases} 0 & \text{for } t > 0, \\ G(t, A) & \text{for } t < 0, \end{cases}, \\ G(+0, A)G(+0, A) = G(+0, A), \\ G(-0, A)G(-0, A) = G(-0, A). \end{gathered}$$

In particular, the above equalities imply

$$G(t, A)h = e^{tA} G(+0, A)h, \quad t > 0.$$

As we know, $\|e^{tA}\| \geqslant e^{-|t|\,\|A\|}$. Therefore,

$$\|G(t, A)\| \geqslant e^{-|t|\,\|A\|} \, \|G(+0, A)h\|, \quad t > 0.$$

Similarly, for $t < 0$

$$G(t, A)h = e^{tA} G(-0, A)h, \quad \|G(t, A)\| \geqslant e^{-|t| \|A\|} \, \|G(-0, A)h\|.$$

Under the above assumption on A, the integrals defining the matrices converge:

$$H_+(A) = H_+^*(A) = \int_0^\infty G^*(t, A)G(t, A)dt,$$

$$H_-(A) = H_-^*(A) = \int_{-\infty}^0 G^*(t, A)G(t, A)dt$$

$$= \int_0^\infty G^*(-t, -A)G(-t, -A)dt = H_+(-A),$$

$$H = H^* = H_+(A) + H_-(A) = \int_{-\infty}^\infty G^*(t, A)G(t, A)dt.$$

For the matrices $H_-(A) = H_+(-A)$, $H_+(A)$, and $H(A) = H(-A)$ we can derive matrix equations which generalize the matrix Lyapunov equation. Let $\tau > 0$ and let

$$J(\tau) = \int_\tau^\infty G^*(t, A)G(t, A)dt = \int_0^\infty G^*(t+\tau, A)G(t+\tau, A)dt.$$

Differentiating these equalities in τ, we obtain

$$\frac{d}{d\tau}J(\tau) = -G^*(\tau, A)G(\tau, A) = A^*\int_0^\infty G^*(t+\tau, A)G(t+\tau, A)dt$$

$$+ \int_0^\infty G^*(t+\tau, A)G(t+\tau, A)dt\, A = A^*J(\tau) + J(\tau)A.$$

Setting $\tau = 0$ and using the equality $J(0) = H_+(A)$, we find

$$H_+(A)A + A^*H_+(A) = -G^*(+0, A)G(+0, A).$$

The equation for $H_-(A)$ follows from this equality and the relations

$$H_-(A) = H_+(-A), \quad G(+0, -A) = -G(-0, +A).$$

It is easy to see that

$$H_-(A)A + A^*H_-(A) = G^*(-0, A)G(-0, A).$$

Adding the equations for H_+ and H_-, we arrive at the equation for $H(A)$:

$$H(A)A + A^*H(A) = G^*(-0, A)G(-0, A) - G^*(+0, A)G(+0, A).$$

The obtained equations

$$H_+(A)A + A^*H_+(A) = -G^*(+0, A)G(+0, A),$$
$$H_-(A)A + A^*H_-(A) = G^*(-0, A)G(-0, A),$$
$$H(A)A + A^*H(A) = G^*(-0, A)G(-0, A) - G^*(+0, A)G(+0, A)$$

can be regarded (cf. [**2**]) as a generalization of the matrix Lyapunov equation

$$HA + A^*H = -I$$

which plays an important role in the study of stability theory. Indeed, if the spectrum of A lies strictly in the left half-plane, then

$$G(t,A)=\begin{cases} e^{tA} & \text{for } t>0,\\ 0 & \text{for } t<0,\end{cases}$$

$$G(+0,A)=I,\quad G(-0,A)=0,$$

$$H_+(A)=\int_0^\infty e^{tA^*}e^{tA}dt,\quad H_-(A)=0,$$

$$H(A)=H_+(A)+H_-(A)=\int_0^\infty e^{tA^*}e^{tA}dt.$$

Therefore,

$$G^*(+0,A)G(+0,A)=I,\quad G^*(-0,A)G(-0,A)=0.$$

Hence for such (Hurwitz) matrices the equation

$$H(A)A+A^*H(A)=G^*(-0,A)G(-0,A)-G^*(+0,A)G(+0,A)$$

is related to the classical matrix Lyapunov equation.

In §2, we introduced the following characteristic $\varkappa(A)$ of the quality of stability for the system $\dot{x}=Ax$:

$$\varkappa(A)=2\,\|A\|\,\|H(A)\|=2\|A\|\left\|\int_0^\infty e^{tA^*}e^{tA}dt\right\|.$$

The matrix integral in the definition of $\varkappa(A)$ converges for Hurwitz matrices. If some eigenvalue of A lies in the right half-plane or on the imaginary axis, then we suggest setting $\varkappa(A)=\infty$. We modify the definition of $\varkappa(A)$ preserving the value of $\varkappa(A)$ for Hurwitz matrices. Namely, we suggest (cf. [**7**]) defining

$$H(A)=\int_{-\infty}^{\infty} G^*(t,A)G(t,A)dt,$$

$$\varkappa(A)=2\,\|A\|\,\|H(A)\|.$$

In the case of Hurwitz matrices, these H and $\varkappa$ coincide with those defined in §2. The integral defining $H(A)$ for a general matrix A converges if the matrix A has no purely imaginary eigenvalues. This effect is most visible if we use the equality

$$H(A)=\int_{-\infty}^{\infty} G^*(t,A)G(t,A)dt=\frac{1}{2\pi}\int_{-\infty}^{\infty}[A^*+i\,\omega I]^{-1}[A^*-i\,\omega I]^{-1}d\omega$$

which was proved in Chapter 1, §11. It is natural to set $\varkappa(A)=\infty$ if A has a purely imaginary eigenvalue.

In the case of a finite $\varkappa(A)$, we can divide the set of all eigenvalues into two subsets. The first subset contains the eigenvalues in the left half-plane and the second one contains the eigenvalues in the right half-plane. If the matrix A has a purely imaginary eigenvalue, such a division is impossible. The possibility of the division of the spectrum into such parts is called the dichotomy of spectrum by the imaginary axis. We may think of the parameter $\varkappa(A)$ as characterizing the quality of dichotomy. It tends to infinity if, in perturbing the matrix, we move at least one eigenvalue to the imaginary axis.

The goal of this section is to prove the inequality

$$\|G(t,A)\| \leqslant \sqrt{\varkappa(A)}\, e^{-|t|\,\|A\|/\varkappa(A)}$$

which gives a local estimate, i.e., the estimate of the Green matrix $G(t,A)$ for any t. We recall that $\varkappa(A)$ is an integral estimate of the Green matrix. It is obvious that the above announced inequality is a generalization of the estimate for the matrix exponential of Hurwitz matrices for $t > 0$:

$$\|e^{tA}\| \leqslant \sqrt{\varkappa(A)}\, e^{-|t|\,\|A\|/\varkappa(A)}.$$

Before proceeding to the proof of the inequality we establish several identities and inequalities which will be used hereinafter.

It is easy to verify the following chain of equalities:

$$G^*(+0,A)H_+(A)G(+0,A) = G^*(+0,A)\Big[\int_0^\infty G^*(t,A)G(t,A)dt\Big]G(+0,A)$$

$$= \int_0^\infty [G(t,A)G(+0,A)]^*[G(t,A)G(+0,A)]dt = \int_0^\infty G^*(t,A)G(t,A)dt = H_+(A).$$

Thus, we have proved that

$$G^*(+0,A)H_+(A)G(+0,A) = H_+(A).$$

In the same way, we can prove the equality

$$G^*(-0,A)H_-(A)G(-0,A) = H_-(A).$$

Hereinafter, if no confusion arises, we will omit "A" in the notation of $G(t,A)$, $H_+(A)$, $H_-(A)$, $H(A)$ and write $G(t)$, H_+, H_-, H for brevity.

For any vector z we have

$$(H_+,z,z) = \Big(\int_0^\infty G^*(t)G(t)dt\, z, z\Big) = \int_0^\infty \|G(t)z\|^2 dt = \int_0^\infty \|G(t)G(+0)z\|^2 dt$$

$$\geqslant \int_0^\infty e^{-2t\,\|A\|}\,\|G(+0)z\|^2 dt = \frac{1}{2\,\|A\|}\,\|G(+0)z\|^2.$$

Thus,

$$(H_+,z,z) \geqslant \frac{\|G(+0)z\|^2}{2\,\|A\|}.$$

We note that the last inequality is also valid if $G(+0) = 0$. In this case, $G(t) = G(t)G(+0) = 0$ for $t > 0$. Consequently,

$$H_+ = \int_0^\infty G^*(t)G(t)\,dt = 0.$$

Let $x(t) = G(t)h$, i.e., $\dfrac{dx}{dt} = Ax$ for $t \neq 0$, $x(+0) - x(-0) = h$, $\|x(t)\| \to 0$ as $t \to \pm\infty$. We consider the quadratic form $(H_+x(t), x(t))$ for $t > 0$ and differentiate it with respect to t:

$$\begin{aligned}\frac{d}{dt}(H_+x(t), x(t)) &= ([H_+A + A^*H_+]x(t), x(t))\\ &= -(G^*(+0)G(+0)x(t), x(t)) = -(G(+0)x(t), G(+0)x(t))\\ &\leqslant -\frac{1}{\|H_+\|}(H_+G(+0)x(t), G(+0)x(t))\\ &= -\frac{1}{\|H_+\|}(G^*(+0)H_+G(+0)x(t), x(t)) = -\frac{1}{\|H_+\|}(H_+x(t), x(t)).\end{aligned}$$

This formula is valid if $\|H_+\| \neq 0$. The obtained inequality implies the estimate

$$\begin{aligned}(H_+x(t), x(t)) &\leqslant e^{-t/\|H_+\|}(H_+x(0), x(0)) = e^{-t/\|H_+\|}(H_+G(+0)h, G(+0)h)\\ &= e^{-t/\|H_+\|}(G^*(+0)H_+G(+0)h, h) = e^{-t/\|H_+\|}(H_+h, h)\\ &\leqslant \|H_+\|\, e^{-t/\|H_+\|}(h, h)\end{aligned}$$

which, together with the inequality

$$\frac{1}{2\,\|A\|}\,\|G(+0)x\|^2 \leqslant 2\,\|A\|\,\|H_+\|\,\|h\|^2 e^{-t/\|H_+\|},$$

yields

$$\|G(+0)x(t)\|^2 \leqslant 2\,\|A\|\,\|H_+\|\,\|h\|^2 e^{-t/\|H_+\|}, \quad t > 0.$$

We recall that for $t > 0$

$$\begin{aligned}x(t) &= G(t)h,\\ G(+0)x(t) &= G(+0)G(t)h = G(t)h,\\ \|H_+\| &\leqslant \|H_+ + H_-\| = \|H\|.\end{aligned}$$

Therefore, for $t > 0$

$$\begin{aligned}\|G(t)h\| &\leqslant \sqrt{2\,\|A\|\,\|H\|}\, e^{-(t\,\|A\|)/(2\,\|A\|\,\|H\|)}\,\|h\|,\\ \|G(t, A)\| \equiv \|G(t)\| &\leqslant \sqrt{2\,\|A\|\,\|H\|}\, e^{-(t\,\|A\|)/(2\,\|A\|\,\|H\|)}\,\|h\|.\end{aligned}$$

We obtained these relations under the assumption $\|H_+\| > 0$. However, if $\|H_+\| = 0$, then for $t > 0$ we have $G(t, A) = 0$, and the last inequality is trivial. Thereby, for $t > 0$ this inequality is proved in all cases.

To derive the estimate of $\|G(t, A)\|$ for $t < 0$ it suffices to use the above equalities $G(t, A) = -G(-t, -A)$, $H(A) = H(-A)$ and their consequence $-G(-t, A) = G(t, A)$. For $t < 0$ we have

$$\begin{aligned}\|G(t, A)\| &= \|G(-t, -A)\| = \|G(|t|, -A)\|\\ &\leqslant \sqrt{2\,\|-A\|\,\|H\|}\, e^{-(|t|\,\|-A\|)/(2\,\|-A\|\,\|H\|)}\\ &= \sqrt{2\,\|A\|\,\|H\|}\, e^{-(|t|\,\|-A\|)/(2\,\|-A\|\,\|H\|)}.\end{aligned}$$

Thus

$$\|G(t, A)\| \leqslant \sqrt{\varkappa(A)}\, e^{-(|t|\,\|A\|)/(\varkappa(A))}$$

for all t, where $\varkappa(A) = 2\|A\|\,\|H\|$. This completes the derivation of the local estimate of the Green matrix in terms of its integral estimate, i.e., the dichotomy parameter $\varkappa(A)$.

§6. Computation of the Green matrix on an infinite line and the determination of the dichotomy parameter

The formulation of the boundary-value problem on an infinite line and the approximation of the solution by a solution to the problem on a bounded interval. The use of the orthogonal sweep method leads to the iteration method based on the successive decompositions of a matrix into the product of a matrix with orthonormal columns and an upper triangular matrix. The estimates that guarantee the validity of the procedure and the estimates of the rate of convergence. The computation of the dichotomy parameter and bases of invariant spaces.

In §4, we sketched a scheme of the computation of the Green matrices appearing in the representation of a solution to the boundary-value problem. The scheme is based on the orthogonal sweep method. In §4, we dealt with problems on finite intervals. We already considered the problem of finding a solution to the vector equation

$$\frac{d}{dt}x(t) = Ax(t) + f(t)$$

that is bounded on the entire line $-\infty < t < \infty$ ($f(t)$ is assumed to be bounded) with the Green matrix $G(t)$ appearing in the representation of the solution

$$x(t) = \int_{-\infty}^{\infty} G(t-s)f(s)ds.$$

This section is devoted to an algorithm for computing the Green matrix $G(t)$.

We recall (cf. §5 and Chapter 1, §7) that $G(t)$ is defined by the conditions

$$\begin{gathered} \frac{d}{dt}G(t) = AG(t) \quad \text{for } t \neq 0, \\ G(+0) - G(-0) = I_N, \\ \|G(t)\| \to 0 \quad \text{as } t \to \pm\infty. \end{gathered}$$

These conditions define $G(t)$ uniquely only if the $N \times N$-matrix A has no purely imaginary eigenvalues, i.e., if $\varkappa(A) = 2\|A\|\,\|H\| < \infty$, where

$$H = \frac{1}{2\pi}\int_{-\infty}^{\infty} (A^* + i\,\omega I)^{-1}(A - i\,\omega I)^{-1}d\omega;$$

moreover, the module of the real part of any eigenvalue of A is at least $\delta = \|A\|/\varkappa(A)$.

We have shown that

$$\|G(t,A)\| \leqslant \sqrt{\varkappa(A)}\,e^{-|t|\,\|A\|/\varkappa(A)},$$

$$\int_{-\infty}^{\infty} G^*(t)G(t)\,dt = H.$$

Under the above assumption the matrix A is represented as

$$A = T^{-1} \begin{bmatrix} B & 0 \\ 0 & C \end{bmatrix} T,$$

where T is a nonsingular matrix and B, C are $N_- \times N_-$-matrices whose eigenvalues lie in the left and right half-planes respectively. As a rule, we assume that $N_+ > 0$ and $N_- > 0$. The proof is simpler for $N_- = 0$ or $N_+ = 0$. We recommend that the reader modify the corresponding constructions for $N_+ = 0$ or $N_- = 0$.

As was mentioned (cf. §§3,5 and Chapter 1, §7),

$$G(t) = \begin{cases} T^{-1} \begin{bmatrix} e^{tB} & 0 \\ 0 & 0 \end{bmatrix} T & \text{for } t > 0, \\ T^{-1} \begin{bmatrix} 0 & 0 \\ 0 & -e^{tC} \end{bmatrix} T & \text{for } t < 0, \end{cases}$$

which can be easily verified. These formulas imply

$$G(+0) = T^{-1} \begin{bmatrix} I_{N_-} & 0 \\ 0 & 0 \end{bmatrix} T,$$

$$G(-0) = -T^{-1} \begin{bmatrix} 0 & 0 \\ 0 & I_{N_+} \end{bmatrix} T.$$

The linear span of the vectors being the columns of $G(+0)$ coincides with the linear span of the columns of the matrix

$$T^{-1} \begin{bmatrix} I_{N_-} \\ 0 \end{bmatrix}.$$

We see that the action of the operator A on the above matrix

$$A\left[T^{-1} \begin{bmatrix} I_{N_-} \\ 0 \end{bmatrix}\right] = T^{-1} \begin{bmatrix} B & 0 \\ 0 & C \end{bmatrix} TT^{-1} \begin{bmatrix} I_{N_-} \\ 0 \end{bmatrix} = T^{-1} \begin{bmatrix} B \\ 0 \end{bmatrix} = \left[T^{-1} \begin{bmatrix} I_{N_-} \\ 0 \end{bmatrix}\right] B,$$

is equivalent to the multiplication from the right of the same matrix by B, i.e., to replacing the columns by their linear combinations. Hence the linear span of the columns of $G(-0)$ is an N_--dimensional A-invariant subspace. If a vector z belongs to the subspace, then it is represented as

$$z = T^{-1} \begin{pmatrix} \xi \\ 0 \end{pmatrix} \quad (\xi \text{ has } N_- \text{ components}),$$

moreover,

$$Az = T^{-1} \begin{bmatrix} B & 0 \\ 0 & C \end{bmatrix} TT^{-1} \begin{pmatrix} \xi \\ 0 \end{pmatrix} = T^{-1} \begin{pmatrix} B\xi \\ 0 \end{pmatrix}.$$

Let z be an eigenvector of A, i.e., $Az = \lambda z$. In addition,

$$\lambda z = T^{-1} \begin{pmatrix} \lambda\xi \\ 0 \end{pmatrix}, \quad Az = T^{-1} \begin{pmatrix} B\xi \\ 0 \end{pmatrix}, \quad B\xi = \lambda\xi.$$

All eigenvectors of A correspond to eigenvalues which coincide with some eigenvalues of B, i.e., they lie in the left half-plane. In other words, the linear span of the columns of $G(+0)$ is an A-invariant subspace corresponding to the eigenvalues in the left half-plane. To construct a basis for this invariant subspace it suffices to take N_- linearly independent columns of $G(+0)$. These columns can be regarded as vectors of the basis. However, the procedure of the choice of basis columns is not easy.

If we have at hand a standard program realizing the singular value decomposition of an arbitrary matrix[1] it is convenient to use it as follows. We find the singular value decomposition

$$G(+0) = U \begin{bmatrix} \sigma_N & & & & & & \\ & \sigma_{N-1} & & & & \mathbf{0} & \\ & & \ddots & & & & \\ & & & \sigma_{N_+ + 1} & & & \\ & & & & 0 & & \\ & \mathbf{0} & & & & \ddots & \\ & & & & & & 0 \end{bmatrix} V^*, \quad U^*U = V^*V = I_N,$$

where $\sigma_N, \sigma_{N-1}, \dots, \sigma_{N_+ + 1}$ are nonzero diagonal elements and the number $N_- = N - N_+$ coincides with the rank of $G(+0)$.

Arguing as above, we can show that the linear span of the columns of $G(-0)$ coincides with the subspace that is invariant under A and corresponds to the eigenvalues in the right half-plane. We recommend that the reader repeat in detail all stages of the above proof for the case of $G(-0)$. The singular value decomposition of $G(-0)$ allows us to construct an orthonormal basis for this subspace.

Discussing properties of $G(+0)$, we depart slightly from our main goal in order to emphasize the importance of the problem (the computation of $G(t)$). We now pass to the method of solution of this problem. Having computed $G(t)$ and, in particular, $G(+0)$ and $G(-0)$, we can effectively distinguish A-invariant subspaces that correspond to eigenvalues of A in the right or left half-plane.

As soon as we begin to compute the Green matrix, we encounter the main difficulty which consists in the fact that the corresponding boundary-value problem is posed on an infinite interval and some part of boundary conditions (namely, conditions imposed at infinity) are given as the requirement that $\|G(t)\|$ tends to zero as $t \to \pm\infty$ and not as relations between components of the considered matrix function. To overcome this difficulty, the original problem is replaced by a" close" problem on an interval which is finite but long enough. Then the difference between solutions of the original problem and the approximate problem is estimated. Based on the estimate, we can find out if it is possible to pass from the original problem to the approximate problem which can be solved by the orthogonal sweep method. The estimate is derived under the assumption that the dichotomy parameter $\varkappa(A)$ of the matrix A is not too large: $\varkappa(A) < \varkappa^*$, where $\varkappa^*$ is a sufficiently large number (e.g., $\varkappa^* = 10^{10}$) so that for $\varkappa(A) > \varkappa^*$ the dichotomy is practically absent.

Realizing the computational algorithm, one can repeatedly verify those consequences of the inequality $\varkappa(A) < \varkappa^*$ that provide the realization of the computational scheme. If these consequences fail, it is necessary to stop the algorithm. In this case, we guarantee that $\varkappa(A) \geqslant \varkappa^*$ and, based on the conclusion, we stop further computation of the Green matrix $G(t)$. If we can complete the computation, then, using the obtained matrix $G(t)$, it is possible to organize an additional

[1]For algorithm of computating singular value decomposition we refer the reader to [**13, 20**].

computation of the parameter

$$\varkappa(A) = 2\|A\|\left\| \int_{-\infty}^{\infty} G^*(t)G(t)\,dt \right\|$$

and verify that $\varkappa(A) < \varkappa^*$; otherwise, we disregard the obtained computation of $G(t)$ and make the conclusion $\varkappa(A) \geqslant \varkappa^*$. This scheme is similar to the scheme which was described in the study of the Lyapunov stability of systems (cf. §2). Therefore, we omit details and consider only the scheme of the computation of $G(t)$ under the assumption $\varkappa(A) < \varkappa^*$. Then we show how to compute $\varkappa(A)$ from $G(+0)$, $G(-0)$, and A.

We begin with some auxiliary constructions. Assume that $\varkappa(A)$ is finite. Using the Green matrix $G(t)$, we form the matrix

$$G_l(t) = \sum_{k=-\infty}^{\infty} G(t + 2kl)$$

depending on parameters l and t. For every t the series representing $G_l(t)$ converges as a geometric sequence. This assertion is a corollary of the inequality

$$\|G(t\| \leqslant \sqrt{\varkappa(A)}\, e^{-(|t|\,\|A\|)/\varkappa(A)}.$$

The matrix $G_l(t)$ is periodic in t with period $2l$, $G_l(t + 2l) = G_l(t)$. In particular, $G_l(l) = G_l(-l)$. It is obvious that $G_l(t)$ is a solution to the following boundary-value problem on the segment $[-l, l]$:

$$\begin{gathered} \frac{d}{dt} G_l(t) = AG_l(t) \qquad \text{for } t \neq 0, \\ G_l(+0) - G_l(-0) = I, \quad G_l(-l) = G_l(l). \end{gathered}$$

The solution of this problem exists (the formula was given) and is unique for any $l > 0$ since the corresponding homogeneous problem

$$\begin{gathered} \frac{d}{dt} \widehat{G}_l(t) = A\widehat{G}_l(t) \qquad \text{for } t \neq 0, \\ \widehat{G}_l(+0) - \widehat{G}_l(-0) = 0, \quad \widehat{G}_l(-l) = \widehat{G}_l(l) \end{gathered}$$

has no nontrivial solutions. Indeed, the continuity of $\widehat{G}_l(t)$ at $t = 0$ ($\widehat{G}_l(+0) - \widehat{G}_l(-0) = 0$) implies that the differential equation holds for all t in the interval $-l \leqslant t \leqslant l$. The boundary conditions at $t = \pm l$ allow us to extend $\widehat{G}_l(t)$ as a periodic function with period $2l$ to the entire line and thereby to obtain a bounded solution to the homogeneous equation

$$\frac{d}{dt} \widehat{G}_l(t) = A\widehat{G}_l(t)$$

on the entire line. As we know (cf. Chapter 1, §6), there is no bounded nonzero solution because $\varkappa(A)$ is finite (this implies that A has no purely imaginary eigenvalues). From the representation

$$G_l(t) = \sum_{k=-\infty}^{\infty} G(t + 2kl)$$

it is obvious that

$$G_l(t) - G(t) = \sum_{k=1}^{\infty}[G(t+2kl) + G(t-2kl)],$$

$$\|G_l(t) - G(t)\| \leqslant \sum_{k=1}^{\infty}[\|G(t+2kl)\| + \|G(t-2kl)\|].$$

For $-l \leqslant t \leqslant l$, $k \geqslant 1$ we have

$$\begin{aligned}\|G(t+2kl)\| &\leqslant \sqrt{\varkappa(A)}\,\exp(-|t+2kl|\,\|A\|/\varkappa(A)) \\ &\leqslant \sqrt{\varkappa(A)}\,\exp(-(2k-1)l\|A\|/\varkappa(A)), \\ \|G(t-2kl)\| &\leqslant \sqrt{\varkappa(A)}\,\exp(-|t-2kl|\,\|A\|/\varkappa(A)) \\ &\leqslant \sqrt{\varkappa(A)}\,\exp(-(2k-1)l\|A\|/\varkappa(A)).\end{aligned}$$

Therefore,

$$\begin{aligned}\|G_l(t) - G(t)\| &\leqslant 2\sqrt{\varkappa(A)}\sum_{k=1}^{\infty}\exp(-(2k-1)l\|A\|/\varkappa(A)) \\ &\leqslant \frac{2\sqrt{\varkappa(A)}\,\exp(-l\,\|A\|/\varkappa(A))}{1-\exp(-2l\|A\|/\varkappa(A))}.\end{aligned}$$

The estimate

$$\|G_l(t) - G(t)\| \leqslant \frac{2\sqrt{\varkappa(A)}\,\exp(-l\,\|A\|/\varkappa(A))}{1-\exp(-2l\|A\|/\varkappa(A))}$$

guarantees that $G_l(t)$ and $G(t)$ are close for l large enough. The commutation of $G_l(t)$ is realistic because it is a solution to the boundary-value problem on the finite interval $-l \leqslant t \leqslant l$. However, this problem is not standard, which causes some inconveniences. Indeed, the boundary conditions are imposed not only at the endpoints of the interval, but at its midpoint $t = 0$; moreover, the conditions imposed at the endpoints $t = \pm l$ connect the values $G_l(\pm l)$ (the boundary conditions on the right and left boundaries were imposed independently in the previous considerations). These inconveniences can be eliminated by a slight transformation of the statement of the boundary-value problem.

For $0 \leqslant t \leqslant l$ we set

$$\begin{aligned}X(t) &= G_l(t-l), \quad -l \leqslant t-l \leqslant 0, \\ Y(t) &= G_l(l-t), \quad 0 \leqslant l-t \leqslant +l.\end{aligned}$$

Then $X(t)$ and $Y(t)$ satisfy the differential equations

$$\frac{d}{dt}X(t) = AX(t), \quad \frac{d}{dt}Y(t) = -AY(t)$$

and the boundary conditions

$$Y(0) - X(0) = 0, \quad Y(l) - X(l) = I_N.$$

The boundary-value problem for $X(t)$ and $Y(t)$ takes a standard form and the boundary conditions imposed on the right and left boundaries are already not connected. Solving this problem, i.e., looking for $X(t)$ and $Y(t)$, we must set

$$G_l(t) = \begin{cases} X(t+l), & -l \leqslant t < 0, \\ Y(l-t), & 0 \leqslant t \leqslant +l, \end{cases}$$
$$G_l(-0) = X(l), \quad G_l(+0) = Y(l).$$

Hereinafter, it is convenient to choose the length l of the interval to be a multiple of $\tau = 1/(2\|A\|)$, i.e., we assume that $l = n/(2\|A\|) = n\tau$. Furthermore,

$$\|G(t) - X\left(t + \frac{n}{2\,\|A\|}\right)\| \leqslant \frac{2\sqrt{\varkappa(A)}\,e^{-n/(2\varkappa(A))}}{1 - e^{-n/\varkappa(A)}}, \qquad -n/(2\,\|A\|) \leqslant t \leqslant 0,$$
$$\|G(t) - Y\left(\frac{n}{2\,\|A\|} - t\right)\| \leqslant \frac{2\sqrt{\varkappa(A)}\,e^{-n/(2\varkappa(A))}}{1 - e^{-n/\varkappa(A)}}, \qquad 0 < t \leqslant n/(2\,\|A\|).$$

In particular,

$$\|G(-0) - X(n\tau)\| \equiv \left\|G(-0) - X\left(\frac{n}{2\,\|A\|}\right)\right\| \leqslant \frac{2\sqrt{\varkappa(A)}\,e^{-n/(2\varkappa(A))}}{1 - e^{-n/\varkappa(A)}},$$
$$\|G(+0) - Y(n\tau)\| \equiv \left\|G(+0) - Y\left(\frac{n}{2\,\|A\|}\right)\right\| \leqslant \frac{2\sqrt{\varkappa(A)}\,e^{-n/(2\varkappa(A))}}{1 - e^{-n/\varkappa(A)}}.$$

Since $X(t)$ and $Y(t)$ satisfy the differential equations

$$\frac{d}{dt}X(t) = AX(t), \quad \frac{d}{dt}Y(t) = -AY(t)$$

as well as the left boundary condition $X(0) = Y(0)$, they can be represented as

$$X(t) = e^{tA}Q_0T_0, \quad Y(t) = e^{-tA}P_0T_0,$$

where $Q_0 = P_0$ is a nonsingular matrix. We set

$$Q_0 = P_0 = \frac{1}{\sqrt{2}}I_N.$$

In particular, for $t = \tau = 1/(2\|A\|)$ we have

$$\begin{bmatrix} X(\tau) \\ Y(\tau) \end{bmatrix} = \begin{bmatrix} e^{\tau A}Q_0 \\ e^{-\tau A}P_0 \end{bmatrix} T_0.$$

We construct an upper triangular $N \times N$-matrix R_1 which realizes the QR-decomposition

$$\begin{bmatrix} e^{\tau A}Q_0 \\ e^{-\tau A}P_0 \end{bmatrix} = \begin{bmatrix} Q_1 \\ P_1 \end{bmatrix} R_1$$

so that $Q_1^*Q_1 + P_1^*P_1 = I_N$. In fact, this equality means that the columns of the matrix $\begin{bmatrix} Q_1 \\ P_1 \end{bmatrix}$ are orthogonal and normalized (orthonormal) in such a manner that the QR-decomposition is a formalized notation of the orthogonalization of $2N$-dimensional vectors, in our case of N columns of the matrix

$$\begin{bmatrix} e^{\tau A}Q_0 \\ e^{-\tau A}P_0 \end{bmatrix}.$$

The orthogonalization can be carried out by the Gram–Schmidt orthogonalization process. However, it is possible to achieve much higher accuracy if the current procedures of linear algebra based on orthogonal reflections are applied. We mentioned this in §4 describing the orthogonal sweep method.

It is obvious that

$$\begin{bmatrix} X(t) \\ Y(t) \end{bmatrix} = \begin{bmatrix} e^{(t-\tau)A}Q_1 \\ e^{-(t-\tau)A}P_1 \end{bmatrix} T_1,$$

where $T_1 = R_1 T_0$. In particular,

$$\begin{bmatrix} X(2\tau) \\ Y(2\tau) \end{bmatrix} = \begin{bmatrix} e^{\tau A}Q_1 \\ e^{-\tau A}P_1 \end{bmatrix} T_1.$$

We again carry out the QR-decomposition ($Q_2^* Q_2 + P_2^* P_2 = I_N$ and R_2 is an upper triangular matrix):

$$\begin{bmatrix} e^{\tau A}Q_1 \\ e^{-\tau A}P_1 \end{bmatrix} T_1 = \begin{bmatrix} Q_2 \\ P_2 \end{bmatrix} R_2.$$

In addition, we obtain the following representation of $X(t)$ and $Y(t)$ ($T_2 = R_2 T_1 = R_2 R_1 T_0$):

$$\begin{bmatrix} X(t) \\ Y(t) \end{bmatrix} = \begin{bmatrix} e^{(t-2\tau)A}Q_2 \\ e^{-(t-2\tau)A}P_2 \end{bmatrix} T_2.$$

We can continue the procedure by setting

$$\begin{bmatrix} e^{\tau A}Q_{j-1} \\ e^{-\tau A}P_{j-1} \end{bmatrix} T_1 = \begin{bmatrix} Q_j \\ P_j \end{bmatrix} R_j$$

and ensuring the upper triangular form of the matrix R_j and the validity of the equality

$$Q_j^* Q_j + P_j^* P_j = I_N.$$

Moreover, the following representations hold:

$$\begin{bmatrix} X(t) \\ Y(t) \end{bmatrix} = \begin{bmatrix} e^{(t-j\tau)A}Q_j \\ e^{-(t-j\tau)A}P_j \end{bmatrix} T_j,$$

which means that the columns of the matrix composed of $X(t)$ and $Y(t)$ are linear combinations of partial solutions which are orthonormal for $t = j\tau$ and, consequently, are almost orthonormal for t close to $j\tau$. In fact, we again gave a description of the procedure used in the orthogonal sweep method in order to construct a basis for the space of solutions to the system

$$\frac{dx}{dt} = -Ax, \quad \frac{dy}{dt} = -Ay$$

satisfying the left boundary condition $x(0) = y(0)$ and adapted the procedure to the equations under consideration. In particular,

$$\begin{pmatrix} X(l) \\ Y(l) \end{pmatrix} \equiv \begin{pmatrix} X(n\tau) \\ Y(n\tau) \end{pmatrix} = \begin{pmatrix} Q_n \\ P_n \end{pmatrix} T_n.$$

The preliminary computation of the matrices $e^{\tau A}$ and $e^{-\tau A}$ appearing in the procedure of the successive determination of Q_j and P_j can be carried out by using, for example, the Taylor formula. The matrix T_n in the representation $X(l) = Q_n T_n$,

$Y(l) = P_n T_n$ can be found from the boundary condition $Y(l) - X(l) = I_N$ which leads to the equation $(-Q_n + P_n)T_n = I_N$ implying

$$T_n = (P_n - Q_n)^{-1}.$$

To show that the matrix $P_n - Q_n$ is invertible we estimate the condition number. As we know,

$$\begin{aligned} &Y(l) = G_l(+0), \quad X(l) = G_l(-0), \quad X(l) = Q_n T_n, \\ &Y(l) = P_n T_n, \quad Q_n^* Q_n + P_n^* P_n = I_N, \end{aligned}$$

where $l = n\tau$. Consequently,

$$\begin{gathered} Q_n^* X(l) + P_n^* Y(l) = Q_n^* Q_n T_n + P_n^* P_n T_n = (Q_n^* Q_n + P_n^* P_n) T_n = T_n, \\ \|T_n\| \leqslant \|Q_n^*\| \, \|X(l)\| + \|P_n^*\| \, \|Y(l)\|. \end{gathered}$$

The norm of any submatrix does not exceed the norm of the whole matrix. Therefore, $\|Q_n^*\| = \|Q_n\|$, $\|P_n^*\| = \|P_n\|$ is at most the norm of the matrix $\begin{bmatrix} Q_n \\ P_n \end{bmatrix}$ with orthonormal columns. The norm of this matrix is equal to 1. Hence

$$\begin{aligned} \|T_n\| &\leqslant \|X(l)\| + \|Y(l)\| = \|G_l(-0)\| + \|G_l(+0)\| \\ &\leqslant \|G(+0)\| + \|G(-0)\| \|G(+0) - G_l(+0)\| + \|G(-0) - G_l(-0)\| \\ &\leqslant 2\sqrt{\varkappa(A)} + 2\sqrt{\varkappa(A)}\, e^{-2n/\varkappa(A)}/[1 - e^{-n/\varkappa(A)}]. \end{aligned}$$

The equality $(P_n - Q_n)T_n = I_N$ implies that the matrix T_n is nonsingular since after the multiplication from the left by $P_n - Q_n$ we obtain the identity matrix. The following estimate is obvious:

$$\|T_n^{-1}\| = \|P_n - Q_n\| \leqslant \|P_n\| + \|Q_n\| \leqslant 2.$$

For the condition number $\mu(T_n) = \mu(P_n - Q_n)$ of the matrices $P_n - Q_n$ and T_n we obtain the estimate

$$\begin{aligned} \mu(P_n - Q_n) &= \|P_n - Q_n\| \, \|(P_n - Q_n)^{-1}\| = \mu(T_n) = \|T_n\| \, \|T_n^{-1}\| \\ &\leqslant 4\sqrt{\varkappa(A)}\left(1 + \frac{e^{-2n/\varkappa(A)}}{1 - e^{-n/\varkappa(A)}}\right) \end{aligned}$$

which shows that $\mu(P_n - Q_n)$ cannot exceed $4\varkappa(A)$ if n large enough.

If we require that $e^{-n/\varkappa(A)} < 3/8$, i.e., n can be greater than $\varkappa(A)$, then

$$4\left(1 + \frac{e^{-2n/\varkappa(A)}}{1 - e^{-m/\varkappa(A)}}\right) < 4\left(1 + \frac{(3/8)^2}{1 - 3/8}\right) = 4\left(1 + \frac{9}{40}\right) < 5.$$

For such n the condition number $\mu(P_n - Q_n)$ is at most $5\varkappa(A)$. If we choose n greater than $\varkappa^*$ and in computing $P_n - Q_n$ it turns out that the condition number of this matrix is greater than $5\varkappa^*$, we can guarantee that $\varkappa(A) > \varkappa^*$. But in this case we did not guarantee the possibility of solving the problem.

Computing T_n, $G_l(+0) = Y(n\tau)$, $G_l(-0) = X(n\tau)$ by the formulas

$$T_n = (P_n - Q_n)^{-1}, \quad G_l(+0) = Y(n\tau) = P_n T_n, \quad G_l(-0) = X(n\tau) = Q_n T_n$$

and using the triangular matrices R_j computed in the orthogonal sweep method, we can carry out the backward sweep and compute

$$T_{n-1}, \quad X[(n-1)\tau], \quad Y[(n-1)\tau], \quad T_{n-2}, \quad X[(n-2)\tau], \ldots$$

from the formulas

$$T_j = R_{j+1}^{-1} T_{j+1}, \quad X[j\tau] = Q_j T_j, \quad Y[j\tau] = P_j T_j.$$

In particular, since $Y[(n-1)\tau] = G_l(\tau)$ and $X[(n-1)\tau] = G_l(-\tau)$, we have

$$G_l(\tau) = P_{n-1} T_{n-1}, \quad T_{n-1} = R_n^{-1} T_n, \quad G_l(-\tau) = Q_{n-1} T_{n-1},$$

where $l = n\tau$. The exact Green matrix $G(t)$ satisfies the equalities

$$\begin{gathered} e^{\tau A} G(-\tau) = G(-0) = G(-\tau) e^{\tau A}, \\ e^{\tau A} G(+\tau) = G(\tau) = G(+0) e^{\tau A}, \\ G(+0)G(-0) = G(-0)G(+0) = 0. \end{gathered}$$

For the matrices $G_{n\tau}(t)$ these equalities hold only approximately and can be used if we verify whether the number of steps used is sufficient to provide the required accuracy $G_{n\tau}(t) \approx G(t)$.

We complete our description of the procedure of computation of the Green matrix. To conclude the section, we explain how to compute the matrix integrals

$$H_+ = \int_0^\infty G^*(t)G(t)\,dt, \quad H_- = \int_0^\infty G^*(-t)g(-t)\,dt,$$

$$H = H_+ + H_- = \int_{-\infty}^{\infty} G^*(t)G(t)\,dt$$

from the obtained $G(\pm 0)$ and $G(\pm\tau)$. In turn, these matrix integrals can be used to find the dichotomy parameter $\varkappa(A)$. We show how to obtain computational formulas for the integral H_+ in detail. The computational formulas for H_- will be given without proof because H_- is considered in the same way as H_+. We recommend that the reader verify this fact.

We use the equalities $G(t+s) = G(t)G(s)$ for $t \geqslant 0$, $s \geqslant 0$ and $G(t) = e^{tA}G(+0)$ for $t > 0$. These equalities are easily derived from the representations of the Green matrix and the matrix exponential for $t > 0$

$$\begin{aligned} e^{tA} &= T^{-1} \begin{bmatrix} e^{tB} & 0 \\ 0 & e^{tC} \end{bmatrix} T, \\ G(t) &= T^{-1} \begin{bmatrix} e^{tB} & 0 \\ 0 & 0 \end{bmatrix} T. \end{aligned}$$

Hence

$$H_+^{(0)} = \int_0^\tau G^*(t)G(t)\,dt = G^*(+0)\left[\int_0^\tau e^{tA^*} e^{tA}\,dt\right] G(+0),$$

$$\int_{(j-1)\tau}^{j\tau} G^*(t)G(t)dt = \int_0^{\tau} G^*[(j-1)\tau + t]G[(j-1)\tau + t]dt$$

$$= [G^*(\tau)]^{j-1}\left[\int_0^{\tau} G^*(t)G(t)dt\right][G(\tau)]^{j-1}$$

$$= [G^*(\tau)]^{j-1}H_+^{(0)}[G(\tau)]^{j-1},$$

$$H_+^{(k)} = \int_0^{2^k} G^*(t)G(t)dt = \sum_{j=1}^{2^k}[G^*(\tau)]^{j-1}H_+^{(0)}[G(\tau)]^{j-1}$$

$$= \sum_{j=1}^{2^{k-1}}[G^*(\tau)]^{j-1}H_+^{(0)}[G(\tau)]^{j-1}$$

$$+ [G^*(\tau)]^{2^{k-1}}\left[\sum_{j=1}^{2^{k-1}}[G^*(\tau)]^{j-1}H_+^{(0)}[G(\tau)]^{j-1}\right][G(\tau)]^{2^{k-1}}$$

$$= H_+^{(k-1)} + [G^*(\tau)]^{2^{k-1}}H_+^{(k-1)}[G(\tau)]^{2^{k-1}}.$$

Introducing the notation $B_+^{(k)} = [G^*(\tau)]^{2^k}$, we get the computational formulas

$$B_+^{(0)} = G(\tau), \quad H_+^{(0)} = G^*(+0)\left[\int_0^{\tau} e^{tA^*}e^{tA}dt\right]G(+0),$$

$$B_+^{(k)} = [B_+^{(k-1)}]^2, \quad H_+^{(k)} = H_+^{(k-1)} + B_+^{(k-1)}H_+^{(k-1)}B_+^{(k-1)^*}$$

which allow us to find the approximations

$$H_+^{(k)} = \int_0^{2^k\tau} G^*(t)G(t)dt$$

to the required integral

$$H_+ = \int_0^{\infty} G^*(t)G(t)dt.$$

From the inequality

$$\|G(t)\| \leqslant \sqrt{\varkappa(A)}\, e^{-t\,\|A\|/\varkappa(A)} \quad (t \geqslant 0)$$

we can estimate the rate of convergence for the described process:

$$\|H_+ - H_+^{(k)}\| \leqslant \frac{\varkappa^2(A)}{2\,\|A\|}\, e^{-2^k/\varkappa(a)},$$

$$\|B_+(k)\| \leqslant \sqrt{\varkappa(A)}\, e^{-2^{k-1}/\varkappa(A)}.$$

Indeed, for $\tau = 1/(2\|A\|)$

$$\|B_+(k)\| = \|G^{2^k}(\tau)\| = \|G(2^k\tau)\| \leqslant \sqrt{\varkappa(A)}\, e^{-2^k\tau\,\|A\|/\varkappa(A)} = \sqrt{\varkappa(A)}\, e^{-2^{k-1}/\varkappa(A)},$$

$$\|H_+ - H_+^{(k)}\| = \left\| \int\limits_{2^k\tau}^{\infty} G^*(t)G(t)dt \right\| \leqslant \int\limits_{2^k\tau}^{\infty} \|G(t)\|^2 dt$$

$$\leqslant \varkappa(A) \int\limits_{2^k\tau}^{\infty} e^{-2t\,\|A\|/\varkappa(A)} dt = \frac{\varkappa^2(A)}{2\,\|A\|} e^{-2^{k+1}\tau\|A\|/\varkappa(A)} = \frac{\varkappa^2(A)}{2\,\|A\|} e^{-2^k/\varkappa(A)}.$$

The computational formulas for H_- are as follows:

$$B_-^{(0)} = G(-\tau), \quad H_-^{(0)} = G^*(-0)\left[\int\limits_0^{\tau} e^{tA^*} e^{tA} dt\right] G(-0),$$

$$B_-^{(k)} = [B_-^{(k-1)}]^2, \quad H_-^{(k)} = H_-^{(k-1)} + B_-^{*(k-1)} H_-^{(k-1)} B_-^{(k-1)}.$$

The rate of convergence is characterized by the following estimates:

$$\|B_-^{(k)}\| \leqslant \sqrt{\varkappa(A)}\, e^{-2^{k-1}/\varkappa(A)},$$

$$\|H_- - H_-^{(k)}\| \leqslant \frac{\varkappa^2(A)}{2\,\|A\|}\, e^{-2^k/\varkappa(A)}.$$

At present, there are more effective methods of computation of the Green matrix and the dichotomy parameter. These methods are based on a new version of the orthogonal sweep method and are adapted to equations with constant coefficients. We already mentioned one version at the end of §3. For applications to the computation of $\varkappa(A)$ we refer to the works of Malyshev [**19, 20**].

This completes the description of procedures used in the study of the dichotomy of the spectrum of A. These procedures must be regarded as basic ones in the analysis of properties of the system $\frac{dx}{dt} = Ax$ and in the study of boundary-value problems on the half-line $0 \leqslant t < \infty$. Using these procedures, one can find the numbers N_- and N_+ of eigenvalues lying in the left and right half-planes, verify the inequalities which provide the Lopatinskii condition (cf. Chapter 1, §13), and compute the Lopatinskii constants. In studying this material, it is useful to write the corresponding programs and to solve sufficiently many numerical problems.

CHAPTER 4

Linear Control Systems

§1. Controllability and observability

Linear control systems. A typical control problem and a solvability criterion. The notion of a controllable matrix pair. The observation problems. The definition of an observable matrix pair. An observability criterion.

We consider an important class of problems appearing in automatic control theory. The right-hand side $g(t)$ of the system

$$\frac{d}{dt}x(t) = Ax(t) + g(t) \tag{1}$$

describes an effect of a control which is at our disposal and has the form

$$g(t) = u(t)b,$$

where b is a fixed vector and $u(t)$ is an arbitrary control. A typical control problem is to find a function $u(t)$, $0 < t < T$, such that a solution $x(t)$ to the system

$$\frac{d}{dt}x(t) = Ax(t) + u(t)b \tag{2}$$

satisfying the boundary condition $x(0) = x^{(0)}$ must also satisfy the condition $x(T) = x^{(1)}$. The problem is not solvable for every vector b. Indeed, let us consider an example.

EXAMPLE 1. Let

$$A = \begin{bmatrix} 1 & 0 \\ 0 & 2 \end{bmatrix}, \quad b = \begin{pmatrix} 0 \\ 1 \end{pmatrix}, \quad x^{(0)} = \begin{pmatrix} 0 \\ 0 \end{pmatrix}, \quad x^{(1)} = \begin{pmatrix} 1 \\ 0 \end{pmatrix}.$$

A general solution to the system

$$\frac{d}{dt}\begin{pmatrix} x_1 \\ x_2 \end{pmatrix} = \begin{bmatrix} 1 & 0 \\ 0 & 2 \end{bmatrix}\begin{pmatrix} x_1 \\ x_2 \end{pmatrix} + \begin{pmatrix} 0 \\ u(t) \end{pmatrix}$$

has the form

$$x_1(t) = e^t x_1(0),$$

$$x_2(t) = e^{2t}x_2(0) + \int_0^t e^{2(t-s)}u(s)ds.$$

Hence the boundary condition at $t = 0$

$$x^{(0)} \equiv \begin{pmatrix} x_1(0) \\ x_2(0) \end{pmatrix} = \begin{pmatrix} 0 \\ 0 \end{pmatrix}$$

implies that $x_1(t) \equiv 0$. Therefore, the boundary condition at $t = T$

$$x^{(1)}(T) \equiv \begin{pmatrix} x_1(T) \\ x_2(T) \end{pmatrix} = \begin{pmatrix} 1 \\ 0 \end{pmatrix}$$

does not hold no matter what $u(t)$ is.

We establish a solvability condition for the problem from Example 1. A solution $x(t)$ to the system (2) satisfying the condition $x(0) = x^{(0)}$ is represented in the form

$$x(t) = e^{tA}x^{(0)} + \int_0^t u(t-s)e^{sA}b\,ds. \tag{3}$$

To satisfy the condition $x(T) = x^{(1)}$ it is necessary to find $u(t)$ from the equality

$$\int_0^T u(T-s)e^{sA}b\,ds = x^{(1)} - e^{tA}x^{(0)}.$$

The solvability condition consists in the possibility to choose $v(t)$ such that

$$\int_0^T v(t)e^{tA}b\,dt = z, \tag{4}$$

where we denote

$$u(T-t) = v(t), \quad x^{(1)} - e^{tA}x^{(0)} = z.$$

DEFINITION 1. A pair A, b is said to be controllable on $0 < t < T$ if for any vector z it is possible to find a control $v(t)$ satisfying (4).

Recall (cf. Chapter 1, §4) that the matrix exponential e^{tA} is represented as a polynomial in A with coefficients depending on t, i.e.,

$$e^{tA} = \omega_1(t)I + \omega_2(t)A + \cdots + \omega_N(t)A^{N-1}. \tag{5}$$

We denote

$$v_j = \int_0^T \omega_j(t)v(t)\,dt, \quad j = 1, 2, \ldots, N. \tag{6}$$

We show that if the pair A, b is controllable, then any vector z is represented as a linear combination

$$z = v_1 b + v_2 Ab + v_3 A^2 b + \cdots + v_N A^{N-1} b.$$

Thus, for controllability it is necessary that the linear span of the vectors b, Ab, $A^2b, \ldots, A^{N-1}b$ coincides with the entire space. Under this condition, to obtain the required control it suffices for any $v_1, v_2, \ldots, v_N$ to find at least one function $v(t)$ such that (6) holds. It turns out that such a function $v(t)$ always exists. We will look for $v(t)$ in the form

$$v(t) = \omega_1(t)\alpha_1 + \omega_2(t)\alpha_2 + \cdots + \omega_N(t)\alpha_N, \tag{7}$$

where α_j, $j = 1, 2, \ldots, N$, satisfy the system

$$\sum_{k=1}^{N} \Big(\int_0^T \omega_j(t)\omega_k(t)dt \Big) \alpha_k = v_j, \quad j = 1, 2, \ldots, N. \tag{8}$$

We denote

$$\int_0^T \omega_j(t)\omega_k(t)dt = g_{jk} \tag{9}$$

and rewrite (8) in the form

$$\sum_{k=1}^{N} g_{jk}\alpha_k = v_j, \quad j = 1, 2, \ldots, N. \tag{10}$$

The determinant of the matrix of coefficients does not vanish (cf. Chapter 1, §4):

$$\det \begin{bmatrix} g_{11} & g_{12} & \cdots & g_{1N} \\ g_{21} & g_{22} & \cdots & g_{2N} \\ \vdots & \vdots & \ddots & \vdots \\ g_{N1} & g_{N2} & \cdots & g_{NN} \end{bmatrix} \neq 0.$$

Therefore, we can find $\alpha_1, \alpha_2, \ldots, \alpha_N$ satisfying (10). We formulate the conclusion as the following theorem.

THEOREM 1. *The pair A, b is controllable on $0 < t < T$ if and only if the linear span of the vectors $b, Ab, A^2b, \ldots, A^{N-1}b$ coincides with the entire space.*

We see that the controllability condition is independent of the choice of the interval $(0, T)$. It is obvious that the linear span of the vectors $b, Ab, A^2b, \ldots, A^{N-1}b$ is contained in any invariant subspace of A that contains the vector b. We show that the linear span itself is an invariant subspace. Indeed, any vector z of the linear span is represented in the form

$$z = \alpha_1 b + \alpha_2 Ab + \cdots + \alpha_N A^{N-1}b;$$

moreover,

$$Az = \alpha_1 Ab + \alpha_2 A^2 b + \cdots + \alpha_{N-1}A^{N-1}b + \alpha_N A^N b.$$

By the Cayley–Hamilton theorem, any matrix satisfies its characteristic equation

$$A^N = -p_1 A^{N-1} - p_2 A^{N-2} - \cdots - p_{N-1}A - p_N I,$$

where the coefficients p_j are defined by the equality

$$\det[\lambda I - A] = \lambda^N + p_1\lambda^{N-1} + \cdots + p_{N-1}\lambda + p_N.$$

Therefore, Az is represented as the linear combination

$$Az = -p_N\alpha_N b + (\alpha_1 - p_{N-1}\alpha_N)Ab + \cdots + (\alpha_{N-1} - p_1\alpha_N)A^{N-1}b,$$

i.e., the linear span of the vectors $b, Ab, A^2b, \ldots, A^{N-1}b$ is an invariant subspace of A. Moreover, it is the minimal invariant subspace containing the vector b, which leads to the following reformulation of Theorem 1.

THEOREM 2. *The pair A, b is controllable on $0 < t < T$ if and only if the minimal invariant subspace of A containing the vector b coincides with the entire space.*

REMARK 1. It is easy to show that in the definition of controllability (cf. Definition 1) the interval $(0, T)$ can be replaced with any interval (T_0, T_1). In this case, we look for a control $u(t)$ such that the problem

$$\dot{x} = Ax + u(t)b, \qquad x(T_0) = x^{(0)}, \quad x(T_1) = x^{(1)}$$

is solvable for any $x^{(0)}$ and $x^{(1)}$. The controllability criterion remains unchanged. Hereinafter, we will often use this fact. The proof is left to the reader as an easy exercise.

The control problem can be formulated in more general form as the problem of finding several controls $u_1(t), \dots, u_m(t)$ such that the system

$$\dot{x}(t) = Ax(t) + u_1(t)b_1 + u_2(t)b_2 + \cdots + u_m(t)b_m$$

has a solution satisfying the boundary conditions $x(0) = x^{(0)}$ and $x(T) = x^{(1)}$ for given $x^{(0}$, $x^{(1)}$, and T.

The vector $\sum\limits_{j=1}^{m} u_j(t)b_j$ runs over the linear span of the vectors $b_1, b_2, \dots, b_m$. This linear span is called the control space.

Introducing the $N \times m$-matrix

$$B = [b_1 \vdots b_2 \vdots b_3 \vdots \dots \vdots b_m]$$

with columns $b_1, b_2, \dots, b_m$ and an m-dimensional vector

$$u(t) = \begin{pmatrix} u_1(t) \\ u_2(t) \\ \vdots \\ u_m(t) \end{pmatrix},$$

it is convenient to write the control system as follows:

$$\frac{d}{dt}x(t) = Ax(t) + Bu(t). \tag{11}$$

The columns $b_1, b_2, \dots, b_m$ of the rectangular matrix B are assumed to be linearly independent. Hence the square symmetric $m \times m$-matrix BB^* is nonsingular.

We say that a control system is given by the pair of matrices A and B.

DEFINITION 2. The matrix pair A, B is said to be controllable if it is possible to choose $u(t)$ so that the vector equation (11) has a solution $x(t)$ satisfying the boundary conditions $x(0) = x^{(0)}$ and $x(T) = x^{(1)}$ for arbitrary $x^{(0)}$ and $x^{(1)}$.

A matrix pair is controllable if and only if for any given vector z it is possible to find functions $v^{(1)}, v^{(2)} \dots, v^{(m)}$ such that

$$\int_0^T \sum_{k=1}^{m} v^{(k)}(t)e^{tA}b_k dt = z \tag{12}$$

or, which is equivalent,

$$z = \sum_{k=1}^{m}\Big(\int_0^T v^{(k)}(t)\omega_1(t)dt\Big)b_k + \sum_{k=1}^{m}\Big(\int_0^T v^{(k)}(t)\omega_2(t)dt\Big)Ab_k + \dots \tag{13}$$
$$+ \sum_{k=1}^{m}\Big(\int_0^T v^{(k)}(t)\omega_N(t)dt\Big)A^{N-1}b_k.$$

To represent an arbitrary z in the form (12), it is necessary that the linear span of the vectors $b_1, b_2, \dots, b_m$, $Ab_1, Ab_2, \dots, Ab_m, \dots, A^{N-1}b_1, \dots, A^{N-1}b_m$ coincide with the entire N-dimensional space. This requirement can be reformulated as follows. The linear span of all those vectors whose components form a column of one of the matrices B, AB, $A^2B, \dots, A^{N-1}B$ has dimension N, i.e., the linear span coincides with the entire space.

This condition is also sufficient. Under this condition, controllability holds. Indeed, for any v_{kj}, $1 \leqslant k \leqslant m$, $1 \leqslant j \leqslant N$, we can construct functions $v^{(k)}(t)$ such that

$$v_{kj} = \int_0^T v^{(k)}(t)\omega_j(t)dt.$$

The last assertion differs from the similar assertion in the case of a single control only by notation.

The linear span of all columns of the matrices $B, AB, \dots, A^{N-1}B$ is the minimal invariant subspace of A containing the control space (i.e., the linear span of columns of B). Therefore, the controllability criterion can be formulated as follows.

THEOREM 3. *The matrix pair A, B is controllable if and only if the minimal invariant subspace containing the control space coincides with the entire space.*

Along with the control problems, the so-called observation problems are often considered. We give a typical example. A solution $x(t)$ to the linear system

$$\frac{d}{dt}x(t) = Ax(t)$$

is represented in the form

$$x(t) = e^{tA}x(0).$$

We want to find $x(0)$ and, consequently, the solution $x(t)$ itself from a given observable vector $y(t) = Cx(t)$. The matrix C is not necessarily square. For example, if a single coordinate of the vector $x(t)$ or some scalar linear function of coordinates of $x(t)$ is observed, then the matrix C has just one row. The question arises if $x(t)$ can be found from $y(t)$ on $0 < t < T$. If the answer is positive we say that the matrix pair A, B is observable. As in the case of controllability, if observability holds on some interval $0 < t < T$, then it holds on any such interval. We recommend that the reader verify that observability on $0 < t < T$ implies observability on $T_0 < t < T_1$ for any T_0 and T_1.

Representing the matrix exponential as a polynomial in A, we find

$$x(t) = [\omega_1(t)I + \omega_2(t)A + \dots + \omega_N(t)A^{N-1}]\,x(0),$$
$$y(t) = Cx(t) = \omega_1(t)y^{(1)} + \omega_2(t)y^{(2)} + \dots + \omega_N(t)y^{(N)},$$

where

$$y^{(1)} = Cx(0),\ y^{(2)} = CAx(0), \ldots, y^{(N)} = CA^{(N-1)}x(0).$$

Since the functions $\omega_1(t), \omega_2, \ldots, \omega_N(t)$ are linearly independent (cf. Chapter 1, §4), it is not hard to restore the components y_j^k of the vectors $y^{(1)}, y^{(2)}, \ldots, y^{(N)}$ from the above equalities provided that the components $y_j(t)$ of the vector-function $y(t)$ are known. Indeed, setting

$$\eta_{jr} = \int_0^T y_j(t)\omega_r(t)dt, \quad g_{kr} = \int_0^T \omega_k(t)\omega_r(t)dt,$$

we establish the equalities

$$\eta_{jr} = \sum_{k=1}^N \Big(\int_0^T \omega_k(t)\omega_r(t)dt \Big) y_j^{(k)} = \sum_{k=1}^N g_{kr} y_j^{(k)}.$$

For fixed j these equalities form the system of equations with respect to N unknowns $y_j^{(1)}, y_j^{(2)}, \ldots, y_j^{(N)}$. The determinant of the matrix with elements g_{kr} does not vanish. Therefore, the system has a unique solution. Solving such systems for $j = 1, 2, \ldots, N$, we find the vectors $y^{(1)}, y^{(2)}, \ldots, y^{(N)}$. The components η_{jr} of the right-hand sides of the systems can be computed with the help of the observable vector $y(t)$.

It is easy to see that the question on the observability of $x(t)$ by $y(t)$ can be replaced by the following equivalent question: Is it possible to restore an N-dimensional vector $x(0)$ from $Cx(0), CAx(0), CA^2x(0), \ldots, CA^{N-1}x(0)$? In other words, the matrix pair A, C is observable if we can restore the vector $x(0)$ from the composite vector

$$\begin{bmatrix} C \\ CA \\ \vdots \\ CA^{N-1} \end{bmatrix} x(0)$$

or, which is equivalent, if the composite matrix

$$\begin{bmatrix} C \\ CA \\ \vdots \\ CA^{N-1} \end{bmatrix}$$

has rank N. Passing to adjoint (transposed) matrices, we find that the matrix composed of all columns of the matrices C^*, A^*C^*, $(A^*)^2C^*, \ldots, (A^*)^{N-1}C^*$ must have rank N.

Thus, the following theorem is true.

THEOREM 4. *The matrix pair A, C is observable if and only if the matrix pair A^*, C^* is controllable.*

§2. The simplest variational control problem

A variational problem. Admissible controls. The minimizing sequence converging in mean. The uniform convergence of trajectories. The optimality condition. The integral representation of an optimal control in terms of an optimal trajectory and a constant vector. The optimal control and auxiliary differential system of equations. The uniqueness of an optimal control. The nonuniqueness of a solution to the auxiliary boundary-value problem. One more controllability criterion. The optimal control using nonhomogeneous equations.

We consider the system

$$\frac{d}{dt}x(t) = Ax(t) + Bu(t), \qquad y(t) = Cx(t), \tag{1}$$

where $u(t)$ is a control and $y(t)$ describes the result of the observation of the trajectory $x(t)$. We assume that the rectangular matrix B has rank at least 1 and all columns of B are linearly independent. If the matrix pair A, B is controllable, it is possible to choose a control $u(t)$ depending continuously on t such that there exists a solution $x(t)$ to (1) satisfying the boundary conditions $x(0) = p$ and $x(T) = q$ for given p and q.

Sometimes it is necessary to consider systems with A, B that are not controllable, i.e., for given $x(0) = p$ the boundary condition $x(T) = q$ is not necessarily valid. For such pairs A, B we will assume that there exists a control $u(t)$ called admissible such that the initial value $x(0) = p$ is transformed to the end value $x(T) = q$. As a rule, we consider controls being continuous or piecewise continuous vector-valued functions with a finite number of discontinuities of the first kind. We also consider control problems on an infinite interval, e.g., $0 < t < \infty$, and admit an infinite number of discontinuities $u(t)$, but under the condition that there is no accumulation point on any finite subinterval.

We consider the problem of finding an admissible control $u(t)$ such that the integral

$$\begin{aligned}\mathcal{J}[u(t), p, q] &= \int_0^T (\|y(t)\|^2 + \|Bu(t)\|^2)dt \\ &= \int_0^T (\|Cx(t)\|^2 + \|Bu(t)\|^2)dt = \int_0^T [(C^*Cx, x) + (B^*Bu, u)]dt\end{aligned}$$

attains the minimal value. It is obvious that $\mathcal{J}[u(t), p, q] \geqslant 0$. Therefore, there exists a lower bound $\mathcal{J}_0[p, q]$ of $\mathcal{J}[u(t), p, q]$ over all admissible controls $u(t)$. The existence of a lower bound implies the existence of a sequence of admissible controls $u^{(k)}(t)$ and the corresponding trajectories $x^{(k)}(t)$ such that

$$\begin{gathered}\frac{d}{dt}x^{(k)}(t) = Ax^{(k)}(t) + Bu^{(k)}(t), \\ x^{(k)}(0) = p, \quad x^{(k)}(T) = q, \\ \lim_{k\to\infty} \mathcal{J}[u^{(k)}(t), p, q] = \mathcal{J}_0[p, q].\end{gathered}$$

We can choose a subsequence $u^{(k_1)}(t), u^{(k_2)}(t), \dots$ of the sequence $u^{(k)}(t)$ such that

$$\mathcal{J}_0[p, q] + \frac{1}{j} \geqslant \mathcal{J}[u^{(k_j)}(t), p, q] \geqslant \mathcal{J}_0[p, q].$$

Hereinafter, we assume that $u^{(k_j)}(t) = u^{[j]}(t)$ are piecewise continuous vector-valued functions and the corresponding trajectories $x^{[j]}(t) \equiv x^{(k_j)}(t)$ are continuous, piecewise continuously differentiable, and satisfy the conditions[1]

$$\frac{d}{dt}x^{[j]}(t) = Ax^{[j]}(t) + Bu^{[j]}(t),$$
$$x^{[j]}(0) = p, \quad x^{[j]}(T) = q,$$
$$\frac{1}{j} \geqslant \int_0^T \Big[(CC^* x^{[j]}(t), x^{[j]}(t)) + (B^*Bu^{[j]}(t), u^{[j]}(t))\Big] dt - \mathcal{J}_0[p,q] \geqslant 0.$$

Let $u^{[j']}(t)$ and $u^{[j'']}(t)$ be admissible controls such that $j' \geqslant j$, $j'' \geqslant j$. In addition,

$$\frac{d}{dt}x^{[j']}(t) = Ax^{[j']}(t) + Bu^{[j']}(t),$$
$$\frac{d}{dt}x^{[j'']}(t) = Ax^{[j'']}(t) + Bu^{[j'']}(t),$$
$$x^{[j']}(0) = x^{[j'']}(0) = p, \quad x^{[j']}(T) = x^{[j'']}(T) = q.$$

It is obvious that

$$\widetilde{u}(t) = \frac{1}{2}[u^{[j']}(t) + u^{[j'']}(t)]$$

is also an admissible control such that the trajectory

$$\widetilde{x}(t) = \frac{1}{2}[x^{[j']}(t) + x^{[j'']}(t)]$$

satisfies the boundary conditions $\widetilde{x}(0) = p$ and $\widetilde{x}(T) = q$. Therefore, $\mathcal{J}[\widetilde{u}(t), p, q] \geqslant \mathcal{J}_0$. It is not hard to verify the equalities

$$\begin{aligned}
\|C\widetilde{x}(t)\|^2 &= \|C\frac{1}{2}[x^{[j']}(t) + x^{[j'']}(t)]\|^2 \\
&= \Big(\frac{1}{2}Cx^{[j']}(t) + \frac{1}{2}Cx^{[j'']}(t), \frac{1}{2}Cx^{[j']}(t) + \frac{1}{2}Cx^{[j'']}(t)\Big) \\
&= \frac{1}{2}\|Cx^{[j']}(t)\|^2 + \frac{1}{2}\|Cx^{[j'']}\|^2 - \frac{1}{4}\|C[x^{[j'']}(t) - x^{[j']}(t)]\|^2, \\
\|B\widetilde{u}(t)\|^2 &= \frac{1}{2}\|Bu^{[j']}(t)\|^2 + \frac{1}{2}\|Bu^{[j'']}(t)\|^2 - \frac{1}{4}\|B[u^{[j'']}(t) - u^{[j']}(t)]\|^2
\end{aligned}$$

which imply that

$$\begin{aligned}
\mathcal{J}[\widetilde{u}(t), p, q] = \frac{1}{2}\mathcal{J}[u^{[j'']}(t), p, q] + \frac{1}{2}\mathcal{J}[u^{[j']}(t), p, q] \\
- \frac{1}{4}\int_0^T (\|C\{x^{[j'']}(t) - x^{[j']}(t)\}\|^2 + \|Bu^{[j'']}(t) - Bu^{[j']}(t)\|^2)dt.
\end{aligned}$$

Since

$$\mathcal{J}[u^{[j'']}(t), p, q] \leqslant \mathcal{J}_0 + \frac{1}{j}, \quad \mathcal{J}[u^{[j']}(t), p, q] \leqslant \mathcal{J}_0 + \frac{1}{j}, \quad \mathcal{J}[\widetilde{u}(t), p, q] \geqslant \mathcal{J}_0,$$

[1] The differential equations are assumed to hold everywhere except points of discontinuity of $Bu^{[j]}$.

we conclude that

$$\mathcal{J}_0 \leqslant \mathcal{J}_0 + \frac{1}{j} - \frac{1}{4}\int\limits_0^T \|Bu^{[j'']}(t) - Bu^{[j']}(t)\|^2 dt,$$

$$\int\limits_0^T \|Bu^{[j'']}(t) - Bu^{[j']}(t)\|^2 dt \leqslant \frac{4}{j}.$$

Thus, if $j' \geqslant j$, $j' \geqslant j$, then

$$\sqrt{\int\limits_0^T \|u^{[j'']}(t) - u^{[j']}(t)\|^2 dt} \leqslant \frac{\sqrt{\int\limits_0^T \|Bu^{[j'']}(t) - Bu^{[j']}(t)\|^2}}{\sqrt{\lambda_1(B^*B)}} \leqslant \frac{2}{\sqrt{j\lambda_1(B^*B)}},$$

where $\lambda_1(B^*B)$ denotes the minimal eigenvalue of the matrix B^*B. We have shown that the sequence $u^{[1]}(t), u^{[2]}(t), \ldots$ minimizing the integral $\mathcal{J}[u(t), p, q]$ is a Cauchy sequence with respect to the mean convergence on the segment $[0, T]$.

It is obvious that the differences $x^{[j'']}(t) - x^{[j']}(t)$ are solutions to the equations

$$\frac{d}{dt}[x^{[j'']}(t) - x^{[j']}(t)] = A[x^{[j'']}(t) - x^{[j']}(t)] + B[u^{[j'']}(t) - u^{[j']}(t)]$$

with the zero initial data $x^{[j'']}(0) - x^{[j']}](0) = 0$. Therefore,

$$x^{[j'']}(t) - x^{[j']}(t) = \int\limits_0^t e^{(t-s)A}[Bu^{[j'']}(s) - Bu^{[j']}(s)]\, ds,$$

$$\|x^{[j'']}(t) - x^{[j']}(t)\| \leqslant e^{T\,\|A\|}\int\limits_0^t \|Bu^{[j'']}(s) - Bu^{[j']}(s)\| ds$$

$$\leqslant e^{T\,\|A\|}\sqrt{\int\limits_0^T 1 ds}\sqrt{\int\limits_0^T \|Bu^{[j'']}(s) - Bu^{[j']}(s)\|^2 ds} \leqslant 2\frac{\sqrt{T}}{j}e^{T\,\|A\|},$$

where the Cauchy–Schwarz–Bunyakovskii inequality was used. The sequence of trajectories $x^{[j]}(t)$ is a Cauchy sequence with respect to the uniform convergence on $[0, T]$. Consequently, setting

$$\widehat{x}(t) = \lim_{j\to\infty} x^{[j]}(t),$$

we can regard $\widehat{x}(t)$ as a continuous vector-valued function of t with the same boundary values at $t = 0$ and $t = T$ as $x^{[j]}(t)$. In other words, $\widehat{x}(0) = p$ and $\widehat{x}(T) = q$. Passing to the limit in the inequality

$$\|x^{[j'']}(t) - x^{[j']}(t)\| \leqslant 2\frac{\sqrt{T}}{j}e^{T\,\|A\|}, \quad j'' \geqslant j,$$

as $j'' \to \infty$, we conclude that for any j and $0 \leqslant t \leqslant T$

$$\|\widehat{x}(t) - x^{[j]}(t)\| \leqslant 2\frac{\sqrt{T}}{j} e^{T\,\|A\|}.$$

It is important to emphasize that we have proved only that $u^{[j]}(t)$ form a Cauchy sequence with respect to the mean convergence. Therefore, we cannot assert that there exists a continuous or piecewise continuous control $\widehat{u}(t)$ with the corresponding trajectory $\widehat{x}(t)$. In fact, this assertion is true ($\widehat{u}(t)$ is continuous and even continuously differentiable), but the proof is not easy. We present the construction on which the proof is based.

From the inequality

$$\sqrt{\int_0^T \|u^{[j_1]}(t) - u^{[j_2]}(t)\|^2 dt} \leqslant \frac{2}{\sqrt{\min(j_1, j_2)}\,\sqrt{\lambda_1(B^*B)}}$$

it follows that there exists a measurable square integrable function $\widehat{u}(t)$ such that

$$\sqrt{\int_0^T \|\widehat{u}(t) - u^{[j]}(t)\|^2} \leqslant \frac{2}{\sqrt{j}\,\sqrt{\lambda_1(B^*B)}}.$$

As was mentioned, so far we cannot conclude that $\widehat{u}(t)$ is continuous. The trajectories

$$x^{[j]}(t) = e^{tA}p + \int_0^t e^{(t-\tau)A}Bu^{[j]}(\tau)d\tau$$

uniformly converge to the limit trajectory

$$\widehat{x}(t) = e^{tA}p + \int_0^t e^{(t-\tau)A}B\widehat{u}(\tau)d\tau \tag{2}$$

as $j \to \infty$. In fact, the formula (2) means that $\widehat{x}(t)$ is a "generalized solution" to the Cauchy problem

$$\frac{d}{dt}\widehat{x}(t) = A\widehat{x}(t) + B\widehat{u}(t), \quad x(0) = p.$$

We recall that the notion of a generalized solution was briefly described in Chapter 1, §9. For the moment, we cannot regard $\widehat{x}(t)$ as a solution because we have no justification of the assumption that $\widehat{x}(t)$ is differentiable and $B\widehat{u}(t)$ is continuous. Later, we will show that $B\widehat{u}(t)$ is not only continuous, but also differentiable, as well as $\widehat{x}(t)$. To prove this fact and to obtain the constructive description of $\widehat{x}(t)$ and $\widehat{u}(t)$, we use the optimality condition which follows from the fact that $\widehat{u}(t)$ yields the minimal value of $\mathcal{J}[u(t), p, q]$. We proceed to the optimality condition.

Let $v(t)$ be a continuous or piecewise continuous control such that if a trajectory $w(t)$ of the control system

$$\frac{d}{dt}w(t) = Aw(t) + Bv(t)$$

starts at the point $w(0) = 0$, then it arrives at the same point $w(T) = 0$. This condition is guaranteed by the equality

$$\int_0^T e^{(T-t)A} Bv(t)dt = 0. \tag{3}$$

Along with the controls $u^{[j]}(t)$, which are elements of the minimizing sequence, we consider the controls $u^{[j]}(t) + \xi v(t)$ and the corresponding trajectories

$$z^{[j]}(t) = x^{[j]}(t) + \xi w(t)$$

which satisfy the same boundary conditions: $z^{[j]}(0) = p$ and $z^{[j]}(T) = q$. It is clear that

$$z^{[j]}(t) \to \widehat{x}(t) + \xi w(t),$$
$$u^{[j]}(t) + \xi v(t) \to \widehat{u}(t) + \xi v(t)$$

as $j \to \infty$; moreover, $z^{[j]}$ uniformly converge to the limit whereas $u^{[j]}(t) + \xi v(t)$ converge in mean:

$$\int_0^T \|[u^{[j]}(t) - \xi v(t)] - [\widehat{u}(t) + \xi v(t)]\|^2 dt = \int_0^T \|u^{[j]}(t) - \widehat{u}(t)\|^2 dt \underset{j\to\infty}{\to} 0.$$

It is clear that

$$\widehat{x} + \xi w(t) = e^{tA}p + \int_0^t e^{(t-\tau)A} B[\widehat{u}(\tau) + \xi v(\tau)]d\tau,$$
$$\mathcal{J}[u^{[j]}(t) + \xi v(t), p, q] \underset{j\to\infty}{\to} \mathcal{J}[\widehat{u}(t) + \xi v(t), p, q],$$

and

$$\mathcal{J}[u^{[j]}(t) + \xi v(t), p, q] \geqslant \mathcal{J}[\widehat{u}(t), p, q].$$

Therefore, for any ξ

$$\mathcal{J}[\widehat{u}(t) + \xi v(t), p, q] \geqslant \mathcal{J}[\widehat{u}(t), p, q].$$

The left-hand side of the last inequality is a quadratic polynomial in ξ:

$$\begin{aligned}
\mathcal{J}[\widehat{u}(t) + \xi v(t), p, q] &= \int_0^T (\|C\widehat{x}(t) + \xi Cw(t)\|^2 + \|B\widehat{u}(t) + \xi Bv(t)\|^2)dt \\
&= \int_0^T (\|C\widehat{x}(t)\|^2 + \|B\widehat{u}(t)\|^2)dt + \xi^2 \int_0^T (\|Cw(t)\|^2 + \|Bv(t)\|^2)dt \\
&\quad + 2\xi \int_0^T [(C\widehat{x}(t), Cw(t)) + (B\widehat{u}(t), Bv(t))]dt \\
&= \mathcal{J}[\widehat{u}(t), p, q] + \xi^2 \mathcal{J}[v(t), 0, 0] + 2\xi \mathcal{I},
\end{aligned}$$

where

$$\mathcal{I} = \int_0^T [(C\widehat{x}(t), Cw(t)) + (B\widehat{u}(t), Bv(t))]dt,$$

$$\mathcal{J}[v(t), 0, 0] = \int_0^T (\|Cw(t)\|^2 + \|Bv(t)\|^2)dt \geqslant \lambda_1(B^*B) \int_0^T \|v(t)\|^2 dt.$$

We have $\mathcal{I} = 0$ since

$$\mathcal{J}[\widehat{u}(t), p, q] + \xi^2 \mathcal{J}[v(t), 0, 0] + 2\xi\mathcal{I} \geqslant \mathcal{J}[\widehat{u}(t), p, q]$$

for all ξ. Thus, for any continuous function $v(t)$ satisfying (3) we have

$$\mathcal{I} = \int_0^T [(C\widehat{x}(t), Cw(t)) + (B\widehat{u}(t), Bv(t))]dt = 0. \tag{4}$$

Before we use (4) to derive the final form of the optimality condition, it is convenient to make some transformation of (4). We apply the relations

$$w(t) = e^{(T-t)A}w(T) - \int_t^T e^{(t-\tau)A}Bv(\tau)d\tau = -\int_t^T e^{(t-\tau)A}Bv(\tau)d\tau,$$

$$\begin{aligned}\int_0^T (C^*C\widehat{x}(t), w(t))dt &= -\int_0^T \left(C^*C\widehat{x}(t), \int_t^T e^{(t-\tau)A}Bv(\tau)d\tau\right)dt \\ &= -\iint\limits_{T\geqslant\tau\geqslant t\geqslant 0} \left(C^*C\widehat{x}(t), e^{(t-\tau)A}Bv(\tau)\right)d\tau dt \\ &= -\int_0^T \left(\int_0^\tau B^*e^{(t-\tau)A^*}C^*C\widehat{x}(t)dt, v(\tau)\right)d\tau.\end{aligned}$$

Interchanging τ and t, we obtain

$$\int_0^T (C^*C\widehat{x}(t), w(t))dt = -\int_0^T \left(\int_0^t B^*e^{(\tau-t)A^*}C^*C\widehat{x}(\tau)d\tau, v(t)\right)dt.$$

By the equality

$$\int_0^T (B\widehat{u}(t), Bv(t))dt = \int_0^T (B^*B\widehat{u}(t), v(t))dt,$$

we conclude

$$\mathcal{I} = \int_0^T [(C^*C\widehat{x}(t), w(t)) + (B^*B\widehat{u}(t), v(t))]dt$$
$$= \int_0^T \Big(B^*B\widehat{u}(t) - \int_0^t B^*e^{(\tau-t)A^*}C^*C\widehat{x}(\tau)d\tau, v(t)\Big)dt = 0.$$

We have justified the optimality condition for the control $\widehat{u}(t)$ and the corresponding trajectory $\widehat{x}(t)$.

OPTIMALITY CONDITION. If $\mathcal{J}[u(t), p, q]$ attains the minimal value at a control $\widehat{u}(t)$, then the equality

$$\int_0^T \Big(B^*B\widehat{u}(t) - \int_0^t B^*e^{(\tau-t)A^*}C^*C\widehat{x}(\tau)d\tau, v(t)\Big)dt = 0$$

holds for any $v(t)$ satisfying (3).

The use of the optimality condition is based on the following lemma.

LEMMA 1. *If*

$$\int_0^T (z(t), v(t))dt = 0$$

for any vector-valued function $v(t)$ satisfying (3) *then $z(t)$ is represented in the form $z(t) = B^*e^{(T-t)A^*}k$ for some constant vector k.*

PROOF. From the formula

$$e^{(T-t)A}B = \begin{bmatrix} n_{11}(t) & n_{12}(t) & \cdots & n_{1m}(t) \\ n_{21}(t) & n_{22}(t) & \cdots & n_{2m}(t) \\ \vdots & \vdots & \cdots & \vdots \\ n_{N1}(t) & n_{N2}(t) & \cdots & n_{Nm}(t) \end{bmatrix},$$

we note that the vector

$$v(t) = \begin{pmatrix} v_1(t) \\ v_2(t) \\ \vdots \\ v_m(t) \end{pmatrix}$$

must satisfy the conditions

$$\int_0^t \sum_{i=1}^m v_i(t)n_{ji}(t)dt = 0, \quad j = 1, 2, \dots, N.$$

In other words, the vector $v(t)$ is orthogonal to every vector whose components form one of the rows of the matrix $e^{(T-t)A}B$ or, which is equivalent, one of the columns of the matrix $B^*e^{(T-t)A^*}$, i.e., $v(t)$ is orthogonal to the linear span of the columns. The orthogonality is understood in the sense of the inner product

$$(f,g) = \sum_i \int_0^T f_i(t)g_i(t)dt.$$

It is conventional to say that in the Hilbert space equipped with the inner product (f,g), the set of vector-valued functions $v(t)$ form the orthogonal complement to the linear span of the columns of the matrix $B^*e^{(t-T)A^*}$.

In the Hilbert space, the orthogonal complement to the orthogonal complement to some finite-dimensional subspace $\mathcal{L}$ coincides with this subspace. In other words, every vector z orthogonal to all those vectors that are orthogonal to $\mathcal{L}$ must lie in the space $\mathcal{L}$.

Indeed, if this were not the case, in the linear span of the basis $l_1, l_2, \dots, l_m$ for the subspace $\mathcal{L}$ and the vector z, it is possible to construct a nonzero vector

$$y = z - \sum_{j=1}^m \alpha_j l_j$$

which is orthogonal to $l_1, l_2, \dots, l_m$. It suffices to find α_j from the system of linear equations

$$(z, l_p) - \sum_{j=1}^m \alpha_j (l_j, l_p) = 0, \quad p = 1, 2, \dots, m,$$

whose determinant, being the determinant of the Gram matrix composed of the basis vectors, does not vanish. By assumption, $(z, l_p) = 0$ for all p. Therefore, $\alpha_1 = \alpha_2 = \dots = \alpha_m = 0$. Consequently, $y = z \neq 0$. Moreover, $(z, y) = \|z\|^2 \neq 0$, which contradicts the assumption that z is orthogonal to all those vectors v that are orthogonal to $\mathcal{L}$.

By the above assertion, if a vector-valued function

$$z(t) = \begin{pmatrix} z_1(t) \\ z_2(t) \\ \vdots \\ z_m(t) \end{pmatrix}$$

is orthogonal to all admissible vectors $v(t)$ that form the orthogonal complement to the linear span of the columns of the matrix $B^*e^{(T-t)A^*}$, then z belongs to this span, i.e.,

$$z(t) = B^*e^{(T-t)A^*}k,$$

where the vector k is composed of the coefficients k_j in decomposition $z(t)$ with respect to the basis formed by the columns of $B^*e^{(T-t)A^*}$. It is obvious that k is a constant vector and its components are independent of t. □

Applying the lemma to the equality

$$\int_0^T \left(B^*B\widehat{u}(t) - \int_0^t B^*e^{(\tau-t)A^*}C^*C\widehat{x}(\tau)d\tau, v(t) \right) dt = 0,$$

we find

$$B^*B\widehat{u}(t) - B^* \int_0^t e^{(\tau-t)A^*} C^*C\widehat{x}(\tau)d\tau = B^*e^{(T-t)A^*}k,$$

which means that the optimal control $\widehat{u}(t)$ has the form

$$\widehat{u}(t) = (B^*B)^{-1}\left(B^*e^{(T-t)A^*}k + B^* \int_0^t e^{(\tau-t)A^*} C^*C\widehat{x}(\tau)d\tau\right).$$

The last formula shows that $\widehat{u}(t)$ is continuously differentiable.

We consider the auxiliary system of equations

$$\frac{d}{dt}\lambda(t) = -A^*\lambda(t) + C^*Cx(t). \tag{5}$$

A general solution $\lambda(t)$ to (5) is given by the formula

$$\lambda(t) = e^{(T-t)A^*}k + \int_0^t e^{(\tau-t)A^*} C^*Cx(\tau)d\tau.$$

The optimal control $\widehat{u}(t)$ is represented in terms of $\lambda(t)$ as follows:

$$\widehat{u}(t) = (B^*B)^{-1}B^*\lambda(t).$$

From the above considerations we obtain the following assertion. If there exists a control $u(t)$ such that the corresponding trajectory $x(t)$ satisfies the system (1) and joins $x(0) = p$ with $x(T) = q$, then there exists a continuously differentiable control that has the same property as $u(t)$ and minimizes the integral

$$\int_0^T (\|Cx(t)\|^2 + \|Bu(t)\|^2)dt.$$

To compute the control $u(t)$ it is necessary to solve the boundary-value problem

$$\frac{d}{dt}\begin{pmatrix} x(t) \\ \lambda(t) \end{pmatrix} = \begin{bmatrix} A & B(B^*B)^{-1}B^* \\ C^*C & -A^* \end{bmatrix} \begin{pmatrix} x(t) \\ \lambda(t) \end{pmatrix},$$
$$x(0) = p, \quad x(T) = q$$

and set

$$u(t) = (B^*B)^{-1}B^*\lambda(t).$$

We will show that an optimal control is unique and can be found by the above computational procedure.[2] To clarify the construction and assertions about its properties, we consider several examples.

---

[2]However, we cannot assert that $\lambda(t)$ is unique; see examples below.

EXAMPLE 1. Let $A = 1$, $B = 1$, $C = 1$ and let $x(t)$, $y(t)$, $\lambda(t)$ be scalar functions, $x(0) = 0$, $x(1) = 1$, $T = 1$. Consider the problem of finding a control $u(t)$ such that a solution $x(t)$ to the equation

$$\frac{dx}{dt} = x + u(t)$$

satisfying the boundary condition $x(0) = 0$ passes through the point $x(1) = 1$; moreover, the control $u(t)$ minimizes the integral

$$\int_0^1 [x^2(t) + u^2(t)]dt.$$

To solve this problem, it is necessary to introduce an additional parameter $\lambda(t)$ so that $x(t)$, $\lambda(t)$ satisfy the system

$$\frac{dx}{dt} = x + \lambda, \quad \frac{d\lambda}{dt} = x - \lambda.$$

A general solution to this system has the form

$$x(t) = ae^{\sqrt{2}t} + be^{-\sqrt{2}t},$$
$$\lambda(t) = (\sqrt{2} - 1)ae^{\sqrt{2}t} - (\sqrt{2} + 1)be^{-\sqrt{2}t}.$$

The constants

$$a = \frac{1}{e^{\sqrt{2}} - e^{-\sqrt{2}}}, \quad b = -\frac{1}{e^{\sqrt{2}} - e^{-\sqrt{2}}}$$

are found from the conditions

$$x(0) = a + b = 0, \quad x(1) = ae^{\sqrt{2}} + be^{-\sqrt{2}} = 1.$$

In addition,

$$x(t) = [e^{\sqrt{2}t} - e^{-\sqrt{2}t}]/[e^{\sqrt{2}} - e^{-\sqrt{2}}],$$
$$\lambda(t) = [(\sqrt{2} - 1)e^{\sqrt{2}t} + (\sqrt{2} + 1)e^{-\sqrt{2}t}]/[e^{\sqrt{2}} - e^{-\sqrt{2}}].$$

In this problem $u(t) = \lambda(t)$.

EXAMPLE 2. It is required to construct a control $u(t)$ such that the value of the integral

$$\int_0^T \|u(t)\|^2 dt$$

is minimal and a solution $x_1(t)$, $x_2(t)$ to the system

$$\frac{d}{dt}\begin{pmatrix} x_1 \\ x_2 \end{pmatrix} = \begin{bmatrix} 1 & 1 \\ 0 & -2 \end{bmatrix}\begin{pmatrix} x_1 \\ x_2 \end{pmatrix} + u(t)\begin{pmatrix} 1 \\ 0 \end{pmatrix}$$

satisfying the boundary condition $x_1(0) = 0$, $x_2(0) = 3$ at $t = 0$ satisfies the boundary condition $x_1(T) = 1$, $x_2(T) = 3e^{-2T}$ at $t = T$. In this example,

$$A = \begin{bmatrix} 1 & 1 \\ 0 & -2 \end{bmatrix}, \quad B = \begin{bmatrix} 1 \\ 0 \end{bmatrix}, \quad C = (0, 0)$$

and the auxiliary system has the form

$$\frac{d}{dt}\begin{pmatrix} x_1 \\ x_2 \\ \lambda_1 \\ \lambda_2 \end{pmatrix} = \begin{bmatrix} 1 & 1 & 1 & 0 \\ 0 & -2 & 0 & 0 \\ 0 & 0 & -1 & 0 \\ 0 & 0 & -1 & 2 \end{bmatrix} \begin{pmatrix} x_1 \\ x_2 \\ \lambda_1 \\ \lambda_2 \end{pmatrix}.$$

A general solution

$$\begin{aligned} x_1 &= a_1 e^t + a_2 e^{-2t} - l_1 e^{-t}, \\ x_2 &= 3a_2 e^{-2t}, \\ \lambda_1 &= 6l_1 e^{-t}, \\ \lambda_2 &= l_2 e^{2t} - 2l_1 e^{-t} \end{aligned}$$

is defined by four constants a_1, a_2, l_1, l_2 which are taken so as to satisfy the boundary conditions at $t = 0$ and $t = T$:

$$\begin{aligned} x_1(0) &= a_1 - a_2 - 3l_1 = 0, \\ x_2(0) &= 3a_2 = 3, \\ x_1(T) &= a_1 e^T + a_2 e^{-2T} - 3l_1 e^{-T} = 0, \\ x_2(T) &= 3a_2 e^{-2T} = 3e^{-2T}. \end{aligned}$$

We obtain

$$a_1 = \frac{e^{-2T} - e^{-T}}{e^T - e^{-T}}, \quad a_2 = 1, \quad l_1 = \frac{e^{-2T} - e^T}{3(e^T - e^{-T})}$$

and conclude that l_2 is not determined by the boundary conditions because these conditions are satisfied by an arbitrary l_2. The control $u(t)$ is found by the formula

$$\begin{aligned} u(t) &= (B^*B)^{-1}B^* \begin{pmatrix} \lambda_1(t) \\ \lambda_2(t) \end{pmatrix} = \left[(1 \vdots 0) \begin{pmatrix} 1 \\ 0 \end{pmatrix} \right]^{-1} (1 \vdots 0) \begin{pmatrix} \lambda_1(t) \\ \lambda_2(t) \end{pmatrix} \\ &= (1 \vdots 0) \begin{pmatrix} \lambda_1(t) \\ \lambda_2(t) \end{pmatrix} = \lambda_1(t) = 2\frac{e^{-2T} - e^T}{e^T - e^{-T}} e^{-t}. \end{aligned}$$

Thus, although we cannot uniquely find $\lambda_2(t)$, the required control $u(t)$ is unique.

REMARK 1. If we modify the boundary conditions at $t = T$ in Example 2 so that $x_2(T) = 0$, whereas $x_1(T) = 1$ remains unchanged, then we cannot choose constants a_1, a_2, l_1, l_2 defining the solution to the extended system since the equalities $x_2(0) = 3$ and $x_2(T) = 0$ are inconsistent with the equality $x_2(t) = a_2 e^{-2t}$ for one of the components of the solution. The last equality is a consequence of the equation $\dot{x}_2(t) = -2x_2(t)$ which is valid for any control $u(t)$. Thus, the above modification of the problem is inadmissible because for any control it is impossible to join the initial point $x_1(0) = 0$, $x_2(0) = 3$ and the end point $x_1(T) = 1$, $x_2(T)$. We recommend that the reader consider the following modification of Example 2. It is required to construct a control $u(t)$ such that a solution $x(t)$ to the system

$$\frac{d}{dt}\begin{pmatrix} x_1 \\ x_2 \end{pmatrix} = \begin{pmatrix} 1 & 1 \\ 0 & -2 \end{pmatrix} \begin{pmatrix} x_1 \\ x_2 \end{pmatrix} + u(t) \begin{pmatrix} 1 \\ 0 \end{pmatrix}$$

satisfies the boundary conditions $x_1(0) = 0$, $x_2(0) = 0$ and $x_1(T) = 1$, $x_2(T) = e^{-2T}$; moreover, the integral

$$\int_0^T [x_1^2(t) + x_2^2(t) + u^2(t)]dt$$

takes the minimal value.

We now prove that the optimal control is unique. Otherwise, if there exist two controls $u^{(1)}(t)$ and $u^{(2)}(t)$ such that the trajectories $x^{(1)}(t)$ and $x^{(2)}(t)$ satisfying

$$\frac{d}{dt}x^{(1)}(t) = Ax^{(1)} + Bu^{(1)}(t),$$
$$\frac{d}{dt}x^{(2)}(t) = Ax^{(2)} + Bu^{(2)}(t)$$

pass through the same points $x^{(1)}(0) = x^{(2)}(0) = p$ and $x^{(1)}(T) = x^{(2)}(T) = q$ and both controls $u^{(1)}(t)$ and $u^{(2)}(t)$ satisfy the equalities

$$\mathcal{J}[u^{(1)}, p, q] = \mathcal{J}[u^{(2)}, p, q] = \mathcal{J}_0(p, q) = \inf_{u(t)} \mathcal{J}[u, p, q],$$

then $\widetilde{x}(t) = \frac{1}{2}x^{(1)}(t) + \frac{1}{2}x^{(2)}(t)$ is the trajectory passing through the points $\widetilde{x}(0) = p$ and $\widetilde{x}(T) = q$ corresponding to the control $\widetilde{u}(t) = \frac{1}{2}[u^{(1)}(t) + u^{(2)}(t)]$.

Acting as in the case of the Cauchy sequence $u^{[j_1]}(t), u^{[j_2]}(t), \ldots$, we show that

$$\begin{aligned}\mathcal{J}[\widetilde{u}(t), p, q] &= \frac{1}{2}\mathcal{J}[u^{(1)}(t), p, q] + \frac{1}{2}\mathcal{J}[u^{(2)}(t), p, q] \\ &\quad - \frac{1}{4}\int_0^T \left(\|C\left\{x^{(1)}(t) - x^{(2)}(t)\right\}\|^2 + \|Bu^{(1)}(t) - Bu^{(2)}(t)\|^2\right)dt \\ &\leqslant \mathcal{J}_0[p, q] - \frac{1}{4}\int_0^T \|B\left\{u^{(1)}(t) - u^{(2)}(t)\right\}\|^2 dt \\ &\leqslant \mathcal{J}_0[p, q] - \frac{\lambda_1(B^*B)}{4}\int_0^T \|u^{(1)}(t) - u^{(2)}(t)\|^2 dt.\end{aligned}$$

Since $\mathcal{J}_0[p, q] \leqslant \mathcal{J}[\widetilde{u}(t), p, q]$, we must have

$$\int_0^T \|u^{(1)}(t) - u^{(2)}(t)\|^2 dt = 0, \quad u^{(1)}(t) = u^{(2)}(t).$$

By the uniqueness theorem in the theory of ordinary differential equations, the fact that the controls and the conditions $x^{(1)}(0) = x^{(2)}(0) = p$ coincide means that the solutions coincide: $x^{(1)}(t) = x^{(2)}(t)$ for all $0 \leqslant t \leqslant T$. Thus, an optimal control is unique.

As we know, the optimal control must satisfy the optimality condition. We show that $\widehat{u}(t)$ is uniquely determined by this condition. We recall that if $\widehat{u}(t)$ is the optimal control,

$$\frac{d}{dt}\widehat{x} = A\widehat{x} + B\widehat{u}, \quad \widehat{x}(0) = p, \quad \widehat{x}(T) = q \tag{6}$$

then for any $v(t)$ satisfying (3) and the solution $w(t)$ to the Cauchy problem

$$w(0) = 0, \quad \frac{dw}{dt} = Aw + Bv(t) \tag{7}$$

taking the value 0 at $t = T$ ($w(T) = 0$ in view of the condition on $v(t)$) the following orthogonality condition holds:

$$\mathcal{I} = \int_0^T [(C\widehat{x}(t), Cw(t)) + (B\widehat{u}(t), Bv(t))]dt = 0. \tag{8}$$

We assume that $u^{(1)}(t)$ and $u^{(2)}(t)$ are two different admissible controls satisfying (6)–(8) and $\widehat{x}^{(1)}$ and $\widehat{x}^{(2)}$ are the corresponding trajectories, i.e., the solutions to the problem (6). Let

$$v(t) = \widehat{u}^{(2)}(t) - \widehat{u}^{(1)}(t), \quad w(t) = \widehat{x}^{(2)}(t) - \widehat{x}^{(1)}(t).$$

It is obvious that $v(t)$ satisfies (3). From (8) it follows that

$$\int_0^T [(C\widehat{x}^{(1)}(t), C[\widehat{x}^{(2)}(t) - x^{(1)}(t)]) + (B\widehat{u}^{(1)}(t), B[u^{(2)}(t) - u^{(1)}(t)])]dt = 0,$$
$$\int_0^T [(C\widehat{x}^{(2)}(t), C[\widehat{x}^{(2)}(t) - x^{(1)}(t)]) + (B\widehat{u}^{(2)}(t), B[u^{(2)}(t) - u^{(1)}(t)])]dt = 0.$$

Subtracting one of these equalities from another, we find

$$0 = \int_0^T \left(\|C[\widehat{x}^{(2)}(t) - \widehat{x}^{(1)}(t)]\|^2 + \|B[\widehat{u}^{(2)}(t) - \widehat{u}^{(1)}(t)]\|^2\right) dt$$
$$\geqslant \lambda_1(B^*B) \int_0^T \|\widehat{u}^{(2)}(t) - \widehat{u}^{(1)}(t)\|^2 dt.$$

Hence $\widehat{u}^{(2)}(t) = \widehat{u}^{(1)}(t)$. It is obvious that $\widehat{x}^{(1)}(t) = x^{(2)}(t)$ in view of the uniqueness theorem for ordinary differential equations.

Thus, if a control $u(t)$ is such that (6)–(8) are satisfied, then it is optimal.

We recall that the condition (8) is equivalent to the existence of a vector k such that the control $\widehat{u}(t)$ is represented as follows:

$$\widehat{u}(t) = (B^*B)^{-1}B^*e^{(T-t)A^*}k + (B^*B)^{-1}B^* \int_0^t e^{(t-\tau)A^*}C^*C\widehat{x}(\tau)d\tau. \tag{9}$$

Therefore, if we find k and the conditions (6), (9) hold, then the control $\widehat{u}(t)$ is optimal and is unique. Consequently, to find $\widehat{u}(t)$ it suffices to solve the auxiliary system

$$\frac{d}{dt}\begin{pmatrix}\widehat{x}(t)\\ \lambda(t)\end{pmatrix}=\begin{bmatrix}A & B(B^*B)^{-1}B^*\\ C^*C & -A^*\end{bmatrix}\begin{pmatrix}\widehat{x}(t)\\ \lambda(t)\end{pmatrix}$$

with the boundary conditions $\widehat{x}(0)=p$, $\widehat{x}(T)=q$ and put

$$\widehat{u}(t)=(B^*B)^{-1}B^*\lambda(t).$$

By the above analysis, $\widehat{u}(t)$ is unique even if the solution to the extended system is not unique (cf. Example 2).

As was mentioned, $\widehat{x}(t)$ is unique. The vector parameter $\lambda(t)$, being a solution to the equation

$$\frac{d\lambda}{dt}=C^*Cx(t)-A^*\lambda(t),$$

is not necessarily unique. If $\lambda'(t)$ and $\lambda''(t)$ are admissible, then their difference satisfies the homogeneous equation

$$\frac{d}{dt}(\lambda''(t)-\lambda'(t))=-A^*[\lambda''(t)-\lambda'(t)],$$

and hence can be represented in the form

$$\lambda''(t)-\lambda'(t)=e^{-(t-t_0)A^*}[\lambda''(t_0)-\lambda'(t_0)].$$

Since $\lambda''(t)$ and $\lambda'(t)$ lead to the same control

$$\widehat{u}(t)=(B^*B)^{-1}B^*\lambda'(t)=(B^*B)^{-1}B^*\lambda''(t),$$

we have

$$(B^*B)^{-1}B^*[\lambda''(t)-\lambda'(t)]=0,$$
$$B^*[\lambda''(t)-\lambda'(t)]=0.$$

The conditions

$$\lambda''(t)-\lambda'(t)=e^{-(t-t_0)A^*}[\lambda''(t_0)-\lambda'(t_0)]\neq 0,$$
$$B^*[\lambda''(t)-\lambda'(t)]=0$$

simultaneously hold if and only if the matrix A^* has a nonempty invariant subspace annihilated by B^*. The vectors $A^*z, (A^*)^2z, \dots (A^*)^{N-1}z$ belong to the same subspace. Therefore,

$$B^*z=0,\ B^*A^*z=0,\ B^*(A^*)^2z=0,\ \dots, B^*(A^*)^{N-1}z=0.$$

These equalities hold for $z\neq 0$ if the rank of the composite matrix

$$\begin{bmatrix}B^*\\ B^*A^*\\ \vdots\\ B^*(A^*)^{N-1}\end{bmatrix}$$

is at most N, i.e., the matrix pair A^*, B^* is not observable or, which is the same, the matrix pair A, B is not controllable. This leads to the following theorem.

THEOREM 1. *If there exists a control $u(t)$ such that the corresponding solution $x(t)$ to the system* (1) *satisfies the boundary conditions $x(0) = p$ and $x(T) = q$ and the matrix pair A, B is controllable, then there exists a unique solution to the equations*

$$\frac{dx}{dt} = Ax + B(B^*B)^{-1}B^*\lambda,$$
$$\frac{d\lambda}{dt} = C^*Cx - A^*\lambda$$

satisfying the boundary conditions $x(0) = p$, $x(T) = q$, and the control $u(t) = (B^*B)^{-1}B^*\lambda(t)$ minimizes the integral

$$\int_0^T (\|Cx(t)\|^2 + \|Bu(t)\|^2)dt. \tag{10}$$

If the matrix pair A, B is not controllable, then the solution $x(t)$, $\lambda(t)$ exists, but $\lambda(t)$ is not unique. However, the optimal control $u(t) = (B^*B)^{-1}B^*\lambda(t)$ and the corresponding trajectory $x(t)$ are unique.

Several easy examples (see below) demonstrate how to apply the formulas describing the optimal control.

EXAMPLE 3. We consider the optimal control $u(t)$ such that a solution $x(t)$ to the system (1) satisfies the boundary conditions $x(0) = 0$, $x(T) = a$ and the integral

$$\int_0^T \|Bu(t)\|^2 dt \tag{11}$$

takes the minimal value. Let the matrix pair A, B be controllable. This problem differs from the above problem only by minimization of the integral (11) instead of (10). Hence the above arguments can be applied in this case too. An optimal control $u(t)$ exists, is unique, and is represented in the form

$$u(t) = (B^*B)^{-1}B^*\lambda(t)$$

where $\lambda(t)$ is a solution to the equation

$$\frac{d\lambda(t)}{dt} = C^*Cx(t) - A^*\lambda(t) = -A^*\lambda(t).$$

Hence

$$\lambda(t) = e^{(T-t)A^*}\lambda(T),$$
$$u(t) = (B^*B)^{-1}B^*e^{(T-t)A^*}\lambda(T),$$
$$x(s) = \left[\int_0^s e^{(s-t)A}B(B^*B)^{-1}B^*e^{(T-t)A^*}dt\right]\lambda(T).$$

We can find $\lambda(T)$ from the condition

$$x(T) \equiv \left[\int_0^T e^{(T-t)A}B(B^*B)^{-1}B^*e^{(T-t)A^*}dt\right]\lambda(T) = a.$$

Since the matrix pair A, B is controllable, from the uniqueness and existence of the optimal control and its representation it follows that $\lambda(T)$ can be found for any a. In other words, if the matrix pair A, B is controllable, then the determinant of the matrix

$$M = \int_0^T e^{(T-t)A} B(B^*B)^{-1}B^* e^{(T-t)A^*}\,dt = \int_0^T e^{tA}B(B^*B)^{-1}B^*e^{tA^*}\,dt$$

does not vanish for any $T > 0$. Conversely, if $\det M \neq 0$ for some $T > 0$, then for any a, computing $\lambda(T)$ from the equality

$$M\lambda(T) \equiv \left[\int_0^T e^{(T-t)A}B(B^*B)^{-1}B^*e^{(T-t)A^*}\,dt\right]\lambda(T) = a,$$

we construct the control

$$u(t) = (B^*B)^{-1}B^*e^{(T-t)A^*}\lambda(T) = (B^*B)^{-1}B^*e^{(T-t)A^*}M^{-1}a$$

such that a solution $x(t)$ to the system (1) satisfying the boundary condition $x(0) = 0$ at $t = 0$ must satisfy the boundary condition $x(T) = a$ at $t = T$. In other words, the condition that the determinant of M differs from zero ($\det M \neq 0$) for some (arbitrary) T is necessary and sufficient for controllability. It is obvious that the matrix M is symmetric and nonnegative definite for any A, B, and T. Therefore, the matrix M is nonsingular if and only if it is positive definite. Since the observability of A, C is equivalent to the controllability of A^*, C^*, the matrix pair A, C is observable if and only if the matrix

$$\int_0^T e^{tA^*}C^*(CC^*)^{-1}Ce^{tA}\,dt$$

is positive definite, where $T > 0$ is arbitrary. It is clear that $\|u(t)\| \leqslant \text{const}\,\|a\|$, where the constant depends only on T, A, and B. For the control problem (1) on the interval (T_1, T_2) with the boundary conditions $x(T_1) = 0$ and $x(T_2) = a$ it is easy to see (the proof is left to the reader) that the optimal control satisfies the same estimate with constant depending only on $T_2 - T_1$, A, and B.

EXAMPLE 4. It is required to find a control $u(t)$ such that a solution $x(t)$ to the nonhomogeneous system

$$\frac{d}{dt}x(t) = Ax(t) + Bu(t) + f(t)$$

satisfies the boundary conditions $x(T_1) = a_1$, $x(T_2) = a_2$ ($T_1 < T_2$) and the integral

$$\int_{T_1}^{T_2} \|Bu(t)\|^2\,dt$$

takes the minimal value. This problem is reduced to the use of the solution of the previous problem. We first construct a partial solution $z(t)$ to the system

$$\frac{d}{dt}z(t) = Az(t) + f(t)$$

satisfying the condition $z(T_1) = a_1$ by using the Cauchy formula

$$z(t) = e^{(t-T_1)A}a_1 + \int_0^{t-T_1} e^{(t-T_1-s)A} f(T_1+s)ds.$$

Hence we arrive at the problem of finding $u(t)$ for the system

$$\frac{d}{dt}(x-z) = A(x-z) + Bu(t)$$

with the boundary conditions

$$x(T_1) - z(T_1) = 0,$$

$$x(T_2) - z(T_2) = a_2 - e^{(T_2-T_1)A}a_1 - \int_0^{T_2-T_1} e^{(T_2-T_1-s)A} f(T_1+s)ds.$$

This problem differs from the above problem only by the interval $(0,T)$ replaced with (T_1,T_2). It is easy to see that

$$\begin{aligned}\|x(T_2)-z(T_2)\| &\leqslant \|a_2\| + e^{(T_2-T_1)\,\|A\|}\,\|a_1\| + e^{(T_2-T_1)\,\|A\|}\int_{T_1}^{T_2}\|f(t)\|dt \\ &\leqslant \|a_2\| + e^{(T_2-T_1)\,\|A\|}\,\|a_1\| + e^{(T_2-T_1)\,\|A\|}\sqrt{T_2-T_1}\sqrt{\int_{T_1}^{T_2}\|f(t)\|^2dt} \\ &\leqslant \text{const}\left(\|a_2\| + \|a_1\| + \sqrt{\int_{T_1}^{T_2}\|f(t)\|^2dt}\right).\end{aligned}$$

Using the solution to the above problem, we obtain the following estimate for the control:

$$\|Bu(t)\| \leqslant \text{const}\left(\|a_2\| + \|a_1\| + \sqrt{\int_{T_1}^{T_2}\|f(t)\|^2dt}\right),$$

where the constant depends only on the length $T_2 - T_1$ of the interval (T_1,T_2) and the controllable matrix pair A, B. This estimate will be used in the sequel.

§3. Stabilizability

Orthogonal transformations and the reduction of a system with noncontrollable matrix pair to the canonical form. The study of an optimal trajectory and an auxiliary boundary-value problem. The definition of stabilizability. Necessary and sufficient conditions for the stabilizability of a matrix pair. The controllability criterion based on the stabilizability criterion.

Let U be an orthogonal matrix (i.e., $U^*U = UU^* = I$). Substituting $x = U^*\widehat{x}$, $\lambda = U^*\widehat{\lambda}$, $A = U^*\widehat{A}U$, $B = U^*\widehat{B}$, $C = \widehat{C}U$ in the equations

$$\begin{aligned}
\frac{d}{dt}x &= Ax + Bu, \\
\frac{d}{dt}x &= Ax + B(B^*B)^{-1}B^*\lambda, \\
\frac{d}{dt}\lambda &= C^*Cx - A^*\lambda,
\end{aligned}$$

we obtain the equivalent system

$$\begin{aligned}
\frac{d}{dt}\widehat{x} &= \widehat{A}\widehat{x} + \widehat{B}u, \\
\frac{d}{dt}\widehat{x} &= \widehat{A}\widehat{x} + \widehat{B}(\widehat{B^*}\widehat{B})^{-1}\widehat{B}^*\widehat{\lambda}, \\
\frac{d}{dt}\widehat{\lambda} &= \widehat{C^*}\widehat{C}\widehat{x} - \widehat{A^*}\widehat{\lambda}.
\end{aligned}$$

We note that the following quantities remain unchanged:

$$\|x(t)\| = \|\widehat{x}(t)\|, \quad \|\lambda(t)\| = \|\widehat{\lambda}(t)\|, \quad \|Bu(t)\| = \|\widehat{B}u(t)\|,$$

$$\int_{T_1}^{T_2} (\|Cx(t)\|^2 + \|Bu(t)\|^2)dt = \int_{T_1}^{T_2} (\|\widehat{C}\widehat{x}(t)\|^2 + \|\widehat{B}u(t)\|^2)dt.$$

This remark allows us to reduce the equations under consideration to the canonical form.

Let the matrix pair A, B be not controllable and let the least invariant subspace of A, being the linear span of columns of the matrices $B, AB, A^2B, \dots, A^{N-1}B$, have dimension $N-k$ ($k > 0$). We choose an orthonormal basis for the N-dimensional linear space so that its first $N-k$ vectors lie in this $(N-k)$-dimensional linear span and its last k vectors are orthogonal to this subspace. We can pass from the initial basis to the new basis by means of an orthogonal transformation. Under the orthogonal transformation, the matrix A goes to the matrix

$$\widehat{A} = \begin{bmatrix} \widehat{A}_{11} & \widehat{A}_{12} \\ 0 & \widehat{A}_{22} \end{bmatrix},$$

where $\widehat{A}_{11}$ and $\widehat{A}_{22}$ are matrix blocks of orders $N-k$ and k respectively. The last k components of each column of B are zero, i.e., the matrix B goes to the matrix

$$\widehat{B} = \begin{pmatrix} \widehat{B}_1 \\ 0 \end{pmatrix} \begin{matrix} \} \, N-k \text{ rows} \\ \} \, k \text{ rows.} \end{matrix}$$

It is convenient to divide the components of $\widehat{x}$ and $\widehat{\lambda}$ into $(N-k)$-dimensional $\widehat{x}_1$, $\widehat{\lambda}_1$ and k-dimensional $\widehat{x}_2$, $\widehat{\lambda}_2$ vectors so that

$$\widehat{x} = \begin{pmatrix} \widehat{x}_1 \\ \widehat{x}_2 \end{pmatrix}, \quad \widehat{\lambda} = \begin{pmatrix} \widehat{\lambda}_1 \\ \widehat{\lambda}_2 \end{pmatrix}.$$

The matrix C is also divided into blocks $C = (\widehat{C}_1 \vdots \widehat{C}_2)$ so that $C\widehat{x} = \widehat{C}_1\widehat{x}_1 + \widehat{C}_2\widehat{x}_2$. Moreover, the system takes the form

$$\begin{aligned}\frac{d\widehat{x}_1}{dt} &= \widehat{A}_{11}\widehat{x}_1 + \widehat{A}_{12}\widehat{x}_2 + \widehat{B}_1 u,\\ \frac{d\widehat{x}_2}{dt} &= \widehat{A}_{22}\widehat{x}_2.\end{aligned}$$

An optimal control and the corresponding trajectory $x(t)$ joining the points $\widehat{x}(T_1) = p$ and $\widehat{x}(T_2) = q$ such that the integral

$$\begin{aligned}\int_{T_1}^{T_2} (\|\widehat{C}\widehat{x}(t)\|^2 + \|\widehat{B}u(t)\|^2)dt = \int_{T_1}^{T_2} \Big[(\widehat{C}_1^*\widehat{C}_1\widehat{x}_1, \widehat{x}_1) + (\widehat{C}_2^*\widehat{C}_2\widehat{x}_2, \widehat{x}_2)\\ + 2(\widehat{C}_1^*\widehat{C}_2\widehat{x}_2, \widehat{x}_1) + \|\widehat{B}_1 u\|^2\Big] dt\end{aligned}$$

takes the minimal value can be found by solving the system

$$\begin{aligned}\frac{d\widehat{x}_1}{dt} &= \widehat{A}_{11}\widehat{x}_1 + \widehat{A}_{12}\widehat{x}_2 + \widehat{B}_1(\widehat{B}_1^*\widehat{B}_1)^{-1}\widehat{B}_1^*\widehat{\lambda}_1,\\ \frac{d\widehat{x}_2}{dt} &= \widehat{A}_{22}\widehat{x}_2,\\ \frac{d\widehat{\lambda}_1}{dt} &= \widehat{C}_1^*\widehat{C}_1\widehat{x}_1 + \widehat{C}_1^*\widehat{C}_2\widehat{x}_2 - \widehat{A}_{11}^*\lambda_1,\\ \frac{d\widehat{\lambda}_2}{dt} &= \widehat{C}_2^*\widehat{C}_1\widehat{x}_1 + \widehat{C}_2^*\widehat{C}_2\widehat{x}_2 - \widehat{A}_{12}^*\widehat{\lambda}_1 - \widehat{A}_{22}^*\widehat{\lambda}_2\end{aligned}$$

from the formula

$$u(t) = (\widehat{B}_1^*\widehat{B}_1)^{-1}\widehat{B}_1^*\widehat{\lambda}_1(t).$$

As we know, $\widehat{x}_1(t)$, $\widehat{x}_2(t)$, and u are uniquely determined along the optimal trajectory. Consequently, the vector-valued function $B_1^*\widehat{\lambda}_1(t)$ is uniquely determined along this trajectory too. Indeed, if $\widehat{\lambda}_1'(t)$ and $\widehat{\lambda}_1''(t) \neq \widehat{\lambda}_1'(t)$ are admissible, then their nonzero difference satisfies the homogeneous equation

$$\frac{d}{dt}(\widehat{\lambda}_1'' - \widehat{\lambda}_1') = -\widehat{A}_{11}^*(\widehat{\lambda}_1'' - \widehat{\lambda}_1');$$

moreover, the equality $B_1^*(\widehat{\lambda}_1'' - \widehat{\lambda}_1') = 0$ holds identically on (T_1, T_2), which contradicts the assumption that the pair $\widehat{A}_{11}^*$, $\widehat{B}_1^*$ is observable and the pair $\widehat{A}_{11}$, $\widehat{B}_1$ is controllable for $\widehat{\lambda}_1' - \widehat{\lambda}_1'' \neq 0$.

The controllability of the pair $\widehat{A}_{11}$, $\widehat{B}_1$ follows from the construction. Indeed, the linear span of columns of the matrices

$$\widehat{B} = \begin{bmatrix}\widehat{B}_1\\ 0\end{bmatrix}, \quad \widehat{A}\widehat{B} = \begin{bmatrix}\widehat{A}_{11}\widehat{B}_1\\ 0\end{bmatrix}, \dots, \widehat{A}^{N-1}\widehat{B} = \begin{bmatrix}\widehat{A}^{N-1}\widehat{B}_1\\ 0\end{bmatrix}$$

coincides with the $(N-k)$-dimensional space of vectors $\widehat{x}$ of the form

$$\widehat{x} = \begin{pmatrix}\widehat{x}_1\\ 0\end{pmatrix},$$

i.e., the rank of the composite matrix

$$[\widehat{B}_1 \vdots \widehat{A}_{11}\widehat{B}_1 \vdots \dots \vdots \widehat{A}_{11}^{N-1}\widehat{B}_1]$$

is $N-k$. By the Cayley–Hamilton theorem, it is easy to see that the matrix

$$[\widehat{B}_1 \vdots \widehat{A}_{11}\widehat{B}_1 \vdots \dots \vdots \widehat{A}_{11}^{N-k}\widehat{B}_1]$$

has the same rank, which means that the matrix pair $\widehat{A}_{11}$, $\widehat{B}_1$ is controllable or the matrix pair $\widehat{A}_{11}^*$, $\widehat{B}_1^*$ is observable.

Thus, $\widehat{x}(t)$, $u(t)$, and $\widehat{\lambda}_1(t)$ are uniquely determined along the optimal trajectory. The vector $\widehat{\lambda}_2(t)$ does not affect the control and is not uniquely determined. The difference of any admissible vectors $\widehat{\lambda}_2'(t)$ and $\widehat{\lambda}_2''(t)$ must be a solution to the homogeneous equation

$$\frac{d}{dt}[\widehat{\lambda}_2'(t) - \widehat{\lambda}_2''(t)] = -A_{22}^*[\widehat{\lambda}_2'(t) - \widehat{\lambda}_2''(t)].$$

We completed the discussion of preliminary material which must facilitate the study of stabilizable controls.

DEFINITION 1. The system

$$\frac{dx}{dt} = Ax + Bu(t)$$

is said to be stabilizable if for any $x(0)$ it is possible to find a control $u(t)$, $0 \leqslant t \leqslant \infty$, such that for the corresponding trajectory the integral

$$\int_0^\infty (\|x(t)\|^2 + \|Bu(t)\|^2)dt \tag{1}$$

converges. In this case, the matrix pair A, B is said to be stabilizable too.

It is clear that if the matrix pair A, B is controllable, then it is stabilizable. Indeed, for an arbitrary $T > 0$ we can construct a trajectory that joins given points $x(0)$ and $x(T) = 0$ (controllability) and compute

$$\int_0^T (\|x(t)\|^2 + \|Bu(t)\|^2)dt.$$

By setting

$$u(t) \equiv 0 \quad \text{if } t > T,$$
$$x(t) \equiv 0 \quad \text{if } t > T,$$

we arrive at one of the possible solutions to the considered problem. Note that $u(t)$ can be discontinuous at $t = T$. In this case, $x(t)$ is regarded as a generalized solution.

It is obvious that if the matrix pair A, B is stabilizable and U is orthogonal, then the matrix pair $\widehat{A} = U^*AU$, $\widehat{B} = U^*B$ is stabilizable. Without loss of generality, we assume that the system under consideration has the form

$$\begin{aligned} \frac{d}{dt}\widehat{x}_1 &= \widehat{A}_{11}\widehat{x}_1 + \widehat{A}_{12}\widehat{x}_2 + \widehat{B}_1 u, \\ \frac{d}{dt}\widehat{x}_2 &= \widehat{A}_{22}\widehat{x}_2 \end{aligned} \tag{2}$$

with controllable matrix pair $\widehat{A}_{11}$, $\widehat{B}_1$.

LEMMA 1. *The system* (2) *is stabilizable if and only if the spectrum of A_{22} lies in the left half-plane.*

PROOF. *Necessity.* If $\widehat{A}_{22}$ has an eigenvector a ($\|a\| \neq 0$) such that $\widehat{A}_{22}a = \lambda a$ and $\operatorname{Re}\lambda \geqslant 0$, then for given initial data

$$\widehat{x}_1(0) = 0, \quad \widehat{x}_2(0) = a$$

we obtain

$$\begin{aligned} \widehat{x}_2(t) &= e^{\lambda t}a, \\ \|\widehat{x}_2(t)\| &= e^{t\operatorname{Re}\lambda}\,\|a\|, \\ \|\widehat{x}(t)\| &= \sqrt{\|\widehat{x}_1(t)\|^2 + \|\widehat{x}_2(t)\|^2} \geqslant e^{t\operatorname{Re}\lambda}\,\|a\|, \end{aligned}$$

which is independent of the choice of the control $u(t)$. Therefore, the integral (1) diverges for every control $u(t)$ provided that the trajectory $\widehat{x}(t)$ starts with a fixed point.

Sufficiency. For any $\widehat{x}_2(0)$ we have $\widehat{x}_2(t) = e^{t\widehat{A}_{22}}\widehat{x}_2(0)$. If every eigenvalue $\lambda_j(\widehat{A}_{22})$ of the matrix $\widehat{A}_{22}$ satisfies the condition $\operatorname{Re}\lambda_j(\widehat{A}_{22}) < -\sigma$, then

$$\begin{aligned} \|e^{t\widehat{A}_{22}}\| &\leqslant \Omega(k)\left(\frac{\|\widehat{A}_{22}\|}{\sigma}\right)^{k-1} e^{-t\sigma/2}, \\ \|\widehat{x}_2(t)\| &\leqslant \Omega(k)\left(\frac{\|A_{22}\|}{\sigma}\right)^{k-1} e^{-t\sigma/2}\,\|\widehat{x}_2(0)\|, \\ \|\widehat{A}_{12}\widehat{x}_2(t)\| &\leqslant \Omega(k)\|\widehat{A}_{12}\|\left(\frac{\|\widehat{A}_{22}\|}{\sigma}\right)^{k-1} e^{-t\sigma/2}\,\|\widehat{x}_2(0)\|. \end{aligned}$$

The variation of $\widehat{x}_1(t)$ is described by the equations

$$\frac{d}{dt}\widehat{x}_1(t) = \widehat{A}_{11}\widehat{x}_1(t) + \widehat{B}_1(t)u(t) + f(t),$$

where the right-hand side $f(t) = \widehat{A}_{12}\widehat{x}_2$ satisfies the estimate

$$\|f(t)\| < Fe^{-\frac{t\sigma}{2}}\,\|\widehat{x}_2(0)\|$$

and F is the constant depending on the matrix $\widehat{A}$.

We will choose a control $u(t)$ later. First we divide the half-axis $t > 0$ into intervals of the length T:

$$sT < t < (s+1)T, \quad s = 0, 1, 2, \ldots.$$

On the first interval ($s = 0$) we choose $u(t)$ so as to satisfy the boundary conditions $\widehat{x}_1(0)$=$\{$the initial value$\}$, and $\widehat{x}_1(T) = 0$.

Such a choice of $u(t)$ is possible since the matrix pair $\widehat{A}_{11}$, $\widehat{B}_{11}$ is controllable. Moreover, as was shown in §2, for the chosen control $u(t)$ and the corresponding trajectory $\widehat{x}_1(t)$ the following estimate holds:

$$\|\widehat{B}_1 u(t)\| \leqslant \text{const}\left(\|\widehat{x}_1(0)\| + \sqrt{\int_0^T \|f(t)\|^2 dt}\right) \leqslant \text{const}\,(\|\widehat{x}_1(0)\| + \|\widehat{x}_2(0)\|).$$

On each subsequent interval ($s \geqslant 1$), we choose a control so as to satisfy the boundary conditions

$$\widehat{x}_1(sT) = 0, \quad \widehat{x}_1[(s+1)T] = 0.$$

This can be done so that the following inequalities hold:

$$\|\widehat{x}_1(t)\| \leqslant \text{const}\sqrt{\int_{sT}^{(s+1)T} \|f(t)\|^2 dt} < \text{const}\, e^{-\frac{\sigma sT}{2}}\|\widehat{x}_2(0)\|,$$

$$\|\widehat{B}_1 u(t)\| < \text{const}\, e^{-\frac{\sigma sT}{2}}\|\widehat{x}_2(0)\|.$$

It is obvious that for such a control the integral converges:

$$\int_0^\infty (\|\widehat{x}_1(t)\|^2 + \|\widehat{B}u(t)\|^2)dt = \int_0^\infty (\|\widehat{x}_1(t)\|^2 + \|\widehat{x}_2(t)\|^2 + \|\widehat{B}_1 u(t)\|^2)dt < \infty. \quad \square$$

We give another formulation of Lemma 1. For stabilizability it is necessary and sufficient that the block $\widehat{A}_{22}$ of the matrix

$$\widehat{A} = \begin{bmatrix} \widehat{A}_{11} & \widehat{A}_{12} \\ 0 & \widehat{A}_{22} \end{bmatrix}$$

has no eigenvalues on the imaginary axis and to the right of it. First, we recall that the subspace of vectors

$$\widehat{x} = \begin{pmatrix} \widehat{x}_1 \\ 0 \end{pmatrix} \begin{matrix} \} \, N-k \text{ components} \\ \} \, k \text{ components} \end{matrix}$$

is the minimal invariant subspace of $\widehat{A}$ containing the control space, being the linear span of columns of $\widehat{B}$. It is obvious that stabilizability is equivalent to the requirement that this minimal subspace contains all root subspaces relative to those eigenvalues which do not lie in the left half-plane. This formulation is independent of the choice of the basis. Therefore, it can be applied not only to $\widehat{A}$, $\widehat{B}$, but to the initial matrices A, B as well. Thus, we have proved the following theorem.

THEOREM 1. *The matrix pair A, B is stabilizable if and only if the linear span of all columns of the composite matrix*

$$[B \,\vdots\, AB \,\vdots\, A^2B \,\vdots\, \dots \,\vdots\, A^{N-1}B]$$

contains those root subspaces of the matrix A that correspond to eigenvalues not lying in the left half-plane.

This theorem immediately leads to the following corollary.

COROLLARY 1. *The matrix pair A, B is controllable if and only if both pairs A, B and $-A$, B are stabilizable.*

§4. Variational approach to the construction of a stabilizing control

A variational functional defined by the integral over an infinite half-line. The differences in the study of minimizing sequences from the case of a finite integration interval in §2. A system of equations for an optimal trajectory and parameters defining the optimal control. An example demonstrating the fact that the optimal control is not necessarily stabilizing.

Considering the system

$$\frac{d}{dt}x(t) = Ax(t) + Bu(t), \quad 0 < t < \infty, \tag{1}$$

with the boundary condition $x(0) = p$, we want to choose, if possible, a control $u(t)$ such that the integral

$$\mathcal{J}[u(t), p] = \int_0^\infty (\|Cx(t)\|^2 + \|Bu(t)\|^2)dt \tag{2}$$

takes the minimal value. This problem is not solvable for each triple A, B, C. Indeed, if $C = I$ and $x(0) = p$, then we can find $u(t)$ such that the integral is finite only if the pair A, B is stabilizable (cf. §3). By arguments below, there exists a control such that the integral (2) is finite. Under this assumption, we show how to derive equations that are satisfied by a control minimizing the integral. In turn, the study of such equations helps to answer the question about the solvability of the stated problem.

There are a number of different controls that provide a finite value of the functional $\mathcal{J}[u(t), p]$. We note that $\mathcal{J}[u(t), p] \geqslant 0$ for any control $u(t)$. Hence there exists a lower bound

$$\mathcal{J}_0(p) = \inf_{u(t)} \int_0^\infty (\|Cx(t)\|^2 + \|Bu(t)\|^2)dt \tag{3}$$

taken over all admissible controls $u(t)$. By admissible controls we mean those controls $u(t)$ for which $\mathcal{J}[u(t), p]$ is finite. We assume that admissible controls exist.

We must also stipulate a function class to which $u(t)$ belongs. It suffices to consider piecewise continuous vector-valued functions $u(t)$ which have only finite number of discontinuities of the first kind on any finite segment of the t-axis.

The existence of a lower bound implies the existence of a sequence of admissible controls $u^{(k)}(t)$ and the corresponding trajectories $x^{(k)}(t)$ of the control system such that

$$\frac{d}{dt}x^{(k)}(t) = Ax^{(k)}(t) + Bu^{(k)}(t),$$

$$x^{(k)}(0) = p,$$

$$\lim_{k\to\infty} \mathcal{J}[u^{(k)}, p] = \mathcal{J}_0(p).$$

From the sequence $u^{(k)}(t)$ it is possible to choose a subsequence

$$u^{(k_1)}(t), u^{(k_2)}(t), \dots, u^{(k_j)}(t), \dots$$

such that

$$\mathcal{J}_0(p) + \frac{1}{j} \geqslant \mathcal{J}(u^{(k)}, p) \geqslant \mathcal{J}_0(p).$$

We will consider sequences with the above property and assume that $u^{(k_j)}(t) \equiv u^{[j]}(t)$ are piecewise continuous and the corresponding trajectories $x^{(k_j)}(t) \equiv x^{[j]}(t)$ are piecewise continuously differentiable; moreover,

$$\frac{d}{dt}x^{[j]}(t) = Ax^{[j]}(t) + Bu^{[j]}(t),$$

$$x^{[j]}(0) = p,$$

$$\frac{1}{j} \geqslant \int_0^\infty [(C^*Cx^{[j]}(t), x^{[j]}(t)) + (B^*Bu^{[j]}(t), u^{[j]}(t))]dt - \mathcal{J}_0(p) \geqslant 0.$$

Then we literally repeat arguments in §2 when we studied minimizing sequences but with the integration limits $0, \infty$ instead of $0, T$. Therefore, we omit details and proceed to formulation of results.

As in §2, we prove that if $j' \geqslant j$ and $j'' \geqslant j$, then

$$\sqrt{\int_0^\infty \|u^{[j'']}(t) - u^{[j']}(t)\|^2 dt} \leqslant \frac{2}{\sqrt{j\lambda_1(B^*B)}}.$$

Thereby, the sequence of controls

$$u^{[1]}(t), u^{[2]}(t), \dots, u^{[j]}(t), \dots$$

is a Cauchy sequence with respect to the convergence in mean on the half-line $0 \leqslant t < \infty$.

As in §2, using the equations

$$\frac{d}{dt}[x^{[j'']}(t) - x^{[j']}(t)] = A[x^{[j'']}(t) - x^{[j']}(t)] + B[u^{[j'']}(t) - u^{[j']}(t)],$$

$$x^{[j'']}(0) - x^{[j']}(0) = 0$$

and the Cauchy formula

$$x^{[j'']}(t) - x^{[j']}(t) = \int_0^t e^{(t-s)A}[Bu^{[j'']}(s) - Bu^{[j']}(s)]ds,$$

we establish the relation

$$\|x^{[j'']}(t) - x^{[j']}(t)\| \leqslant \sqrt{T}e^{T\,\|A\|}\sqrt{\int_0^\infty \|Bu^{[j'']}(t) - Bu^{[j']}(t)\|^2 dt}$$

$$\leqslant \frac{2\,\|B\|}{\lambda_1(B^*B)}\sqrt{\frac{T}{j}}e^{T\,\|A\|}$$

for any $T > 0$ and $0 \leqslant t \leqslant T$. As in §2, the sequence of trajectories $x^{[j]}(t)$ is a Cauchy sequence with respect to the uniform convergence on $[0, T]$. Therefore, there exists the limit $\widehat{x}(t) = \lim_{j\to\infty} x^{[j]}(t)$ which is a continuous function on $[0, T]$. Since T is arbitrary, we can assume that the continuous trajectory $\widehat{x}(t)$ is defined on the entire half-line $0 < t < \infty$, but we cannot regard the convergence $x^{[j]}(t) \to x(t)$ as uniform. The uniform convergence holds only on each finite segment of the half-line $0 < t < \infty$. We have the representation

$$\widehat{x}(t) = e^{tA}p + \int_0^t e^{(t-s)A}B\widehat{u}(s)ds,$$

where $\widehat{u}(s)$ is the limit of the sequence $u^{[j]}(s)$ in the sense of the mean convergence on $(0, \infty)$.

The sequence $Cx^j(t)$ also converges in mean on $[0, \infty)$ in view of the inequalities

$$\mathcal{J}[u^{[j'']}, p] \leqslant \mathcal{J}_0 + \frac{1}{j}, \quad \mathcal{J}[u^{[j']}, p] \leqslant \mathcal{J}_0 + \frac{1}{j},$$

$$\mathcal{J}_0 \leqslant \frac{1}{2}\mathcal{J}[u^{[j'']}, p] + \frac{1}{2}\mathcal{J}[u^{[j']}, p]$$
$$-\frac{1}{4}\int_0^\infty \left(\|Cx^{[j'']}(t) - Cx^{[j']}(t)\|^2 + \|Bu^{[j'']}(t) - Bu^{[j']}(t)\|^2\right)dt$$

for $j' \geqslant j$, $j'' \geqslant j$. Therefore,

$$\int_0^\infty \|Cx^{[j'']}(t) - Cx^{[j']}(t)\|^2\, dt \leqslant \frac{4}{j}.$$

This inequality follows from the properties of the minimizing sequence in the same way as the similar inequality was obtained in §2.

It is clear that the sequence $Cx^{[j]}(t)$ is a Cauchy sequence in mean on any finite segment $[0, T]$ and uniformly converges to $C\widehat{x}(t)$ there.

It is obvious that a continuous function $C\widehat{x}(t)$ is also the limit of $Cx^{[j]}(t)$ with respect to the mean convergence on $[0, T]$. Therefore,

$$\int_0^T \|C\widehat{x}(t)\|^2 dt = \lim_{j\to\infty}\int_0^T \|Cx^{[j]}(t)\|^2 dt \leqslant \lim_{j\to\infty}\int_0^\infty \|Cx^{[j]}(t)\|^2 dt,$$

$$\int_0^\infty \|C\widehat{x}(t)\|^2 dt \leqslant \lim_{j\to\infty}\int_0^\infty \|Cx^{[j]}(t)\|^2 dt \leqslant \lim_{j\to\infty}\int_0^\infty (\|Cx^{[j]}(t)\|^2 + \|Bu^{[j]}(t)\|^2)dt.$$

Hence $C\widehat{x}(t)$ is continuous and square integrable on $0 < t < \infty$. It is an easy exercise in analysis to prove that $Cx^{[j]}(t)$ converges to $C\widehat{x}(t)$ in mean on the entire half-line. The proof is left to the reader.

Thus, we have

$$\lim_{j\to\infty}\int_0^\infty \|u^{[j]}(t)-\widehat{u}(t)\|^2dt=0,$$

$$\lim_{j\to\infty}\int_0^\infty \|Cx^{[j]}(t)-C\widehat{x}(t)\|^2dt=0.$$

Therefore,

$$\mathcal{J}_0(p)=\lim_{j\to\infty}\int_0^\infty\Big(\|Cx^{[j]}(t)\|^2+\|Bu^{[j]}(t)\|^2\Big)dt=\int_0^\infty(\|C\widehat{x}(t)\|^2+\|B\widehat{u}(t)\|^2)dt,$$

i.e., the existence of the optimal control $\widehat{u}(t)$ and the corresponding continuous trajectory $\widehat{x}(t)$ is proved.

As in §2, we can prove the uniqueness of the optimal control.

We fix $t=T_1$, $t=T_2$ $(0\leqslant T_1<T_2<\infty)$ and points $x(T_1)$ and $x(T_2)$ of the optimal trajectory on the entire half-line. It is clear that on the segment $0\leqslant T_1<t<T_2$ the optimal control which provides the passage of the trajectory $\widehat{x}(t)$ through the points $\widehat{x}(T_1)$, $\widehat{x}(T_2)$ coincides with $\widehat{u}(t)$ constructed for the entire half-line. On this segment, $\widehat{x}(t)$ and $\widehat{u}(t)$ satisfy the differential equations

$$\begin{aligned}\frac{d\widehat{x}(t)}{dt}&=A\widehat{x}(t)+B(B^*B)^{-1}B^*\lambda(t),\\ \frac{d}{dt}\lambda(t)&=C^*C\widehat{x}(t)-A^*\lambda(t),\\ \widehat{u}(t)&=(B^*B)^{-1}B^*\lambda(t),\end{aligned}$$

which imply the differentiability of $\widehat{x}(t)$, $\widehat{u}(t)$.

As we know, the optimal trajectory $\widehat{x}(t)$ is uniquely determined for all t $(0<t<\infty)$. Consequently, $C^*C\widehat{x}(t)$ is uniquely determined. We see (cf. §2) that $\lambda(t)$ may be not completely determined but the difference $\lambda^{(1)}(t)-\lambda^{(2)}(t)$ of two admissible $\lambda(t)$ vanishes under the action of

$$B^*(\lambda^{(1)}(t)-\lambda^{(2)}(t))=0$$

and satisfy the homogeneous system

$$\frac{d}{dt}(\lambda^{(1)}(t)-\lambda^{(2)}(t))=-A^*(\lambda^{(1)}(t)-\lambda^{(2)}(t)).$$

If we choose some admissible $\lambda(t)$, $0\leqslant t<T$, then we can extend it on the entire half-line $0\leqslant t<\infty$ as a solution to the vector equation

$$\frac{d}{dt}\lambda(t)=C^*C\widehat{x}(t)-A^*\lambda(t).$$

Such an extension exists in view of the existence and uniqueness theorems for the Cauchy problem (cf. Chapter 1, §9). The vector-valued function $\widehat{u}(t)=(B^*B)^{-1}B^*\lambda(t)$ is independent of the choice of $\lambda(t)$.

EXAMPLE 1. It is required to find a control $u(t)$ such that a solution $x(t)$ to the system

$$\frac{d}{dt}\begin{pmatrix} x_1 \\ x_2 \end{pmatrix} = \begin{bmatrix} 1 & 0 \\ 0 & 2 \end{bmatrix}\begin{pmatrix} x_1 \\ x_2 \end{pmatrix} + u(t)\begin{pmatrix} 0 \\ 1 \end{pmatrix}$$

with the boundary conditions $x_1(0) = p_1$ $x_2(0) = p_2$ and the integral

$$\int_0^\infty [x_2^2(t) + u^2(t)]dt$$

takes the minimal value. In this example,

$$A = \begin{bmatrix} 1 & 0 \\ 0 & 2 \end{bmatrix}, \quad B = \begin{bmatrix} 0 \\ 1 \end{bmatrix}, \quad C = (0 \vdots 1),$$

$$B(B^*B)^{-1}B^* = \begin{bmatrix} 0 & 0 \\ 0 & 1 \end{bmatrix}, \quad C^*C = \begin{bmatrix} 0 & 0 \\ 0 & 1 \end{bmatrix}$$

and the equations for the optimal trajectory $x_1(t)$, $x_2(t)$ and parameters $\lambda_1(t)$, $\lambda_2(t)$ take the form

$$\frac{dx_1}{dt} = x_1, \quad \frac{d\lambda_1}{dt} = -\lambda_1,$$

$$\frac{dx_2}{dt} = 2x_2 + \lambda_2(t), \quad \frac{d\lambda_2}{dt} = x_2 - 2\lambda_2.$$

In addition, $u(t) = \lambda_2(t)$. The equations for x_2, λ_2 can be separated as the system

$$\frac{dx_2}{dt} = 2x_2 + \lambda_2,$$

$$\frac{d\lambda_2}{dt} = x_2 - 2\lambda_2.$$

A general solution to this system is as follows:

$$x_2 = ae^{\sqrt{5}t} + \alpha e^{-\sqrt{5}t},$$

$$\lambda_2 = (\sqrt{5} - 2)ae^{\sqrt{5}t} - \alpha e^{-\sqrt{5}t}(\sqrt{5} + 2) = u(t).$$

The integral

$$\int_0^\infty [x_2^2(t) + u^2(t)]dt$$

will be finite if we consider only solutions with the coefficient $a = 0$:

$$x_2 = \alpha e^{-\sqrt{5}t},$$

$$\lambda_2 = -\alpha(\sqrt{5} + 2)e^{-\sqrt{5}t} = u(t).$$

The boundary condition $x_2(0) = p_2$ allows us to find a:

$$x_2 = p_2 e^{-\sqrt{5}t}, \quad u(t) = \lambda_2 = -p_2(\sqrt{5} + 2)e^{-\sqrt{5}t}.$$

From the equation for $x_1(t)$ with the boundary condition $x_1(0) = p_1$ we derive that $x_1(t) = p_1 e^t$.

The parameter $\lambda_1(t)$ can be found from the differential equation up to a constant, i.e., $\lambda_1 = \beta e^{-t}$. Thus, we have established that

$$\begin{pmatrix} x_1(t) \\ x_2(t) \end{pmatrix} = \begin{pmatrix} p_1 e^t \\ p_2 e^{-\sqrt{5}t} \end{pmatrix}$$

and if the optimal control $u(t)$ exists, it has the form $u(t) = -p_2(\sqrt{5}+2)e^{-\sqrt{5}t}$; moreover, $x_2(t) = p_2 e^{-\sqrt{5}t}$ along the optimal trajectory. However, we do not know if there exists at least one control $u(t)$ such that the integral

$$\int_0^\infty [x_2^2(t) + u^2(t)]dt$$

is finite. But if it exists, then there exists a unique optimal control which coincides with the constructed control. For the constructed control the integral is finite:

$$\int_0^\infty [x_2^2(t) + u^2(t)]dt = \int_0^\infty p_2^2[1 + (\sqrt{5}+2)^2]e^{-2\sqrt{5}t}dt = \frac{1 + (\sqrt{5}+2)^2}{2\sqrt{5}} p_2^2.$$

It is clear that the problem is solvable and we have found its solution

$$x_1(t) = p_1 e^t, \quad x_2(t) = p_2 e^{-\sqrt{5}t},$$
$$u(t) = -p_2(\sqrt{5}+2)e^{-\sqrt{5}t}.$$

We note that for $p \neq 0$ the component $x_1(t)$ unboundedly increases as $t \to \infty$. However, it does not prevent the convergence of the minimized integral since the latter does not contain $x_1(t)$. The same is true with respect to the indefiniteness of the parameter $\lambda_1(t)$ which contains an arbitrary coefficient β.

In what follows, we will discuss how to recognize whether or not the matrix pair A, B is stabilizable and controllable by studying properties of solutions to the equations with the optimal control $u(t)$.

§5. Hamiltonian systems of equations and their use in the study of stabilizability

Hamiltonian systems of ordinary differential equations. The Poincaré–Lyapunov theorem on characteristic roots for a Hamiltonian matrix. A Hamiltonian matrix $\mathcal{H}$ without purely imaginary eigenvalues and the structure of a basis for the invariant subspace of $\mathcal{H}$ that corresponds to the eigenvalues of $\mathcal{H}$ with negative real part. A Hamiltonian system constructed with the help of the stabilizable matrix pair A, B. The possibility of choosing a solution decreasing as $t \to \infty$. The formulation of the problem on a half-line. Stabilizability is equivalent to the Lopatinskii condition. A controllability criterion. One more necessary and sufficient condition for stabilizability.

Let p and q be N-dimensional vectors and let $H = H(p,q)$ be a function of $2N$ variables $p_1, p_2, \ldots p_N$; $q_1, q_2, \ldots, q_N$. We consider the case where $H(p,q)$ is the quadratic form

$$H = \frac{1}{2}\Big(\sum_{i,j=1}^N H_{p_i p_j} p_i p_j + \sum_{i,j=1}^N H_{p_i q_j} p_i q_j + \sum_{i,j=1}^N H_{q_j p_i} q_j p_i + \sum_{i,j=1}^N H_{q_i q_j} q_i q_j \Big)$$

and the partial derivatives $H_{p_ip_j}$, $H_{p_iq_j}$, $H_{q_ip_j}$, $H_{q_iq_j}$ are constant. In various problems of mechanics and calculus of variations, one deals with equations of the form

$$\frac{dp_i}{dt} = \frac{\partial H}{\partial q_i}, \quad \frac{dq_i}{dt} = -\frac{\partial H}{\partial p_i}$$

or, in detailed notation,

$$\begin{aligned}\frac{dp_i}{dt} &= \sum_{j=1}^{N} \frac{\partial^2 H}{\partial q_i \partial p_j} p_j + \sum_{j=1}^{N} \frac{\partial^2 H}{\partial q_i \partial q_j} q_j, \\ \frac{dq_i}{dt} &= -\sum_{j=1}^{N} \frac{\partial^2 H}{\partial p_i \partial p_j} p_j - \sum_{j=1}^{N} \frac{\partial^2 H}{\partial p_i \partial q_j} q_j.\end{aligned}$$

Such systems are called Hamiltonian systems. We note that the matrices

$$\Phi = \frac{\partial^2 H}{\partial q_i \partial q_j}, \quad \Psi = \frac{\partial^2 H}{\partial p_i \partial p_j}$$

are symmetric ($\Phi = \Phi^*$, $\Psi = \Psi^*$) and the matrices

$$A = \frac{\partial^2 H}{\partial q_i \partial p_j}, \quad -A^* = -\frac{\partial^2 H}{\partial p_i \partial q_j}$$

are obtained each from other by transposition and changing the sign of every element. This fact allows us to write the Hamiltonian system in the form

$$\begin{aligned}\frac{dp}{dt} &= Ap + \Phi q, \\ \frac{dq}{dt} &= \Psi p - A^* q.\end{aligned}$$

We recall that in §§2,4 we already considered such equations

$$\begin{aligned}\frac{d}{dt} x &= Ax + B(B^*B)^{-1}B^*\lambda, \\ \frac{d}{dt}\lambda &= C^*Cx - A^*\lambda\end{aligned}$$

in order to describe optimal controls, but we used another notation: x, λ instead of p, q, and $\Psi = C^*C$, $\Phi = B(B^*B)^{-1}B^*$.

Hamiltonian systems have many remarkable properties, some of which will be useful later. A matrix of coefficients of a Hamiltonian system, i.e., a real matrix of the form

$$\begin{bmatrix} A & \Phi \\ \Psi & -A^* \end{bmatrix}, \quad \Phi = \Phi^*, \quad \Psi = \Psi^*$$

is said to be Hamiltonian.

It is easy to check that the characteristic polynomial of a Hamiltonian matrix

$$\det \begin{bmatrix} A - \mu I & \Phi \\ \Psi & -A^* - \mu I \end{bmatrix}$$

does not change if μ is replaced with $-\mu$. Indeed,

$$\det\begin{bmatrix} A-\mu I & \Phi \\ \Psi & -A^*-\mu I \end{bmatrix} = (-1)^N \det\begin{bmatrix} \Phi & A-\mu I \\ -A^*-\mu I & \Psi \end{bmatrix}$$
$$= \det\begin{bmatrix} -\Phi & A-\mu I \\ A^*+\mu I & \Psi \end{bmatrix} = \det\begin{bmatrix} -\Phi & A-\mu I \\ A^*+\mu I & \Psi \end{bmatrix}^*$$
$$= \det\begin{bmatrix} -\Phi & A+\mu I \\ A^*-\mu I & \Psi \end{bmatrix} = \det\begin{bmatrix} A+\mu I & \Phi \\ \Psi & -A^*+\mu I \end{bmatrix}.$$

In other words, the characteristic polynomial of a Hamiltonian matrix contains only even powers of μ. Hence if the characteristic equation

$$\det\begin{bmatrix} A-\mu I & \Phi \\ \Psi & -A^*-\mu I \end{bmatrix} = 0$$

has a root μ_0 of multiplicity k, then it also has the root $-\mu_0$ of the same multiplicity k. This assertion is known as the Poincaré–Lyapunov theorem. We note that if $\mu = 0$ is a characteristic root, then its multiplicity is even.

The above arguments mean that if N_1 roots (counting their multiplicity) lie in the left half-plane, then N_1 roots lie in the right half-plane and $2(N - N_1)$ roots are purely imaginary. Consequently, $N_1 \leqslant N$.

We assume that for any initial vector $x(0) = a$ we can find a vector $\lambda(0)$ such that a solution $x(t)$, $\lambda(t)$ to the Hamiltonian system

$$\frac{d}{dt}x(t) = Ax(t) + \Phi\lambda(t),$$
$$\frac{d}{dt}\lambda(t) = \Psi x(t) - A^*\lambda(t)$$

with the initial data $x(0)$, $\lambda(0)$ tends to zero as $t \to \infty$:

$$\|x(t)\| \to 0, \quad \|\lambda(t)\| \to 0.$$

There exist at least N linearly independent vectors

$$\begin{pmatrix} 1 \\ 0 \\ 0 \\ \vdots \\ 0 \\ \lambda^{(1)}(0) \end{pmatrix}, \begin{pmatrix} 0 \\ 1 \\ 0 \\ \vdots \\ 0 \\ \lambda^{(2)}(0) \end{pmatrix}, \dots, \begin{pmatrix} 0 \\ 0 \\ 0 \\ \vdots \\ 1 \\ \lambda^{(N)}(0) \end{pmatrix} \begin{matrix} \left.\vphantom{\begin{matrix}0\\0\\0\\0\\0\end{matrix}}\right\} N \text{ components} \\ \} \; N \text{ components} \end{matrix}$$

such that taking them for the initial data we obtain a solution decreasing in norm as $t \to \infty$. Any invariant subspace of a Hamiltonian matrix containing these vectors has dimension at least N. We consider the minimal invariant subspace. It is easy to see that any solution with the initial data from the minimal invariant subspace tends to zero as $t \to \infty$. Thereby, all characteristic roots corresponding to this subspace lie in the left half-plane. The number of such roots coincides with the dimension of the subspace and cannot be less than N. But it cannot be greater than N. Thus, a Hamiltonian matrix cannot have purely imaginary eigenvalues (including zero). Furthermore, N eigenvalues lie in the left half-plane and N eigenvalues lie in the right half-plane.

Consider the $2N \times N$-matrix composed of two square blocks

$$\begin{bmatrix} X_0 \\ \Lambda_0 \end{bmatrix}$$

so that its columns constitute a basis for the invariant subspace corresponding to all eigenvalues lying in the left half-plane. Note that the square matrix X_0 is nonsingular ($\det X_0 \neq 0$). Indeed, by assumption, any N-dimensional vector a can be expressed as a linear combination of columns of X_0. If R is a nonsingular matrix, then the columns of the matrix

$$\begin{bmatrix} X_0 \\ \Lambda_0 \end{bmatrix} R = \begin{bmatrix} X_0 R \\ \Lambda_0 R \end{bmatrix}$$

form the basis for the same space. In particular, taking $R = X_0^{-1}$, we obtain the basis formed by columns of the matrix with the identity upper block:

$$\begin{bmatrix} I \\ \Lambda \end{bmatrix} = \begin{bmatrix} I \\ \Lambda_0 X_0^{-1} \end{bmatrix}.$$

This helps us to study an optimal control for the matrix stabilizable pair A, B in detail. If the matrix pair A, B is stabilizable, then for any $x(0)$ there exists a control $u(t)$ such that for a solution to the equation

$$\frac{d}{dt}x(t) = Ax(t) + Bu(t)$$

with the initial value $x(0)$, the integral

$$\int_0^\infty (\|x(t)\|^2 + \|Bu(t)\|^2)dt \tag{1}$$

is finite. In this case, we can choose a control $u(t)$ (cf. §4) such that the integral (1) attains the minimal value. The optimal control can be constructed by the formula $u(t) = (B^*B)^{-1}B^*\lambda(t)$ by solving the Hamiltonian system

$$\begin{aligned} &\frac{d}{dt}x(t) = Ax(t) + B(B^*B)^{-1}B^*\lambda(t), \\ &\frac{d}{dt}\lambda(t) = x(t) - A^*\lambda(t) \end{aligned} \tag{2}$$

with the initial value $x(0)$. As we know, $u(t) \to 0$ as $t \to \infty$. We can ask whether or not these conditions uniquely determine a solution to the Hamiltonian system and, thereby, whether it is possible to find a control $u(t)$ by solving the boundary-value problem (2) on the half-line $0 \leqslant t < \infty$. It turns out that for a given stabilizable matrix pair A, B these conditions determine a solution to the Hamiltonian system, but not in a unique way. However, the nonuniqueness does not affect the choice of $u(t)$. We can impose some additional conditions on $\lambda(t)$ under which the Hamiltonian system is solvable and the solution is unique. We now justify this assertion.

As in §3, it is convenient to make the orthogonal transformation U ($U^*U = I$) of variables x and λ and the corresponding transformation of the matrices A and B as follows:

$$x = U^*\widehat{x}, \quad \lambda = U^*\widehat{\lambda},$$
$$A = U^* \begin{bmatrix} \widehat{A}_{11} & \widehat{A}_{12} \\ 0 & \widehat{A}_{22} \end{bmatrix} U, \quad B = U^* \begin{pmatrix} \widehat{B}_1 \\ 0 \end{pmatrix}.$$

The Hamiltonian system takes the form

$$\begin{aligned} \frac{d}{dt}\widehat{x}_1 &= \widehat{A}_{11}\widehat{x}_1 + \widehat{A}_{12}\widehat{x}_2 + \widehat{B}_1(\widehat{B}_1^*\widehat{B}_1)\widehat{B}_1^*\widehat{\lambda}_1, \\ \frac{d}{dt}\widehat{x}_2 &= \widehat{A}_{22}\widehat{x}_2, \\ \frac{d}{dt}\widehat{\lambda}_1 &= \widehat{x}_1 - \widehat{A}_{11}^*\widehat{\lambda}_1, \\ \frac{d}{dt}\widehat{\lambda}_2 &= \widehat{x}_2 - \widehat{A}_{12}^*\widehat{\lambda}_1 - \widehat{A}_{22}^*\widehat{\lambda}_2. \end{aligned} \tag{3}$$

We recall that we choose this canonical form in order to provide the controllability of the matrix pair $\widehat{A}_{11}$, $\widehat{B}_1$ and, assuming that the matrix pair A, B is stabilizable, we deduce that the spectrum of $\widehat{A}_{22}$ lies in the left half-plane. Moreover, the expression $(\widehat{B}_1^*\widehat{B}_1)^{-1}$ makes sense. The optimal control $u(t)$ is found from $\widehat{\lambda}_1(t)$ by the formula

$$u(t) = (\widehat{B}_1^*\widehat{B}_1)^{-1}\widehat{B}_1^*\widehat{\lambda}_1(t).$$

Conversely, if $u(t)$ is known, then $\widehat{\lambda}_1(t)$ can be found from the linear system of equations

$$\begin{aligned} &\widehat{B}_1^*\widehat{\lambda}_1(t) = \widehat{B}_1^*\widehat{B}_1 u(t), \\ &\widehat{A}_{11}^*\widehat{B}_1^*\widehat{\lambda}_1(t) = \widehat{A}_{11}^*\widehat{B}_1^*\widehat{B}_1 u(t), \\ &(\widehat{A}_{11}^*)^2\widehat{B}_1^*\widehat{\lambda}_1 = (\widehat{A}_{11}^*)^2\widehat{B}_1^*\widehat{B}_1 u(t), \\ &\dots\dots\dots\dots\dots\dots\dots\dots\dots\dots\dots \\ &(\widehat{A}_{11}^*)^{N_1-1}\widehat{B}_1^*\widehat{\lambda}_1(t) = (\widehat{A}_{11}^*)^{N-1}\widehat{B}_1^*\widehat{B}_1 u(t), \end{aligned}$$

where N_1 denotes the order of $\widehat{A}_{11}$. These equations are solvable since the matrix pair $\widehat{A}_{11}^*$, $\widehat{B}_1^*$ is observable (the matrix pair $\widehat{A}_{11}$, $\widehat{B}_1$ is controllable). Consequently, the rank of the composite matrix

$$\begin{bmatrix} \widehat{B}_1^* \\ \widehat{A}_{11}^*\widehat{B}_1^* \\ \vdots \\ (A_{11}^*)^{N_1-1}\widehat{B}_1^* \end{bmatrix}$$

is N_1 (cf. §1). Moreover, $\widehat{\lambda}_1(t)$ is uniquely determined by a given $u(t)$. It is obvious that if $u(t) \to 0$ as $t \to \infty$, then $\widehat{\lambda}_1(t) \to 0$ as $t \to +\infty$.

We rewrite the Hamiltonian system, rearranging the equations:

$$\begin{aligned}
\frac{d}{dt}\widehat{\lambda}_2 &= -\widehat{A}_{22}^*\lambda_2 - \widehat{A}_{12}^*\widehat{\lambda}_1 + \widehat{x}_2,\\
\frac{d}{dt}\widehat{x}_1 &= \widehat{A}_{11}\widehat{x}_1 + \widehat{B}_1(\widehat{B}_1^*\widehat{B}_1)^{-1}\widehat{B}_1^*\widehat{\lambda}_1 + \widehat{A}_{12}\widehat{x}_2,\\
\frac{d}{dt}\widehat{\lambda}_1 &= \widehat{x}_1 - \widehat{A}_{11}^*\widehat{\lambda}_1,\\
\frac{d}{dt}\widehat{x}_2 &= \widehat{A}_{22}\widehat{x}_2.
\end{aligned}$$

As we know, the optimal control $u(t)$ ($u(t) \to 0$ as $t \to \infty$) exists and is unique and also there exists the unique trajectory $\widehat{x}_1(t)$, $\widehat{x}_2(t)$. Therefore, there exists a solution $\widehat{x}_1(t)$, $\widehat{x}_2(t)$, $\widehat{\lambda}_1(t)$ to the last three equations for any given $\widehat{x}_1(0)$, $\widehat{x}_2(0)$ such that the integrals

$$\int_0^\infty \|\widehat{x}_1(t)\|^2 dt, \quad \int_0^\infty \|\widehat{x}_2(t)\|^2 dt, \quad \int_0^\infty \|\widehat{\lambda}_1(t)\|^2 dt$$

converge. Using the results from Chapter 1, §5 it is easy to show that $\|\widehat{x}_i(t)\|$ and $\|\widehat{\lambda}_1(t)\|$ exponentially converge to zero as $t \to \infty$:

$$\|\widehat{x}_1(t)\| < Ke^{-\sigma t}, \quad \|\widehat{x}_2(t)\| < Ke^{-\sigma t}, \quad \|\widehat{\lambda}_1(t)\| < Ke^{-\sigma t}.$$

Therefore, $\varphi(t) = -\widehat{A}_{12}^*\widehat{\lambda}_1 + \widehat{x}_2$ exponentially decreases:

$$\|\varphi(t)\| \leqslant K'e^{-\sigma t}.$$

We consider the equation

$$\frac{d}{dt}\widehat{\lambda}_2 = -\widehat{A}_{22}^*\widehat{\lambda}_2 - \widehat{A}_{12}^*\widehat{\lambda}_1 + \widehat{x}_2(t) = -\widehat{A}_{22}^*\widehat{\lambda}_2 + \varphi(t) \tag{4}$$

and take a solution

$$\widehat{\lambda}_{20}(t) = \int_t^\infty e^{(s-t)\widehat{A}_{22}^*}\varphi(s)ds.$$

The verification of the convergence of the integral is based on the fact that the spectrum of $\widehat{A}_{22}$, and consequently the spectrum of $\widehat{A}_{22}^*$, lies in the left half-plane, which yields

$$\begin{gathered}
\|e^{(s-t)\widehat{A}_{22}^*}\| \leqslant K''e^{-\sigma''(s-t)} \quad (s \geqslant t),\\
\|e^{(s-t)\widehat{A}_{22}^*}\varphi(s)\| \leqslant K'K''e^{-(\sigma''+\sigma)(s-t)}e^{-\sigma t},\\
\|\widehat{\lambda}_2(t)\| \leqslant \frac{K'K''}{\sigma+\sigma''}e^{-\sigma t}.
\end{gathered}$$

The constructed solution tends to zero as $t \to \infty$ and is unique since all eigenvalues of $-\widehat{A}_{22}^*$ (cf. (4)) lie in the right half-plane.

Thus, if the Hamiltonian system (3) is stabilizable, then for any $\widehat{x}_1(0)$, $\widehat{x}_2(0)$ there exists a solution $\widehat{x}_1(t)$, $\widehat{x}_2(t)$, $\widehat{\lambda}_1(t)$, $\widehat{\lambda}_2(t)$ such that $\|\widehat{x}_i(t)\| \to 0$ and $\|\widehat{\lambda}_j(t)\| \to 0$ as $t \to +\infty$. It is obvious that the similar assertion is true for the original Hamiltonian system. Thus, we have proved the following lemma.

LEMMA 1. *If the matrix pair A, B is stabilizable, then for any $x(0)$ there exists a solution $x(t)$, $\lambda(t)$ to the system*

$$\frac{d}{dt}x(t) = Ax(t) + B(B^*B)^{-1}B^*\lambda(t),$$
$$\frac{d}{dt}\lambda(t) = x(t) - A^*\lambda(t)$$

defined on the half-line $0 \leqslant t < +\infty$ and such that $\|x(t)\| \to 0$, $\|\lambda(t)\| \to \infty$ as $t \to \infty$.

By Lemma 1 and the properties of Hamiltonian matrices (cf. the beginning of this section), the matrix

$$\begin{bmatrix} A & B(B^*B)^{-1}B^* \\ I & -A^* \end{bmatrix}$$

has exactly N eigenvalues in the left half-plane, N eigenvalues in the right half-plane, and no eigenvalues (including zero) on the imaginary axis. For a basis of the invariant subspace corresponding to eigenvalues in the left half-plane we can take columns of the rectangular matrix $\begin{bmatrix} I \\ \Lambda \end{bmatrix}$. A solution to the Hamiltonian system decreasing as $t \to \infty$ has the form

$$\begin{pmatrix} x(t) \\ \lambda(t) \end{pmatrix} = \exp\left[t\begin{bmatrix} A & B(B^*B)^{-1}B^* \\ I & -A^* \end{bmatrix}\right]\begin{bmatrix} I \\ \Lambda \end{bmatrix} x(0)$$

and, in view of the last formulas, is uniquely determined by $x(0)$.

Thus, if the matrix pair A, B is stabilizable, then the boundary-value problem

$$x(0) = a, \quad \|x(t)\| \to 0, \quad \|\lambda(t)\| \to 0 \text{ as } t \to \infty,$$
$$\frac{d}{dt}\begin{pmatrix} x(t) \\ \lambda(t) \end{pmatrix} = \begin{bmatrix} A & B(B^*B)^{-1}B^* \\ I & -A^* \end{bmatrix}\begin{pmatrix} x(t) \\ \lambda(t) \end{pmatrix}$$

on the half-line $0 \leqslant t < +\infty$ has a unique solution for any a. In other words, the Lopatinskii condition holds (cf. Chapter 1, §13). Conversely, by the Lopatinskii condition, for any $x(0) = a$ the Hamiltonian system has a unique solution $x(t)$, $\lambda(t)$ exponentially decreasing as $t \to +\infty$. The control $u(t) = (B^*B)^{-1}B^*\lambda(t)$ also exponentially decreases. A solution $x(t)$ to the system

$$\frac{d}{dt}x(t) = Ax(t) + Bu(t) \quad (x(0) = a)$$

with such a control coincides with the solution $x(t)$ to the Hamiltonian system and exponentially decreases too. Therefore, the integral

$$\int_0^\infty (\|x(t)\|^2 + \|Bu(t)\|^2)dt$$

converges, which means that the pair A, B is stabilizable.

Thus, for the Hamiltonian system under consideration the Lopatinskii condition is equivalent to the stabilizability of the pair A, B.

Recall that the stabilizability of both pairs A, B and $-A, B$ is equivalent to the unique solvability for any a of the boundary-value problem

$$\frac{d}{dt}\begin{pmatrix} x(t) \\ \lambda(t) \end{pmatrix} = \begin{bmatrix} A & B(B^*B)^{-1}B^* \\ I & -A^* \end{bmatrix}\begin{pmatrix} x(t) \\ \lambda(t) \end{pmatrix} \quad \text{for } t \neq 0,$$
$$\|x(t)\| \to 0, \quad \|\lambda(t)\| \to 0 \quad \text{as } t \to \pm\infty, \quad x(0) = a$$

on the entire line $-\infty < t < \infty$. The accurate verification of this assertion is left to the reader.

We give one more necessary and sufficient condition for the stabilizability of the pair A, B.

PROPOSITION 1. *The matrix pair A, B is stabilizable if and only if there is a Hurwitz matrix of the form $A + BM$.*

PROOF. *Sufficiency.* Since $A + BM$ is a Hurwitz matrix, for any $x(0)$ the corresponding solution $x(t)$ to the equation

$$\frac{d}{dt}x = Ax + BMx$$

exponentially decreases as $t \to \infty$. Setting $u(t) = Mx(t)$, we obtain a control such that for a solution $x(t)$ to the system

$$\frac{d}{dt}x = Ax + Bu(t)$$

with the initial data $x(0)$ the integral

$$\int_0^\infty (\|x(t)\|^2 + \|Bu(t)\|^2)dt = \int_0^\infty (\|x(t)\|^2 + \|BMx(t)\|^2)dt$$

is finite.

Necessity. As we know, the stabilizability of A, B implies the existence of the N-dimensional invariant subspace of the Hamiltonian matrix

$$\mathcal{H} = \begin{bmatrix} A & B(B^*B)^{-1}B^* \\ I & -A^* \end{bmatrix}$$

such that vectors of the form

$$\begin{pmatrix} x \\ \lambda \end{pmatrix} = \begin{bmatrix} I \\ \Lambda \end{bmatrix} x = \begin{pmatrix} x \\ \Lambda x \end{pmatrix}$$

are contained in this subspace and for any $x(0)$, $\lambda(0)$ in the subspace ($\lambda(0) = \Lambda x(0)$) the solution $x(t)$, $\lambda(t)$ to the Hamiltonian system

$$\frac{d}{dt}\begin{pmatrix} x(t) \\ \lambda(t) \end{pmatrix} = \mathcal{H}\begin{pmatrix} x(t) \\ \lambda(t) \end{pmatrix}$$

with the initial data $x(0)$, $\lambda(0)$ belongs to this subspace ($\lambda(t) = \Lambda x(t)$) for all $t > 0$. Moreover, $\|x(t)\| \to 0$ as $t \to \infty$. For such solutions we have

$$\frac{d}{dt}x = Ax(t) + B(B^*B)^{-1}B^*\lambda(t) = Ax(t) + B(B^*B)^{-1}B^*\Lambda x(t)$$
$$= [A + B(B^*B)^{-1}B^*\Lambda]x(t).$$

Setting $M = (B^*B)^{-1}B^*\Lambda$, we can assert that any solution to the vector equation

$$\frac{d}{dt}x(t) = (A + BM)x(t)$$

tends to zero as $t \to \infty$, which may be only if the spectrum of the matrix $A + BM$ lies in the left half-plane. □

§6. Further study of variational problems on a half-line. The notion of detectability

A functional with not necessarily positive definite quadratic form in $x(t)$. The existence of an optimal control for a stabilizable pair A, B such that a solution $\lambda(t)$ to the Hamiltonian system tends to zero as $t \to \infty$. Examples. The definition of detectability. The properties of the boundary-value problem with stabilizable matrix pair A, B and detectable matrix pair A, C. The duality of notions of detectability and stabilizability.

Let the matrix pair A, B be stabilizable, i.e., there exists a control $u(t)$ such that for a solution $x(t)$ to the Cauchy problem

$$\frac{d}{dt}x(t) = Ax(t) + Bu(t)$$

with a given $x(0)$ the integral

$$\mathcal{J}_I = \int_0^\infty (\|x(t)\|^2 + \|Bu(t)\|^2)dt$$

converges and takes a finite limit value. For any matrix C the integral

$$\mathcal{J}_C = \int_0^\infty (\|Cx(t)\|^2 + \|Bu(t)\|^2)dt$$

is finite. We assume that the rows of the matrix C are linearly independent. This assumption is similar to the above assumption that columns of B are linearly independent.

Under these assumptions, there exists a unique optimal control $u(t)$ such that the integral $\mathcal{J}_C$ takes the minimal value. The optimal control $u(t)$ and the corresponding trajectory $x(t)$ satisfy the equations

$$\begin{aligned}
&\frac{d}{dt}x(t) = Ax(t) + B(B^*B)^{-1}B^*\lambda(t),\\
&\frac{d}{dt}\lambda(t) = C^*Cx(t) - A^*\lambda(t),\\
&u(t) = (B^*B)^{-1}B^*\lambda(t)
\end{aligned}$$

and the condition that the trajectory $x(t)$ passes through a given point $x(0)$ at $t = 0$. Moreover, $\|u(t)\| \to 0$ as $t \to \infty$. As was mentioned, it does not mean that $\lambda(t) \to 0$ as $t \to \infty$. If the matrix pair A, B is only stabilizable, but not controllable, then $\lambda(t)$ is not uniquely determined. However, this does not affect the values of $u(t)$.

It is possible to choose an admissible $\lambda(t)$ such that $\|\lambda(t)\| \to 0$ as $t \to \infty$ and this condition uniquely determines $\lambda(t)$. The proof of this fact repeats similar arguments from §5 where this problem is considered with $C = I$. Therefore, we omit details.

As in §§3,5, we make the orthogonal transformation

$$x = U^* \begin{pmatrix} \widehat{x}_1 \\ \widehat{x}_2 \end{pmatrix}, \quad \lambda = U^* \begin{pmatrix} \widehat{\lambda}_1 \\ \widehat{\lambda}_2 \end{pmatrix},$$

$$A = U^* \begin{bmatrix} \widehat{A}_{11} & \widehat{A}_{12} \\ 0 & \widehat{A}_{22} \end{bmatrix} U, \quad B = U^* \begin{bmatrix} \widehat{B}_1 \\ 0 \end{bmatrix}, \quad C = \left(\widehat{C}_1 \vdots \widehat{C}_2\right) U,$$

replace the original Hamiltonian system for x, λ by the Hamiltonian system

$$\begin{aligned}
\frac{d}{dt}\widehat{x}_1 &= \widehat{A}_{11}\widehat{x}_1 + \widehat{A}_{12}\widehat{x}_2 + \widehat{B}_1(\widehat{B}_1^*\widehat{B}_1)^{-1}\widehat{B}_1^*\widehat{\lambda}_1, \\
\frac{d}{dt}\widehat{x}_2 &= \widehat{A}_{22}\widehat{x}_2, \\
\frac{d}{dt}\widehat{\lambda}_1 &= \widehat{C}_1^*(\widehat{C}_1\widehat{x}_1 + \widehat{C}_2\widehat{x}_2) - \widehat{A}_{11}^*\widehat{\lambda}_1, \\
\frac{d}{dt}\widehat{\lambda}_2 &= \widehat{C}_2^*(\widehat{C}_1\widehat{x}_1 + \widehat{C}_2\widehat{x}_2) - \widehat{A}_{12}^*\widehat{\lambda}_1 - \widehat{A}_{22}^*\widehat{\lambda}_2,
\end{aligned}$$

where the matrix pair $\widehat{A}_{11}$, $\widehat{B}_1$ is controllable and all eigenvalues of $\widehat{A}_{22}$ lie in the left half-plane. The control $u(t)$ is connected with $\widehat{\lambda}_1(t)$ by the equality

$$u(t) = (\widehat{B}_1^*\widehat{B}_1)^{-1}\widehat{B}_1^*\widehat{\lambda}_1(t)$$

which allows us (cf. §5) to restore uniquely $\widehat{\lambda}_1(t)$ from $u(t)$ since the matrix pair $\widehat{A}_{11}$, $\widehat{B}_1$ is controllable. Moreover,

$$\|Cx\|^2 = \|\widehat{C}_1\widehat{x}_1 + \widehat{C}_2\widehat{x}_2\|^2.$$

Thus, for the optimal trajectory $\widehat{\lambda}_1(t)$ we see that $\widehat{C}_1\widehat{x}_1(t) + \widehat{C}_2\widehat{x}_2(t)$ tends to zero as $t \to \infty$ and the system of linear differential equations with constant coefficients

$$\begin{aligned}
\frac{d}{dt}\widehat{x}_1 &= \widehat{A}_{11}\widehat{x}_1 + \widehat{A}_{12}\widehat{x}_2 + \widehat{B}_1(\widehat{B}_1^*\widehat{B}_1)^{-1}\widehat{B}_1^*\widehat{\lambda}_1, \\
\frac{d}{dt}\widehat{x}_2 &= \widehat{A}_{22}\widehat{x}_2, \\
\frac{d}{dt}\widehat{\lambda}_1 &= \widehat{C}_1^*[\widehat{C}_1\widehat{x}_1 + \widehat{C}_2\widehat{x}_2] - \widehat{A}_{11}^*\widehat{\lambda}_1
\end{aligned}$$

is satisfied. This implies (cf. Chapter 1, §5) that $\widehat{\lambda}_1(t)$ and $\widehat{C}_1x_1(t) + \widehat{C}_2x_2(t)$ exponentially tend to zero:

$$\|\widehat{C}_1\widehat{x}_1(t) + \widehat{C}_2\widehat{x}_2(t)\| < Ke^{-\sigma t}, \quad \|\widehat{\lambda}_1\| < Ke^{-\sigma t},$$

where $K > 0$ and $\sigma > 0$. Therefore,

$$\|\widehat{C}_2^*[\widehat{C}_1\widehat{x}_1(t) + \widehat{C}_2\widehat{x}_2(t)] - \widehat{A}_{12}^*\widehat{\lambda}_1(t)\| < K'e^{-\sigma t}.$$

For $\widehat{\lambda}_2(t)$ we have the equation

$$\frac{d}{dt}\widehat{\lambda}_2 + \widehat{A}_{22}^*\widehat{\lambda}_2 = \varphi(t) \equiv \widehat{C}_2^*[\widehat{C}_1\widehat{x}_1 + \widehat{C}_2\widehat{x}_2(t)] - \widehat{A}_{12}^*\widehat{\lambda}_1(t),$$
$$\|\varphi(t)\| < K'e^{-\sigma t}.$$

Since $\varphi(t)$ exponentially decreases, there exists (cf. Chapter 1, §9) at least one solution $\widehat{\lambda}_2(t)$ such that $\|\lambda_2(t)\|$ exponentially decreases as $t \to \infty$. As we know, all eigenvalues of $\widehat{A}_{22}$, and consequently, of A_{22}^* lie in the left half-plane. Hence there is only one solution $\widehat{\lambda}_2$ to the equation

$$\frac{d}{dt}\widehat{\lambda}_2 + \widehat{A}_{22}^*\widehat{\lambda}_2 = \varphi(t),$$

which tends to zero as $t \to \infty$. This decreasing solution (it exists and exponentially decreases) is taken for $\widehat{\lambda}_2(t)$. The exponential decrease of $\|\widehat{\lambda}_1(t)\|$ and $\|\widehat{\lambda}_2(t)\|$ means the exponential decrease of the norm

$$\|\widehat{\lambda}(t)\| = \sqrt{\|\widehat{\lambda}_1(t)\|^2 + \|\widehat{\lambda}_2(t)\|^2}.$$

As in §5, we can prove that the norm

$$\|u(t)\| = \|(\widehat{B}_1^*\widehat{B}_1)^{-1}\widehat{B}_1^*\widehat{\lambda}_1(t)\|$$

of the control $u(t)$ exponentially decreases. Thus, for the system written in the canonical form we have proved that the solution can be restored from $\widehat{\lambda}_1(t)$, $\widehat{\lambda}_2(t)$, $\widehat{C}_1\widehat{x}_1(t) + \widehat{C}_2\widehat{x}_2(t)$ such that

$$\|\widehat{C}_1\widehat{x}_1(t) + \widehat{C}_2\widehat{x}_2(t)\| \to 0, \quad \|\widehat{\lambda}_1(t)\| \to 0, \quad \|\widehat{\lambda}_2(t)\| \to 0$$

as $t \to \infty$. Now, it is obvious that for the stabilizable pair A, B the original Hamiltonian system

$$\frac{d}{dt}x(t) = Ax(t) + B(B^*B)^{-1}B^*\lambda(t),$$
$$\frac{d}{dt}\lambda(t) = C^*Cx(t) - A^*\lambda(t)$$

has a solution corresponding to the control

$$u(t) = (B^*B)^{-1}B^*\lambda(t)$$

such that the integral

$$\mathcal{J}_C = \int_0^\infty (\|Cx(t)\|^2 + \|Bu(t)\|^2)dt$$

takes the minimal value and

$$\|Cx(t)\| \to 0, \quad \|\lambda(t)\| \to 0, \quad \|u(t)\| \to 0$$

as $t \to \infty$; moreover, $\|Cx(t)\|$, $\|Bu(t)\|$ exponentially decrease with some exponent $\sigma > 0$:

$$\|Cx(t)\| < Ke^{-\sigma t}, \quad \|Bu(t)\| < \|B\|\, Me^{-\sigma t}.$$

It is important to emphasize that we prove the convergence to zero only for $\|Cx(t)\|$ but not for $\|x(t)\|$, although the stabilizability of the matrix pair A, B

implies the existence of a control which provides the convergence of $\|x(t)\|$ to zero (we have used this fact in the proof above). The fact is that among all controls we try to find a control which provides the minimal value of $\mathcal{J}_C$, but on this optimal control $\|x(t)\|$ does not necessarily tend to zero.

EXAMPLE 1. Let

$$A = \begin{bmatrix} 1 & 0 \\ 0 & 2 \end{bmatrix}, \quad B = \begin{bmatrix} 1 \\ 1 \end{bmatrix}, \quad C = [1 \vdots 0].$$

The matrix pair A, B is controllable, and consequently, is stabilizable. Therefore, for any $x_1(0)$, $x_2(0)$ it is possible to construct a control $u(t)$ such that for a solution to the system

$$\frac{dx_1}{dt} = x_1 + u,$$
$$\frac{dx_2}{dt} = 2x_2 + u$$

the integral

$$\mathcal{J} = \int_0^\infty (x_1^2 + u^2)dt$$

takes the minimal value. The formula

$$u(t) = (B^*B)^{-1}B^*\lambda(t) \equiv \frac{1}{2}\lambda_1(t) + \frac{1}{2}\lambda_2(t)$$

connects the optimal control with some solution to the Hamiltonian system

$$\frac{d}{dt}\begin{pmatrix} x_1 \\ x_2 \end{pmatrix} = \begin{bmatrix} 1 & 0 \\ 0 & 2 \end{bmatrix}\begin{pmatrix} x_1 \\ x_2 \end{pmatrix} + \frac{1}{2}\begin{bmatrix} 1 & 1 \\ 1 & 1 \end{bmatrix}\begin{pmatrix} \lambda_1 \\ \lambda_2 \end{pmatrix},$$
$$\frac{d}{dt}\begin{pmatrix} \lambda_1 \\ \lambda_2 \end{pmatrix} = \begin{bmatrix} 1 & 0 \\ 0 & 0 \end{bmatrix}\begin{pmatrix} x_1 \\ x_2 \end{pmatrix} - \begin{bmatrix} 1 & 0 \\ 0 & 2 \end{bmatrix}\begin{pmatrix} \lambda_1 \\ \lambda_2 \end{pmatrix}.$$

A general solution to this system has the form

$$x_1(t) = \frac{\alpha}{\sqrt{6}-2}\, e^{t\sqrt{3/2}} + \frac{\beta}{\sqrt{6}+2} e^{-t\sqrt{3/2}} - \frac{\gamma}{5}\, e^{-2t},$$
$$x_2(t) = -\frac{\alpha}{4-\sqrt{6}}\, e^{t\sqrt{3/2}} + \frac{\beta}{4+\sqrt{6}}\, e^{-t\sqrt{3/2}} - \frac{3\gamma}{20}\, e^{-2t} + \delta e^{2t},$$
$$\lambda_1(t) = \alpha e^{t\sqrt{3/2}} - \beta\, e^{-t\sqrt{3/2}} + \frac{\gamma}{5}\, e^{-2t},$$
$$\lambda_2(t) = \gamma e^{-2t}.$$

The corresponding control can be expressed as

$$u(t) = \frac{1}{2}[\lambda_1(t) + \lambda_2(t)] = \frac{\alpha}{2}\, e^{t\sqrt{3/2}} - \frac{\beta}{2}\, e^{-t\sqrt{3/2}} + \frac{3\gamma}{5}\, e^{-2t}.$$

It is easy to see that the condition $\alpha = 0$ is necessary for the convergence of the integral

$$\mathcal{J} = \int_0^\infty [x_1^2(t) + u^2(t)]dt.$$

In this case, x_1, λ_1, λ_2, u tends to zero as $t \to \infty$ for any β, γ, δ. In view of the initial conditions $x_1(0)$, $x_2(0)$ the coefficients β, γ, and δ are connected by the formula

$$\frac{\beta}{\sqrt{6}+2} - \frac{\gamma}{5} = x_1(0), \qquad \frac{\beta}{4+\sqrt{6}} - \frac{3\gamma}{20} + \delta = x_2(0).$$

The integral $\mathcal{J}$ is expressed in terms of β, γ as follows:

$$\begin{aligned}\mathcal{J} &= \int_0^\infty \left\{ \left[\frac{\beta}{\sqrt{6}+2} e^{-\sqrt{3/2}t} - \frac{\gamma}{5} e^{-2t} \right]^2 + \left[-\frac{\beta}{2} e^{-t\sqrt{3/2}} + \frac{3\gamma}{5} e^{-2t} \right]^2 \right\} dt \\ &= \beta^2 \left[\frac{1}{(\sqrt{6}+2)^2} + \frac{1}{4} \right] \int_0^\infty e^{-\sqrt{6}t} dt \\ &\quad - 2\beta\gamma \left[\frac{1}{5(\sqrt{6}+2)} + \frac{3}{10} \right] \int_0^\infty e^{-(2+\sqrt{3/2})t} dt + \gamma^2 \left(\frac{1}{25} + \frac{9}{25} \right) \int_0^\infty e^{-4t} dt \\ &= \beta^2 \frac{1}{\sqrt{6}} \left[\frac{1}{(\sqrt{6}+2)^2} + \frac{1}{4} \right] - 2\beta\gamma \frac{1}{2+\sqrt{3/2}} \left[\frac{1}{5(\sqrt{6}+2)} + \frac{3}{10} \right] + \frac{1}{10}\gamma^2.\end{aligned}$$

We see that for coefficients β, γ, δ we have only two equations from which these coefficients cannot be found in a unique way. With each possible choice of these coefficients we can associate the value of $\mathcal{J}$ which can be expressed in terms of β and γ only. Since $x_1(0)$ is given, for a given γ we find

$$\beta = \frac{2(x_1(0) + \gamma/5}{\sqrt{6}-2}$$

and express $\mathcal{J}$ as a function (quadratic polynomial) of γ:

$$\mathcal{J} = m_0 + m_1\gamma + m_2\gamma^2 = m_2 \left(\gamma + \frac{m_1}{2m_2} \right)^2 + \frac{4m_0m_2 - m_1^2}{m_2}.$$

To avoid cumbersome expressions, we do not compute coefficients. The functional $\mathcal{J}$ attains the minimal value at

$$\gamma = -\frac{m_1}{2m_2}.$$

Now we can find β. We recall that α was established above ($\alpha = 0$). It remains to determine δ. To do it we use the initial value $x_2(0)$ and the formula

$$x_2(t) = -\frac{\alpha}{4-\sqrt{6}} e^{\sqrt{3/2}t} + \frac{\beta}{4+\sqrt{6}} e^{-\sqrt{3/2}t} - \frac{3\gamma}{20} e^{-2t} + \delta e^{2t}$$

which implies that ($\alpha = 0$)

$$\delta = x_2(0) - \frac{\beta}{4+\sqrt{6}} + \frac{3\gamma}{20}.$$

Since $x_1(0)$, $x_2(0)$ are arbitrary, we cannot guarantee that $\delta = 0$. Hence $x_2(t)$ does not decrease as $t \to \infty$, but unboundedly increases.

In Example 1 we have shown that even if the matrix pair A, B is stabilizable, it is not necessary that $\|x(t)\| \to 0$ as $t \to \infty$ along the optimal trajectory.

Now we show that the above phenomena takes place for some initial data if C annihilates an invariant subspace of A which contains at least one eigenvector corresponding to an eigenvalue lying on the imaginary axis or to the right of it. The matrix pair A, B is assumed to be stabilizable.

Indeed, let z be an eigenvector of A ($Az = \mu z$) corresponding to an eigenvalue μ lying in the right half-plane or on the imaginary axis (i.e., $\operatorname{Re}\mu \geqslant 0$) and let $Cz = 0$. It is obvious that

$$e^{tA}z = e^{t\mu}z, \quad \|e^{tA}z\| = e^{t\operatorname{Re}\mu}\|z\| \geqslant \|z\|,$$
$$Ce^{tA}z = C[e^{t\mu}z] = e^{t\mu}Cz = 0.$$

For the control $u(t) = 0$ the system

$$\frac{d}{dt}x(t) = Ax(t) + Bu(t)$$

with the initial condition $x(0) = z$ has the solution $x(t) = e^{tA}z$. On this solution we have

$$\mathcal{J}_C = \int\limits_0^\infty (\|Cx(t)\|^2 + \|Bu(t)\|^2)dt = 0.$$

It is clear that $\mathcal{J}_C \geqslant 0$ for any control. Therefore, the control $u(t) = 0$ is optimal. It yields the minimal value of the integral $\mathcal{J}_C$. Since an optimal control and the corresponding trajectory are unique (cf. §4), for the initial condition $x(0) = z$ the optimal trajectory and control are as follows:

$$x(t) = e^{tA}z = e^{\mu t}z, \quad u(t) = 0;$$

moreover, $\|x(t)\| \geqslant \|z\|$ does not converge to zero as $t \to \infty$.

DEFINITION 1. The matrix pair A, C is said to be detectable if either there are no invariant subspaces of A that are annihilated by C or each invariant subspace of A annihilated by C contains only those eigenvectors the corresponding eigenvalues of which lie in the left half-plane.

This definition can be interpreted as follows. If the matrix pair A, C is detectable and the nonzero vector x lies in an invariant subspace of A corresponding to eigenvalues lying in the right half-plane and on the imaginary axis, then there are nonzero vectors among $Cx, CAx, \dots, CA^{N-1}x$. This means that the composite matrix

$$\begin{bmatrix} C \\ CA \\ \vdots \\ CA^{N-1} \end{bmatrix}$$

does not send a nonzero vector of this invariant subspace to zero. In other words, the operator corresponding to this composite matrix is nonsingular on this invariant subspace.

We have proved that if the matrix pair A, C is not detectable and the matrix pair A, B is stabilizable, then for the system

$$\frac{d}{dt}x(t) = Ax(t) + Bu(t)$$

the optimal control $u(t)$ minimizing the integral

$$\int_0^\infty (\|Cx(t)\|^2 + \|Bu(t)\|^2)dt$$

yields (at least for some initial values $x(0)$) the trajectory $x(t)$ which does not vanish as $t \to \infty$ (the existence and uniqueness of the optimal control follow from §4). The above example illustrates this fact. We continue to consider the example in order to emphasize one more difficulty which can arise in the case of a nondetectable matrix pair.

Studying the system

$$\frac{d}{dt}\begin{pmatrix} x_1(t) \\ x_2(t) \end{pmatrix} = \begin{bmatrix} 1 & 0 \\ 0 & 2 \end{bmatrix}\begin{pmatrix} x_1(t) \\ x_2(t) \end{pmatrix} + \begin{pmatrix} 1 \\ 1 \end{pmatrix} u(t), \quad t > 0,$$

with the initial values $x_1(0)$, $x_2(0)$, we have established that the optimal control minimizing the integral

$$\mathcal{J} = \int_0^\infty (x_1^2 + u^2)dt$$

can be computed by the formula

$$u(t) = \frac{1}{2}\lambda_1(t) + \frac{1}{2}\lambda_2(t)$$

in terms of a solution to the Hamiltonian system

$$\begin{aligned} \frac{d}{dt}\begin{pmatrix} x_1(t) \\ x_2(t) \end{pmatrix} &= \begin{bmatrix} 1 & 0 \\ 0 & 2 \end{bmatrix}\begin{pmatrix} x_1 \\ x_2 \end{pmatrix} + \begin{bmatrix} 0,5 & 0,5 \\ 0,5 & 0,5 \end{bmatrix}\begin{pmatrix} \lambda_1 \\ \lambda_2 \end{pmatrix}, \\ \frac{d}{dt}\begin{pmatrix} \lambda_1(t) \\ \lambda_2(t) \end{pmatrix} &= \begin{bmatrix} 1 & 0 \\ 0 & 0 \end{bmatrix}\begin{pmatrix} x_1 \\ x_2 \end{pmatrix} - \begin{bmatrix} 1 & 0 \\ 0 & 2 \end{bmatrix}\begin{pmatrix} \lambda_1 \\ \lambda_2 \end{pmatrix} \end{aligned}$$

such that $x_1(0)$ and $x_2(0)$ coincide with given values and $|\lambda_1(t)| \to 0$, $|\lambda_2(t)| \to 0$, $|x_1(t)| \to 0$ as $t \to \infty$. All solutions such that $|\lambda_1(t)|$, $|\lambda_2(t)|$, $|x_1(t)|$ decrease can be described by the formulas

$$\begin{aligned} x_1(t) &= \frac{\beta}{\sqrt{6}+2}\, e^{-t\sqrt{3/2}} - \frac{\gamma}{5}\, e^{-2t}, \\ x_2(t) &= \frac{\beta}{4+\sqrt{6}}\, e^{-t\sqrt{3/2}} - \frac{3\gamma}{20}\, e^{-2t} + \delta e^{2t}, \\ \lambda_1(t) &= -\beta\, e^{-t\sqrt{3/2}} + \frac{\gamma}{5}\, e^{-2t}, \\ \lambda_2(t) &= \gamma e^{-2t}. \end{aligned}$$

In addition, we require that $|x_2(t)| \to 0$ as $t \to \infty$. This requirement is satisfied if $\delta = 0$. The initial conditions are satisfied if β and γ satisfy the system

$$\begin{aligned} \frac{\beta}{\sqrt{6}+2} - \frac{\gamma}{5} &= x_1(0), \\ \frac{\beta}{4+\sqrt{6}} - \frac{3\gamma}{20} &= x_2(0). \end{aligned}$$

This system is compatible and uniquely solvable. Thus, for any $x_1(0)$, $x_2(0)$ the Hamiltonian system

$$\frac{d}{dt}\begin{pmatrix} x_1(t) \\ x_2(t) \end{pmatrix} = \begin{bmatrix} 1 & 0 \\ 0 & 2 \end{bmatrix}\begin{pmatrix} x_1 \\ x_2 \end{pmatrix} + \begin{bmatrix} 0,5 & 0,5 \\ 0,5 & 0,5 \end{bmatrix}\begin{pmatrix} \lambda_1 \\ \lambda_2 \end{pmatrix},$$
$$\frac{d}{dt}\begin{pmatrix} \lambda_1(t) \\ \lambda_2(t) \end{pmatrix} = \begin{bmatrix} 1 & 0 \\ 0 & 0 \end{bmatrix}\begin{pmatrix} x_1 \\ x_2 \end{pmatrix} - \begin{bmatrix} 1 & 0 \\ 0 & 2 \end{bmatrix}\begin{pmatrix} \lambda_1 \\ \lambda_2 \end{pmatrix}$$

has a unique solution such that $\|\lambda(t)\| \to 0$ and $\|x(t)\| \to 0$ as $t \to \infty$. For the control system

$$\frac{d}{dt}\begin{pmatrix} x_1 \\ x_2 \end{pmatrix} = \begin{bmatrix} 1 & 0 \\ 0 & 2 \end{bmatrix}\begin{pmatrix} x_1 \\ x_2 \end{pmatrix} + \begin{pmatrix} 1 \\ 1 \end{pmatrix} u(t)$$

the control

$$u(t) = \frac{1}{2}\lambda_1(t) + \frac{1}{2}\lambda_2(t)$$

guarantees that $\|x(t)\| \to 0$ as $t \to \infty$ and the integral

$$\mathcal{J} = \int_0^\infty (x_1^2 + u^2)\,dt$$

is finite.

The constructed control is stabilizing. However, as a rule, it is not optimal. The optimal control was constructed above. In our example, for this optimal control we have $\|x(t)\| = \sqrt{x_1^2 + x_2^2} \to \infty$ as $t \to \infty$.

EXAMPLE 2. Let

$$A = \begin{bmatrix} 0 & -1 \\ 1 & 0 \end{bmatrix}, \quad B = \begin{pmatrix} 0 \\ 1 \end{pmatrix}, \quad C = (0 \vdots 0).$$

Since the matrix C is zero, the matrix pair A, C is not detectable and, even more so, is not observable. The matrix pair A, B is controllable, hence is stabilizable. Indeed, the vectors

$$\begin{pmatrix} 0 \\ 1 \end{pmatrix}, \quad A\begin{pmatrix} 0 \\ 1 \end{pmatrix} = \begin{bmatrix} 0 & -1 \\ 1 & 0 \end{bmatrix}\begin{pmatrix} 0 \\ 1 \end{pmatrix} = \begin{pmatrix} -1 \\ 0 \end{pmatrix}$$

are linearly independent. We formally construct a control minimizing the integral

$$\int_0^\infty (\|Cx\|^2 + \|Bu\|^2)\,dt = \int_0^\infty \|Bu(t)\|^2\,dt$$

for given $x_1(0)$, $x_2(0)$ and try to find optimal trajectory $x(t)$ and control $u(t)$ by solving the Hamiltonian system

$$\frac{d}{dt}\begin{pmatrix} x_1 \\ x_2 \end{pmatrix} = \begin{bmatrix} 0 & -1 \\ 1 & 0 \end{bmatrix}\begin{pmatrix} x_1 \\ x_2 \end{pmatrix} + \begin{bmatrix} 0 & 0 \\ 0 & 1 \end{bmatrix}\begin{pmatrix} \lambda_1 \\ \lambda_2 \end{pmatrix},$$
$$\frac{d}{dt}\begin{pmatrix} \lambda_1 \\ \lambda_2 \end{pmatrix} = \begin{bmatrix} 0 & 1 \\ -1 & 0 \end{bmatrix}\begin{pmatrix} \lambda_1 \\ \lambda_2 \end{pmatrix}.$$

The characteristic equation for this system

$$\det \begin{bmatrix} -\mu & -1 & 0 & 0 \\ 1 & -\mu & 0 & 1 \\ 0 & 0 & -\mu & 1 \\ 0 & 0 & -1 & -\mu \end{bmatrix} \equiv (\mu^2+1)^2 = 0$$

has only purely imaginary roots ($\mu_{1,2} = i$, $\mu_{3,4} = -i$). Therefore, we cannot find solutions decreasing as $t \to \infty$, so that this Hamiltonian system cannot be used for the construction of stabilizing controls. The reason is that the matrix pair A, C is not detectable. In this example, the integral does not contain $x(t)$, hence we cannot hope that the optimal trajectory $x(t)$ tends to zero as $t \to \infty$. Furthermore, setting

$$Bu = B(B^*B)^{-1}B^* \begin{pmatrix} \lambda_1 \\ \lambda_2 \end{pmatrix} = \begin{bmatrix} 0 & 0 \\ 0 & 1 \end{bmatrix} \begin{pmatrix} \lambda_1 \\ \lambda_2 \end{pmatrix} = \begin{pmatrix} 0 \\ \lambda_2 \end{pmatrix}$$

and taking into account that λ_1, λ_2 from the Hamiltonian system must be of the form

$$\lambda_1 = K\cos(t-t_0),$$
$$\lambda_2 = K\sin(t-t_0),$$

we can establish that the optimal control

$$Bu = \begin{pmatrix} 0 \\ K\sin(t-t_0) \end{pmatrix}$$

yields a finite integral only if $K = 0$, i.e., if $Bu = 0$, $\lambda_1(t) = 0$, $\lambda_2(t) = 0$. Moreover,

$$\frac{dx_1}{dt} = -x_2, \quad \frac{dx_2}{dt} = x_1,$$
$$x_1(t) = x_1(0)\cos t - x_2(0)\sin t,$$
$$x_2(t) = x_1(0)\sin t + x_2(0)\cos t,$$
$$\|x(t)\| = \sqrt{x_1^2(t) + x_2^2(t)} = \sqrt{x_1^2(0) + x_2^2(0)} = \|x(0)\|.$$

The norm $\|x(t)\|$ is constant for optimal trajectories and is equal to $\|x(0)\|$.

THEOREM 1. *If the matrix pair A, B is stabilizable and the matrix pair A, C is detectable, then on the optimal control $u(t)$ for the system*

$$\frac{d}{dt}x(t) = Ax(t) + Bu(t)$$

with the initial value which minimizes the integral

$$\mathcal{J}_C = \int_0^\infty (\|Cx(t)\|^2 + \|Bu(t)\|^2)dt$$

for a given $x(0)$, the norm $\|x(t)\|$ tends to zero as $t \to \infty$.

To prove the theorem, we recall that under the assumption that the matrix pair A, B is stabilizable for the optimal control $u(t)$ it was proved (cf. the beginning of this section) that the norms $\|Bu(t)\|$ and $\|Cx(t)\|$ exponentially decrease as $t \to \infty$. We apply the following lemma.

LEMMA 1. *If the matrix pair A, C is detectable and $\|f(t)\| < M'e^{-\sigma t}$, then from the inequality $\|Cx(t)\| < M''e^{-\sigma t}$ for the solution $x(t)$, $0 \leqslant t < \infty$, to the system $\frac{d}{dt}x(t) = Ax(t) + f(t)$ it follows that $x(t) \to 0$ as $t \to \infty$.*

PROOF. As was shown in Chapter 1, §9, if the inequality $\|f(t)\| < M'e^{-\sigma t}$ holds for $0 \leqslant t < +\infty$, then the nonhomogeneous system has at least one solution that decreases exponentially. The general solution $x(t)$ to the system

$$\frac{d}{dt}x(t) = Ax(t) + Bu(t)$$

can be represented as the sum $x(t) = y(t) + \xi(t)$, where $\xi(t)$ is a solution decreasing exponentially whereas $y(t)$ is a solution to the homogeneous system

$$\frac{d}{dt}y(t) = Ay(t).$$

In addition, $\|Cy(t)\|$ decreases as t increases because

$$\|Cy(t)\| = \|Cx(t) - C\xi(t)\| \leqslant \|Cx(t)\| + \|C\xi(t)\|.$$

The decrease is exponential:

$$\|Cy(t)\| \leqslant \text{const}\, e^{-\sigma_1 t} \quad (\sigma_1 > 0).$$

We choose R so that

$$y(t) = R\begin{pmatrix} z_1(t) \\ z_2(t) \end{pmatrix}$$

and the equation for $y(t)$ is reduced to the canonical form

$$\frac{d}{dt}\begin{pmatrix} z_1(t) \\ z_2(t) \end{pmatrix} = \begin{bmatrix} A_1 & 0 \\ 0 & A_2 \end{bmatrix}\begin{pmatrix} z_1(t) \\ z_2(t) \end{pmatrix},$$

where all eigenvalues of $N_1 \times N_2$-matrix A_1 lie in the left half-plane and all eigenvalues of $(N - N_1) \times (N - N_1)$-matrix A_2 lie in the right half-plane and, perhaps, on the imaginary axis. Since the spectrum of A_1 lie in the left half-plane, we have $\|z_1(t)\| \to 0$ as $t \to \infty$. If $\|z_2(t)\|$ also decreases, then

$$\|y(t)\| \leqslant \|R^{-1}\|\sqrt{\|z_1(t)\|^2 + \|z_2(t)\|^2}.$$

Consequently, we obtain the required assertion that $\|x(t)\|$ decreases since

$$\|x(t)\| = \|y(t) + \xi(t)\| \leqslant \|y(t)\| + \|\xi(t)\|.$$

In the proof we must use the fact that the matrix pair A, C is detectable, i.e., C does not annihilate any nonzero vector of an invariant subspace of A corresponding to an eigenvalue lying in the right half-plane or on the imaginary axis. It is obvious that any vector in these subspaces, being represented in the form

$$y = R\begin{pmatrix} z_1 \\ z_2 \end{pmatrix},$$

has $z_1 = 0$, i.e., admits the representation

$$y = R\begin{pmatrix} 0 \\ z_2 \end{pmatrix};$$

moreover, $Cy = CR\begin{pmatrix} 0 \\ z_2 \end{pmatrix} \neq 0$ if $z_2 \neq 0$. We consider

$$\inf_{z_2\neq 0} \frac{\|Cy\|}{\|z_2\|} = \inf_{\|z_2\|=1} \|Cy\|.$$

Since the sphere $\|z_2\| = 1$ is finite-dimensional hence compact, there exists a vector z_{20} ($\|z_{20}\| = 1$) on which the lower bound is attained. Setting

$$y_0 = CR\begin{bmatrix} 0 \\ z_{20} \end{bmatrix}$$

we have

$$\inf_{z_2\neq 0} \frac{\|Cy\|}{\|z_2\|} = \frac{\|Cy_0\|}{\|z_{20}\|} = \rho > 0.$$

Consequently,

$$\|Cy\| = \left\| CR\begin{pmatrix} 0 \\ z_2 \end{pmatrix} \right\| \geqslant \rho\|z_2\|.$$

Let y be arbitrary, i.e., admits the representation

$$y = R\begin{pmatrix} 0 \\ z_2 \end{pmatrix} + R\begin{pmatrix} z_1 \\ 0 \end{pmatrix}.$$

In this case,

$$\|Cy\| \geqslant \left\| CR\begin{pmatrix} 0 \\ z_2 \end{pmatrix} \right\| - \left\| CR\begin{pmatrix} z_1 \\ 0 \end{pmatrix} \right\| \geqslant \rho\,\|z_2\| - \|C\|\,\|R\|\,\|z_1\|.$$

Let $y(t)$ be a solution to the homogeneous equation

$$\frac{d}{dt}y(t) = Ay(t).$$

It can be represented in the form

$$y(t) = R\begin{pmatrix} z_1(t) \\ z_2(t) \end{pmatrix} = R\begin{pmatrix} e^{tA_1}z_1(0) \\ e^{tA_2}z_2(0) \end{pmatrix}.$$

Therefore,

$$\begin{aligned} \|Cy(t)\| &\geqslant \rho\,\|e^{tA_2}z_2(0)\| - \|C\|\,\|R\|\,\|e^{tA_1}z_1(0)\|, \\ \|e^{tA_2}z_2(0)\| &\leqslant \frac{1}{\rho}[\,\|Cy(t)\| + \|C\|\,\|R\|\,\|e^{tA_1}z_1(0)\|\,], \\ \|z_2(0)\| = \|e^{-tA_2}e^{tA_2}z_2(0)\| &\leqslant \|e^{-tA_2}\|\,\|e^{tA_2}z_2(0)\| \\ &\leqslant \|e^{-tA_2}\|\frac{1}{\rho}[\|Cy(t)\| + \|C\|\,\|R\|\,\|e^{tA_1}z_1(0)\|]. \end{aligned}$$

The assumptions on spectra of A_1, A_2 and the assumption that $\|Cy(t)\|$ exponentially decreases lead to the following inequalities satisfied for $t > 0$:

$$\begin{aligned} \|e^{tA_1}\| &\leqslant \Omega(N_1, \sigma_3)\,e^{-\sigma_3 t}, \\ \|Cy(t)\| &\leqslant \text{const}\,e^{-\sigma_1 t}, \\ \|e^{-tA_2}\| &\leqslant \Omega'(N - N_1, \sigma_2)\,e^{\sigma_2 t}, \end{aligned}$$

where σ_2 is an arbitrary positive number that does not exceed σ_1, σ_3 ($\sigma_2 < \sigma_1$, $\sigma_2 < \sigma_3$). Therefore,

$$\begin{aligned}\|z_2(0)\| &\leqslant \|e^{-tA_2}\|\frac{1}{\rho}[\|Cy(t)\| + \|C\|\,\|R\|\,\|e^{tA_1}z_1(0)\|] \\ &\leqslant \text{const}\, e^{\sigma_2 t}[e^{-\sigma_1 t} + e^{-\sigma_3 t}].\end{aligned}$$

The constant on the right-hand side of this inequality depends on $\|z_1(0)\|$ and the choice of σ_2. It is clear that for

$$0 < \sigma_2 < \min\left\{\frac{1}{2}\sigma_1, \frac{1}{2}\sigma_3\right\}$$

the right-hand side decreases as t increases whereas the left-hand side $\|z_2(0)\|$ is constant. This can happen only if

$$z_2(0) = 0, \quad z_2(t) = e^{tA_2}z_2(0) = 0.$$

We have proved that not only $\|z_2(t)\| \to 0$, but even $\|z_2(t)\| \equiv 0$. □

As was mentioned, Lemma 1 implies Theorem 1. The optimal trajectory can be found by solving the Hamiltonian system

$$\begin{aligned}\frac{d}{dt}x(t) &= Ax(t) + B(B^*B)^{-1}B^*\lambda(t), \\ \frac{d}{dt}\lambda(t) &= C^*Cx(t) - A^*\lambda(t)\end{aligned}$$

such that $x(0)$ is given and $\|x(t)\| \to 0$, $\|\lambda(t)\| \to 0$ as $t \to \infty$.

Under the assumption that the matrix pair A, B is stabilizable and the matrix pair A, C is detectable, the optimal trajectory and, consequently, the solution to the Hamiltonian system exists for any $x(0)$. Hence the space of solutions decreasing as $t \to \infty$ is at least N-dimensional. This space coincides with the invariant subspace of the Hamiltonian matrix

$$\begin{bmatrix} A & B(B^*B)^{-1}B^* \\ C^*C & -A^* \end{bmatrix}$$

which corresponds to eigenvalues lying in the left half-plane.

By the Poincaré–Lyapunov theorem, the space of decreasing solutions cannot have dimension greater than N. Comparing these assertions, we conclude that the dimension is N and that the solution to the Hamiltonian system which tends to zero as $t \to 0$ ($\|x(t)\| \to 0$, $\|\lambda(t)\| \to 0$) is uniquely determined by the value $x(0)$. This conclusion is obtained under the assumption that the matrix pair A, B is stabilizable and the matrix pair A, C is detectable. Under this assumption, the boundary-value problem for the Hamiltonian system with the initial value $x(0)$ and $\|x(t)\| \to 0$, $\|\lambda(t)\| \to 0$ as $t \to \infty$ is uniquely solvable for any $x(0)$, i.e., the Lopatinskii condition holds. The optimal control $u(t)$ minimizing the integral

$$\int_0^\infty (\|Cx(t)\|^2 + \|Bu(t)\|^2)dt$$

for the system

$$\frac{d}{dt}x(t) = Ax(t) + Bu(t)$$

can be found from $\lambda(t)$ by solving the above boundary-value problem according to the formula

$$u(t) = (B^*B)^{-1}B^*\lambda(t).$$

We now discuss one more point of view on the notion of detectability. Let the eigenvalues $\tau_1, \tau_2, \dots, \tau_{N_0}$ of A lie in the left half-plane,

$$\operatorname{Re}\tau_j < 0 \text{ for } j = 1, 2, \dots, N_0,$$

and the eigenvalues $\tau_{N_0+1}, \tau_{N_0+2}, \dots, \tau_{N-1}, \tau_N$ lie in the right half-plane or on the imaginary axis,

$$\operatorname{Re}\tau_j \geqslant 0 \text{ for } N_0 + 1 \leqslant j \leqslant N.$$

The polynomials

$$p_-(\tau) = (\tau - \tau_1)(\tau - \tau_2)\dots(\tau - \tau_{N_0}),$$
$$p_+(\tau) = (\tau - \tau_{N_0+1})(\tau - \tau_{N_0+2})\dots(\tau - \tau_N)$$

are relatively prime. Therefore, there exist polynomials $q_-(\tau)$ and $q_+(\tau)$ such that

$$q_-(\tau)p_-(\tau) + q_+(\tau)p_+(\tau) = 1.$$

This algebraic fact we already used in Chapter 1, §4. We denote

$$q_-(\tau)p_-(\tau) = \pi_-(\tau), \quad q_+(\tau)p_+(\tau) = \pi_+(\tau)$$

and form the matrix polynomials

$$\pi_-(A) = q_-(A)p_-(A), \quad \pi_+(A) = q_+(A)p_+(A).$$

We claim that their sum is the identity matrix, $\pi_+(A) + \pi_-(A) = I$, and their product vanishes, $\pi_+(A)\pi_-(A) = \pi_-(A)\pi_+(A) = 0$. Indeed,

$$\begin{aligned}\pi_+(A)\pi_-(A) &= q_+(A)q_-(A)p_-(A)p_+(A)\\ &= q_+(A)q_-(A)(A - \tau_1 I)\dots(A - \tau_{N_0}I)(A - \tau_{N_0+1}I)\dots(A - \tau_N I) = 0.\end{aligned}$$

Here we use the Cayley–Hamilton theorem, which claims that every matrix satisfies its characteristic equation. Any vector x can be represented as the sum

$$x = \pi_+(A)x + \pi_-(A)x,$$

where the first term vanishes after the application of the operator $p_-(A)$ whereas the second term vanishes after the application of the operator $p_+(A)$:

$$p_-(A)\pi_+(A) = 0, \quad p_+(A)\pi_-(A) = 0.$$

This assertion also follows from the Cayley–Hamilton theorem.

In other words, $\pi_+(A)x$ and $\pi_-(A)x$ lie in the invariant subspaces of A that correspond to eigenvalues in the left half-plane and in the right half-plane or on the imaginary axis respectively. It is clear that if x runs over the entire N-dimensional space, then $\pi_-(A)x$ and $\pi_-(A)x$ run over all vectors of the corresponding spaces. The detectability condition consists in the requirement that each vector of the form $\pi_-(A)x$ can be restored from the vector

$$\begin{pmatrix} C\pi_-(A)x \\ CA\pi_-(A)x \\ \vdots \\ CA^{N-1}\pi_-(A)x \end{pmatrix}.$$

In other words, the rank of the matrix

$$\begin{bmatrix} C\pi_-(A) \\ CA\pi_-(A) \\ \vdots \\ CA^{N-1}\pi_-(A) \end{bmatrix}$$

must coincide with the dimension of the invariant subspace corresponding to eigenvalues in the closed right half-plane, i.e., the rank is $N - N_0$.

Since the rank of a matrix coincides with that of transposed matrix, the detectability criterion can be formulated as follows: the rank of the matrix

$$\begin{aligned} &[\pi_-(A^*)C^* \vdots \pi_-(A^*)A^*C^* \vdots \dots \vdots \pi_-(A^*)(A^*)^{N-1}C^*] \\ &\equiv \pi_-(A^*)[C^* \vdots A^*C^* \vdots \dots \vdots (A^*)^{N-1}C^*] \end{aligned} \tag{1}$$

is $N - N_0$.

Since the characteristic polynomials of the matrices A and A^* coincide, the decomposition $z = \pi_+(A^*)z + \pi_-(A^*)z$ represents z as the sum of vectors belonging to invariant subspaces of A^* corresponding to eigenvalues such that $\operatorname{Re}\tau_j(A^*) < 0$ and $\operatorname{Re}\tau_j(A^*) \geqslant 0$ respectively. The dimension of the subspace formed by vectors $\pi_-(A^*)z$ is $N - N_0$.

Thus, the condition that the rank of the matrix (1) is $N - N_0$ can be formulated as the requirement that the linear span of columns of the matrix

$$[C^* \vdots A^*C^* \vdots \dots \vdots (A^*)^{N-1}C^*]$$

contains the entire invariant subspace of the matrix A^* corresponding to eigenvalues in the right half-plane and on the imaginary axis. In other words, the matrix pair A^*, C^* must be stabilizable.

THEOREM 2. *The matrix pair A, C is detectable if and only if the matrix pair A^*, C^* is stabilizable.*

We recall that the matrix pair A, B is stabilizable if and only if it is possible to choose a matrix M such that all eigenvalues of $A + BM$ lie in the left half-plane. From this assertion and Theorem 2 we obtain the following corollary.

COROLLARY 1. *The matrix pair A, C is detectable if and only if there exists a matrix M such that $A + MC$ is a Hurwitz matrix.*

§7. The Lur′e–Riccati matrix equation

Invariant subspaces of a Hamiltonian matrix and bases. The matrix notation of a basis and the relations between matrix blocks. An integral relation. The Lur′e–Riccati equation. Theorems on nonnegative definite and positive definite solutions. Stabilizability criteria formulated by means of solutions to the Lur′e–Riccati equation.

We consider the boundary-value problem for the system

$$\begin{aligned} \frac{d}{dt}x(t) &= Ax(t) + B(B^*B)^{-1}B^*\lambda(t), \\ \frac{d}{dt}\lambda(t) &= C^*Cx(t) - A^*\lambda(t) \end{aligned} \tag{1}$$

which consists in finding vector-valued functions $x(t)$, $\lambda(t)$, $0 \leqslant t < \infty$, such that $\|x(t)\| \to 0$, $\|\lambda(t)\| \to 0$ as $t \to \infty$ and the vector-valued function $x(t)$ takes the value $x(0)$ at $t = 0$. We suppose that the problem satisfies the Lopatinskii condition (i.e., it is well posed). In this case, half of all eigenvalues of the Hamiltonian $2N \times 2N$-matrix

$$\begin{bmatrix} A & B(B^*B)^{-1}B^* \\ C^*C & -A^* \end{bmatrix}$$

lie in right half-plane and the other half in the left half-plane. The N-dimensional invariant subspace of this Hamiltonian matrix corresponding to the part of the spectrum which lies in the left half-plane can be described as the basis composed of vectors whose components form the $2N \times N$-matrix of the form[3]

$$\begin{bmatrix} I \\ -\Lambda \end{bmatrix}.$$

The columns of this matrix form a basis for the invariant subspace if and only if the following equality holds:

$$\begin{bmatrix} A & B(B^*B)^{-1}B^* \\ C^*C & -A^* \end{bmatrix} \begin{bmatrix} I \\ -\Lambda \end{bmatrix} = \begin{bmatrix} I \\ -\Lambda \end{bmatrix} K,$$

where K is an $N \times N$-matrix K. It is obvious that

$$\begin{bmatrix} A & B(B^*B)^{-1}B^* \\ C^*C & -A^* \end{bmatrix}^m \begin{bmatrix} I \\ -\Lambda \end{bmatrix} = \begin{bmatrix} I \\ -\Lambda \end{bmatrix} K^m,$$

$$\left\{\sum_{m=0}^{\infty} \frac{t^m}{m!} \begin{bmatrix} A & B(B^*B)^{-1}B^* \\ C^*C & -A^* \end{bmatrix}^m\right\} \begin{bmatrix} I \\ -\Lambda \end{bmatrix} = \begin{bmatrix} I \\ -\Lambda \end{bmatrix} \left\{\sum_{m=0}^{\infty} \frac{t^m}{m!} K^m\right\},$$

$$\left\{\exp\left[t\begin{pmatrix} A & B(B^*B)^{-1}B^* \\ C^*C & -A^* \end{pmatrix}\right]\right\} \begin{bmatrix} I \\ -\Lambda \end{bmatrix} = \begin{bmatrix} I \\ -\Lambda \end{bmatrix} e^{tK} = \begin{bmatrix} e^{tK} \\ -\Lambda e^{tK} \end{bmatrix}.$$

Since the invariant subspace under consideration corresponds to eigenvalues in the left half-plane, we have

$$\left\|\exp\left\{t\begin{bmatrix} A & B(B^*B)^{-1}B^* \\ C^*C & -A^* \end{bmatrix}\right\} \begin{bmatrix} I \\ -\Lambda \end{bmatrix}\right\| \to 0 \text{ as } t \to \infty.$$

Consequently, $\|e^{tK}\| \to 0$ as $t \to \infty$. Hence all eigenvalues of K lie in the left half-plane (i.e., K is a Hurwitz matrix).

Thus, if $x(t)$, $\lambda(t)$ is a decreasing (as $t \to \infty$) solution to the Hamiltonian system (1), we have $x(t) = e^{tK}x(0)$ and $\lambda(t) = -\Lambda e^{tK}x(0)$. Let $y(t) = e^{tK}y(0)$ and $\mu(t) = -\Lambda e^{tK}y(0)$ be another solution. Take the inner product of both sides of the equation

$$\frac{d}{dt}x(t) = Ax(t) + B(B^*B)^{-1}B^*\lambda(t)$$

with $\mu(t)$ and of both sides of the equation

$$\frac{d}{dt}\mu(t) = C^*Cy(t) - A^*\mu(t)$$

[3]It is convenient to distinguish the sign in the notation of the lower block. This notation differs from the similar notation used in previous sections by only this sign.

with $x(t)$, and sum up the results:

$$\begin{aligned}\left(\frac{d}{dt}x(t),\mu(t)\right)+\left(x(t),\frac{d}{dt}\mu(t)\right)&\equiv\frac{d}{dt}\big(x(t),\mu(t)\big)\\ =\big(Ax(t),\mu(t)\big)+\big(B(B^*B)^{-1}B^*\lambda(t),\mu(t)\big)&+\big(x(t),C^*Cy(t)\big)-\big(x(t),A^*\mu(t)\big)\\ \equiv\big(Ax(t),\mu(t)\big)+\big(B(B^*B)^{-1}B^*\lambda(t),\mu(t)\big)&+\big(C^*Cx(t),y(t)\big)-\big(Ax(t),\mu(t)\big)\\ &\equiv\big(B(B^*B)^{-1}B^*\lambda(t),\mu(t)\big)+\big(C^*Cx(t),y(t)\big).\end{aligned}$$

We obtain

$$\frac{d}{dt}\big(x(t),\mu(t)\big)=\big(B(B^*B)^{-1}B^*\lambda(t),\mu(t)\big)+\big(C^*Cx(t),y(t)\big).$$

We use the relations

$$\begin{aligned}x(t)&=e^{tK}x(0),\quad &\mu(t)&=-\Lambda e^{tK}y(0),\\ y(t)&=e^{tK}y(0),\quad &\lambda(t)&=-\Lambda e^{tK}x(0)\end{aligned}$$

to obtain the equality

$$\begin{aligned}&-\frac{d}{dt}\big(e^{tK}x(0),\Lambda e^{tK}y(0)\big)\\ &\qquad=\big(B(B^*B)^{-1}B^*\Lambda e^{tK}x(0),\Lambda e^{tK}y(0)\big)+\big(C^*Ce^{tK}x(0),e^{tK}y(0)\big),\end{aligned}$$

which can be rewritten as follows:

$$\begin{aligned}&-\frac{d}{dt}\big(e^{tK^*}\Lambda e^{tK}y(0),x(0)\big)\\ &\qquad=\big(e^{tK^*}\Lambda^*B(B^*B)^{-1}B^*\Lambda e^{tK}y(0),x(0)\big)+\big(e^{tK^*}C^*Ce^{tK}y(0),x(0)\big).\end{aligned}$$

We integrate the last equality with respect to t from 0 to ∞ and, taking into account that K is a Hurwitz matrix, find

$$\begin{aligned}\big(\Lambda y(0),x(0)\big)&=\left(\left\{\int_0^\infty e^{tK^*}\Lambda^*B(B^*B)^{-1}B^*\Lambda e^{tK}\,dt\right\}y(0),x(0)\right)\\ &\quad+\left(\left\{\int_0^\infty e^{tK^*}C^*Ce^{tK}\,dt\right\}y(0),x(0)\right).\end{aligned}$$

Since $y(0)$ and $x(0)$ are arbitrary, we obtain the matrix equality

$$\Lambda=\int_0^\infty e^{tK^*}\Lambda^*B(B^*B)^{-1}B^*\Lambda e^{tK}\,dt+\int_0^\infty e^{tK^*}C^*Ce^{tK}\,dt,$$

where the right-hand side is symmetric and positive definite. Therefore, $\Lambda=\Lambda^*$ and $(\Lambda x,x)\geqslant 0$. Now we show that Λ satisfies a matrix equality. The equality

$$\begin{bmatrix}A & B(B^*B)^{-1}B^*\\ C^*C & -A^*\end{bmatrix}\begin{bmatrix}I\\ -\Lambda\end{bmatrix}=\begin{bmatrix}I\\ -\Lambda\end{bmatrix}K$$

implies that

$$\begin{aligned}A-B(B^*B)^{-1}B^*\Lambda&=K,\\ C^*C+A^*\Lambda&=-\Lambda K.\end{aligned}$$

Adding the second and the third relations, we obtain the equation for Λ which does not contain K:

$$\Lambda A + A^*\Lambda = \Lambda B(B^*B)^{-1}B^*\Lambda - C^*C. \tag{2}$$

This "quadratic" equation for Λ was obtained by Lur′e in 1951 and, independently, by Kalman in 1960 who called it the matrix Riccati equation because of analogy with the well-known Riccati differential equation. This term was adopted in literature. However, in the Russian mathematical texts the equation is usually called the Lur′e equation. It is obvious that the equation can be rewritten as follows:

$$\begin{aligned}\Lambda[A - B(B^*B)^{-1}B^*\Lambda] + [A - B(B^*B)^{-1}B^*\Lambda]^*\Lambda \\ = -\Lambda B(B^*B)^{-1}B^*\Lambda - C^*C.\end{aligned}$$

Recalling that $A - B(B^*B)^{-1}B^*\Lambda = K$, we obtain the following form of our quadratic equation:

$$\Lambda K + K^*\Lambda = -\Lambda^* B(B^*B)^{-1}B^*\Lambda - C^*C,$$

which, being added to the equality connecting K and A, is equivalent to the Lur′e–Riccati equation.

We suppose that for some K we have found a solution $\Lambda = \Lambda^* \geqslant 0$ to the above equations and the matrix pair A, C is detectable. We prove that all eigenvalues of K lie in the left half-plane. Thereby, we will establish the following theorem.

THEOREM 1. *If there is a nonnegative definite solution $\Lambda = \Lambda^* \geqslant 0$ to the Lur′e–Riccati equation* (2) *and the matrix pair A, C is detectable, then all eigenvalues of the matrix $K = A - B(B^*B)^{-1}B^*\Lambda$ lie in the left half-plane. Hence the matrix pair A, B is stabilizable.*

Before the proof of Theorem 1 we establish the lemma.

LEMMA 1. *Let the matrix pair A, C be detectable. Then $K = A + F$ is a Hurwitz matrix if and only if the norm $\|S(t)\|$ of the matrix*

$$S(t) = \int_0^t e^{sK^*}(C^*C + F^*F)e^{sK}\,ds$$

is bounded as $t \to +\infty$

PROOF. *Necessity.* We prove that if all eigenvalues of K lie in the left half-plane, then $\|S(t)\|$ is bounded. Indeed, for a Hurwitz matrix K the parameter

$$\varkappa(K) = 2\|K\|\left\|\int_0^\infty e^{sK^*}e^{sK}\,ds\right\|, \tag{3}$$

characterizing the stability quality is finite and the matrix exponentials e^{tK} and e^{tK^*} can be estimated in terms of this parameter:

$$\|e^{tK}\| = \|e^{tK^*}\| \leqslant \sqrt{\varkappa(K)}\, e^{-\frac{t\|K\|}{\varkappa(K)}}.$$

Therefore,

$$\|S(t)\| \leqslant (\|C\|^2 + \|F\|^2) \int_0^t \|e^{sK^*}\| \cdot \|e^{sK}\| \, ds \leqslant (\|C\|^2 + \|F\|^2) \int_0^\infty \|e^{sK^*}\| \cdot \|e^{sK}\| \, ds$$

$$\leqslant (\|C\|^2 + \|F\|^2) \varkappa(K) \int_0^\infty e^{-\frac{2s\|K\|}{\varkappa(K)}} \, ds = \frac{(\|C\|^2 + \|F\|^2)\varkappa^2(K)}{2\|K\|}.$$

Sufficiency. We must prove that if $S(t)$ is bounded, then there is no eigenvalue of $K = A + F$ in the right half-plane or on the imaginary axis. Assume that there exists a nonzero eigenvector x ($\|x\| \neq 0$) of $A + F$ such that $(A + F)x = \mu x$, $\operatorname{Re} \mu \geqslant 0$; moreover, ($s \geqslant 0$)

$$\begin{aligned} e^{s(A+F)}x &= e^{\mu s}x, \\ Ce^{s(A+F)}x &= Ce^{\mu s}x = e^{\mu s}Cx, \\ Fe^{s(A+F)}x &= Fe^{\mu s}x = e^{\mu s}Fx, \end{aligned}$$

$$\begin{aligned} \big(e^{s(A^*+F^*)}[C^*C + F^*F]e^{s(A+F)}x, x\big) &= \|Ce^{s(A+F)x}\|^2 + \|Fe^{s(a+F)}x\|^2 \\ &= \|e^{\mu s}Cx\|^2 + \|e^{\mu s}Fx\|^2 \\ &= e^{2s \operatorname{Re} \mu}(\|Cx\|^2 + \|Fx\|^2) \geqslant \|Cx\|^2 + \|Fx\|^2. \end{aligned}$$

Therefore,

$$\begin{aligned} \big(S(t)x, x\big) &= \int_0^t \big(e^{sK^*}[C^*C + F^*F]e^{sK}x, x\big) \, ds \\ &\geqslant \int_0^t (\|Cx\|^2) + \|Fx\|^2) \, ds = t(\|Cx\|^2) + \|Fx\|^2). \end{aligned}$$

It is clear that the boundedness of $\|S(t)\|$ and consequently the boundedness of $(S(t)x, x)$ as $t \to \infty$ holds only if the equalities $Cx = 0$ and $Fx = 0$ hold simultaneously. Moreover, $(A + F)x = Ax = \mu x$ ($\operatorname{Re} \mu \geqslant 0$, $\|x\| \neq 0$). We have established that C annihilates the eigenvector x of A corresponding to the eigenvalue μ in the right half-plane or on the imaginary axis. For the detectable matrix pair A, C there is no such vector. □

Proof of Theorem 1. We prove that K is a Hurwitz matrix. Let $K = A+F$, $F = -B(B^*B)^{-1}B^*\Lambda$, where $\Lambda = \Lambda^*$ is a solution to the equation (2). We note that

$$\Lambda B(B^*B)^{-1}B^*\Lambda = [\Lambda^*B(B^*B)^{-1}B^*][B(B^*B)^{-1}B^*\Lambda] = F^*F,$$

hence the equation for Λ can be written as follows:

$$\Lambda K + K^*\Lambda = -F^*F - C^*C.$$

We form the matrix integral

$$M(t) = \int_0^t e^{sK^*}[F^*F + C^*C]e^{sK} \, ds$$

and establish that

$$\frac{d}{dt}M(t) = e^{tK^*}[F^*F + C^*C]e^{tK} = -e^{tK^*}(K^*\Lambda + \Lambda K)e^{tK}$$

$$= -e^{tK^*}K^*\Lambda e^{tK} - e^{tK^*}\Lambda K e^{tK} = -\frac{d}{dt}[e^{tK^*}\Lambda e^{tK}],$$

$$\frac{d}{dt}[M(t) + e^{tK^*}\Lambda e^{tK}] = 0,$$

$$M(t) + e^{tK^*}\Lambda e^{tK} = M(0) + e^{0\cdot K^*}\Lambda e^{0\cdot K} = \Lambda.$$

Hence

$$\Lambda = e^{tK^*}\Lambda e^{tK} + \int_0^t e^{sK^*}[F^*F + C^*C]e^{sK}\,ds, \quad t > 0.$$

By the assumption of the theorem, $\Lambda = \Lambda^* \geqslant 0$. Hence both terms on the right-hand side are symmetric nonnegative definite matrices. Consequently, for any $t > 0$

$$\Lambda \geqslant \int_0^t e^{sK^*}[F^*F + C^*C]e^{sK}\,ds,$$

$$\left\| \int_0^t e^{sK^*}[F^*F + C^*C]e^{sK}\,ds \right\| \leqslant \|\Lambda\|.$$

Applying Lemma 1 ($K = A + F$, the pair A, C is detectable), we conclude that all eigenvalues of the matrix

$$K = A + F = A - B(B^*B)^{-1}B^*\Lambda$$

lie in the left half-plane. □

The equality defining K and the matrix equation (2) are equivalent to the relations

$$A - B(B^*B)^{-1}B^*\Lambda = K$$
$$C^*C + A^*\Lambda = -\Lambda K,$$

which admit the matrix notation

$$\begin{bmatrix} A & B(B^*B)^{-1}B^* \\ C^*C & -A^* \end{bmatrix}\begin{bmatrix} I \\ -\Lambda \end{bmatrix} = \begin{bmatrix} I \\ -\Lambda \end{bmatrix} K.$$

The last expression means that the columns of the rectangular matrix $\begin{bmatrix} I \\ -\Lambda \end{bmatrix}$ form a basis for the N-dimensional invariant subspace of the matrix of coefficients of the Hamiltonian system

$$\frac{d}{dt}\begin{pmatrix} x(t) \\ \lambda(t) \end{pmatrix} = \begin{bmatrix} A & B(B^*B)^{-1}B^* \\ C^*C & -A^* \end{bmatrix}\begin{pmatrix} x(t) \\ \lambda(t) \end{pmatrix}.$$

Solutions lying in the invariant subspace have the form

$$x(t) = e^{tK}x(0),$$
$$\lambda(t) = -\Lambda e^{tK}x(0),$$

tend to zero as $t \to \infty$ since $\|e^{tK}\| \to 0$ and are uniquely defined by the value $x(0)$.

The uniqueness of the invariant subspace means that Λ is uniquely determined by A, B, and C. Hence we have proved the following theorem.

THEOREM 2. *If the matrix pair A, C is detectable, then a nonnegative definite solution $\Lambda = \Lambda^* \geqslant 0$ to the equation* (2) *is unique. The existence of such a solution implies that the matrix pair A, B is stabilizable.*

Under more restrictive assumptions on A, B, C, Theorem 2 admits an important generalization. Namely, if instead of detectability of the matrix pair A, C we require this pair to be observable and keep the requirement of the stabilizability of the matrix pair A, B, then it is possible to prove that the solution (which exists and is unique under these assumptions) to the equation (2) is strictly positive definite.

As was mentioned, by setting

$$F = -B(B^*B)^{-1}B^*\Lambda, \quad K = A + F,$$

we can assert that K is a Hurwitz matrix and $\Lambda = \Lambda^*$ can be regarded as a solution to the Lyapunov equation

$$\Lambda K + K^*\Lambda = -F^*F - C^*C.$$

Therefore, we have the representation

$$\Lambda = \int_0^\infty e^{tK^*}(F^*F + C^*C)e^{tK}\,dt,$$

which implies that

$$(\Lambda x, x) = \int_0^\infty \|Fe^{tK}x\|^2\,dt + \int_0^\infty \|Ce^{tK}x\|^2\,dt.$$

Let $x \neq 0$ and let

$$\int_0^\infty \|Fe^{tK}x\|^2\,dt = 0.$$

The last equality holds only if

$$Fe^{tK}x \equiv 0, \quad 0 \leqslant t < \infty.$$

Differentiating this identity with respect to t, we obtain for $0 \leqslant t < \infty$

$$\begin{aligned} &FKe^{tK}x = 0, \\ &FK^2e^{tK}x = 0, \\ &\dots\dots\dots\dots \\ &FK^me^{tK}x = 0\ . \end{aligned}$$

Setting $t = 0$ and taking into account the equality $K = A + F$, we conclude that

$$\begin{aligned}
&Fx = 0,\\
&F(A+F)x = 0,\\
&F(A+F)^2x = 0,\\
&\dots\dots\dots\dots\\
&F(A+F)^m x = 0,\\
&\dots\dots\dots\dots.
\end{aligned}$$

These relations lead to the equalities

$$\begin{aligned}
(A+F)x &= Ax + Fx = Ax,\\
(A+F)^2x &= (A+F)(Ax+Fx) = (A+F)Ax = A^2x + FAx\\
&= A^2 + FAx + FFx = A^2x + (F(A+F)x = A^2x.
\end{aligned}$$

We prove by induction that

$$(A+F)^m x = A^m x.$$

Indeed, if the equality

$$(A+F)^{m-1}x = A^{m-1}x$$

is already proved, we use the relation $F(A+F)^{m-1}x = 0$ and obtain

$$(A+F)^m x = A(A+F)^{m-1}x + F(A+F)^{m-1}x = AA^{m-1}x = A^m x.$$

Hence for any t we have

$$\begin{aligned}
e^{tK}x = e^{t(A+F)}x &= x + \frac{t}{1!}(A+F)x + \frac{t^2}{2!}(A+F)^2x + \dots\\
&= x + \frac{t}{1!}Ax + \frac{t^2}{2!}A^2x + \dots = e^{tA}x.
\end{aligned}$$

Thus, if $x \neq 0$ and

$$\int_0^\infty \|Fe^{tK}x\|^2\,dt = 0,$$

then $e^{tK}x = e^{tA}x$; therefore,

$$\int_0^\infty \|Ce^{tK}x\|^2\,dt = \int_0^\infty \|Ce^{tA}x\|^2\,dt \geqslant \int_0^1 \|Ce^{tA}x\|^2\,dt.$$

In §2 it was shown that if the matrix pair A, C is observable, then for any $T > 0$ the matrix

$$\int_0^T e^{tA^*}C^*Ce^{tA}\,dt$$

is strictly positive definite. Hence for $x \neq 0$ such that

$$\int_0^\infty \|Fe^{tK}x\|^2 dt = 0$$

the following inequalities hold:

$$\int_0^1 \|Ce^{tA}x\|^2\,dt > 0, \quad \int_0^\infty \|Ce^{tK}x\|^2\,dt > 0.$$

Thus, we have shown that if the matrix pair A, C is detectable and $x \neq 0$, then either

$$\int_0^\infty \|Fe^{tK}x\|^2\,dt > 0$$

or

$$\int_0^\infty \|Ce^{tK}x\|^2\,dt > 0.$$

Therefore, the quadratic form

$$(\Lambda x, x) = \int_0^\infty \|Fe^{tK}x\|^2\,dt + \int_0^\infty \|Ce^{tK}x\|^2\,dt$$

is positive definite. This completes the proof of the assertion that for the stabilizable matrix pair A, B and observable matrix pair A, C a solution Λ to the matrix equation (2) is positive definite. As we know, if the matrix pair A, C is detectable, then the existence of a nonnegative solution Λ to the equation (2) is equivalent to the stabilizability of the matrix pair A, B. Now we can specify this assertion.

THEOREM 3. *If the matrix pair A, C is observable, then the existence of a nonnegative definite solution $\Lambda = \Lambda^*$ to the matrix equation* (2) *is equivalent to the stabilizability of the matrix pair A, B. In addition, the matrix Λ is positive definite ($\Lambda = \Lambda^* > 0$). Another solution $\Lambda = \Lambda^*$, if it exists, has at least one strictly negative eigenvalue.*

COROLLARY 1. *The existence of a nonnegative definite solution Λ to the equation*

$$\Lambda A + A^*\Lambda = \Lambda B(B^*B)^{-1}B^*\Lambda - I$$

is a necessary and sufficient condition for stabilizability of the matrix pair A, B. In this case, a nonnegative definite solution is unique and positive definite.

§8. The use of variational problems in the study of control systems

The controllability criterion (reminder). The verification of the stabilizability criterion based on the study of the boundary-value problem and the solution to the matrix Lur′e–Riccati equations. The verification of controllability, detectability, and observability of matrix pairs. The scheme of solution and analysis of a general variational control problem. Remark about the form of the Hamiltonian system for extremals of the functionals corresponding to arbitrary quadratic forms and the derivation of the Hamiltonian system by the formalism of calculus of variation with Lagrange multipliers. The example of the application of the theory described in this chapter. The dynamic compensation method.

The previous sections of this chapter were devoted to detailed description of the notions of controllability, observability, stabilizability, and detectability as well as the solvability conditions for variational problems formulated with the help of the introduced notions. In this section, we sum up the results of previous sections and give some rules for studying systems of the form

$$\frac{d}{dt}x(t) = Ax(t) + Bu(t). \tag{1}$$

We assume that the observation for behavior of such systems is realized by studying the vector-valued function $y(t) = Cx(t)$. The obtained results lead to the constructions of some controllability and observation laws.

CRITERION 1. *The matrix pair A, B is controllable if the rank of the composite matrix*

$$[B \vdots AB \vdots \dots \vdots A^{N-1}B]$$

coincides with the number N of rows, i.e., with the dimension N of the space on which A acts.

The study of a matrix pair by this criterion is reduced to the study of a classical problem of linear algebra. However, some difficulties often appear in practical solution of this problem.

In §§2–4, we discussed the relationships between main notions of control theory and properties of variational problems. By solving variational problems, due to these relationships we are able to make certain conclusions about whether the matrix pair A, B is controllable or stabilizable and the matrix pair A, C is observable or detectable.

We begin with the description of the procedure that allows us to verify if the matrix pair A, B is stabilizable. This procedure was justified in §5. Considering the system (1), we established that the necessary and sufficient condition for stabilizability of the matrix pair A, B is unique solvability of the boundary-value problem for the Hamiltonian system

$$\frac{d}{dt}\begin{pmatrix} x(t) \\ \lambda(t) \end{pmatrix} = \mathcal{H}\begin{pmatrix} x(t) \\ \lambda(t) \end{pmatrix}, \quad \mathcal{H} = \begin{bmatrix} A & B(B^*B)^{-1}B^* \\ I & -A^* \end{bmatrix}, \quad t \geqslant 0, \tag{2}$$

for any $x(0)$. This problem is to find a solution vanishing at infinity ($\|x(t)\| \to 0$, $\|\lambda(t)\| \to 0$ as $t \to \infty$) for a given $x(0)$. In other words, the stabilizability of the matrix pair A, B is equivalent to the Lopatinskii condition for the boundary-value problem.

From Chapter 1, §13 it follows that the Lopatinskii condition implies the requirement that the dichotomy parameter $\varkappa(\mathcal{H})$ is finite.

We note that there is an effective algorithm which distinguishes the desired invariant subspace and computes $\varkappa(\mathcal{H})$ at the same time. This algorithm does not work only if the dichotomy of the spectrum by the imaginary axis does not hold.[4] If $\varkappa(\mathcal{H}) \geqslant \varkappa^*$, then the matrix pair A, B is regarded as noncontrollable. If $\varkappa(\mathcal{H}) < \varkappa^*$, then there is an N-dimensional invariant subspace $\mathcal{H}$ corresponding to eigenvalues in the left half-plane. We choose the orthonormal basis in this subspace. We form the $2N \times N$-matrix with columns containing coordinates of the basis vectors. This

[4] In this case, the algorithm tells that $\varkappa(\mathcal{H}) \geqslant \varkappa^*$, where $\varkappa^*$ is a very large number so that it is regarded as infinity from the computational point of view.

matrix is represented as the matrix composed of two square blocks

$$\begin{bmatrix} X_0 \\ \Lambda_0 \end{bmatrix}.$$

As was mentioned in §5, in this representation X_0 is nonsingular only if the matrix pair A, B is stabilizable. The nonsingularity follows from the solvability of the boundary-value problem

$$x(0) = a, \quad \|x(t)\| \to 0, \quad \|\lambda(t)\| \to 0 \text{ as } t \to +\infty,$$
$$\frac{d}{dt}\begin{pmatrix} x(t) \\ \lambda(t) \end{pmatrix} = \mathcal{H}\begin{pmatrix} x(t) \\ \lambda(t) \end{pmatrix}, \quad 0 \leqslant t < +\infty$$

for any a. Indeed, any solution to the Hamiltonian system decreasing as $t \to \infty$ has the form

$$\begin{pmatrix} x(t) \\ \lambda(t) \end{pmatrix} = e^{t\mathcal{H}}\begin{bmatrix} X_0 \\ \Lambda_0 \end{bmatrix}\xi,$$

where ξ is an N-dimensional vector. On such a solution we have $x(0) = X_0\xi$. The solvability of the boundary-value problem implies the solvability of the equation $X_0\xi = a$ with respect to ξ for any a, i.e., the matrix X_0 is nonsingular. It is clear that the columns of the matrix

$$\begin{bmatrix} X_0 \\ \Lambda_0 \end{bmatrix} X_0^{-1} = \begin{bmatrix} I \\ \Lambda_0 X_0^{-1} \end{bmatrix}$$

can be assumed to be composed of coordinates of vectors that form the basis for the invariant subspace of $\mathcal{H}$ corresponding to the eigenvalues in the left half-plane. Thus, having computed X_0 and Λ_0 with the help of the singular value decomposition of P and setting $\Lambda = -\Lambda_0 X_0^{-1}$, we arrive at the representation of the basis by the composite matrix

$$\begin{bmatrix} I \\ -\Lambda \end{bmatrix}.$$

If the matrix X_0 is singular, we can assert that the matrix pair A, B is not stabilizable. In this case, we write $\|\Lambda\| = \infty$.

If the matrix X_0 is nonsingular, and we can compute Λ, then, in view of the results in §7, we can claim that Λ is positive definite and it is a unique positive definite solution to the matrix Lur′e–Riccati equation

$$\Lambda A + A^*\Lambda = \Lambda B(B^*B)^{-1}B^*\Lambda - I.$$

Also, we can claim that $A + BM$, where $M = -(B^*B)^{-1}B^*\Lambda$ is a Hurwitz matrix (i.e., its spectrum lies to the left of the imaginary axis), and the optimal control minimizing the integral

$$\int_0^\infty (\|x(t)\|^2 + \|Bu(t)\|^2)dt$$

over solutions to the system (1) with given $x(0)$ have the form $u(t) = Mx(t)$. For such a control we have

$$\int_0^\infty (\|x(t)\|^2 + \|Bu(t)\|^2)dt = (\Lambda x(0), x(0)).$$

Thus, the verification of the stabilizability of the matrix pair A, B is reduced to the verification of the Lopatinskii condition which guarantees the well-posedness of the boundary-value problem

$$x(0) = a, \quad \sqrt{\|x(t)\|^2 + \|\lambda(t)\|^2} \to 0 \text{ as } t \to +\infty;$$
$$\frac{d}{dt}\begin{pmatrix} x(t) \\ \lambda(t) \end{pmatrix} = \begin{bmatrix} A & B(B^*B)^{-1}B^* \\ I & -A^* \end{bmatrix}\begin{pmatrix} x(t) \\ \lambda(t) \end{pmatrix}$$

on the half-line $0 \leqslant t \leqslant \infty$. These conditions hold if $\varkappa(\mathcal{H}) < \infty$ and for any vector $x(0)$ it is possible to construct the initial data such that a solution to the Hamiltonian system

$$\begin{pmatrix} x(0) \\ \lambda(0) \end{pmatrix} = \begin{pmatrix} a \\ -\Lambda a \end{pmatrix}$$

with these initial data tends to zero as $t \to \infty$.

The Lopatinskii condition fails if $\varkappa(\mathcal{H}) = \infty$ (practically) or $\varkappa(\mathcal{H})$ is finite but we cannot find Λ of finite norm. Consequently, $\varkappa(\mathcal{H})$ and $\|\Lambda\|$ allow us to make conclusions about the stabilizability of the matrix pair A, B. For the stabilizability of the matrix pair A, B it is necessary and sufficient that both these quantities be finite. Note that some numerical characteristics of "stabilizability degree" can be combined by these quantities.

This completes the review of computational procedures that help us to clarify whether or not the matrix pair A, B is stabilizable and, if the answer is positive, to compute a matrix M such that $A + BM$ is a Hurwitz matrix.

These procedures can be also used if we want to find out whether or not the matrix pair A, C is detectable and, if the answer is positive, to compute a matrix M such that $A + MC$ is a Hurwitz matrix. Indeed, instead of A, C we consider the matrices A^*, C^* and verify whether they form a stabilizable matrix pair. If the answer is positive, then the matrix pair A, C is detectable. In this case, we construct a rectangular matrix $\widetilde{M}$ such that $A^* + C^*\widetilde{M}$ is a Hurwitz matrix and set $M = \widetilde{M}^*$. It is obvious that the real matrix

$$A + MC = (A^* + C^*M^*)^* = (A^* + C^*\widetilde{M})^*$$

has the same eigenvalues as the matrix $A^* + C^*\widetilde{M}$. Since all eigenvalues of $A^* + C^*\widetilde{M}$ lie in the left half-plane, the same assertion is true for the matrix $A + MC$.

The verification of the controllability of the matrix pair A, B and the observability of the matrix pair A, C can be also reduced to the above procedures. Indeed, as was shown in §3, the matrix pair A, B is controllable if and only if both matrix pairs A, B and $-A$, B are stabilizable. Similarly, the matrix pair A, C is observable if and only if the matrix pairs A, C and $-A$, C are detectable.

Now we consider the question about the computation of $u = Mx$ which defines a control $u(t)$ such that every trajectory $x(t)$ of the system (1) approaches the point $x = 0$ as $t \to \infty$ ($\|x(t)\| \to 0$ as $t \to \infty$); moreover, for a given $x(0)$ it minimizes the integral

$$\int_0^\infty (\|Cx(t)\|^2 + \|Bu(t)\|^2)dt.$$

As was shown in §6, this problem is not always solvable. For its unique solvability it suffices to require that the matrix pair A, B be stabilizable and the matrix pair A, C be detectable. Therefore, it is necessary to verify these conditions before solving the problem. It suffices to verify that the matrix pair A, C is detectable and after that to analyze and solve the problem. For this, we consider the Hamiltonian system

$$\frac{d}{dt}\begin{pmatrix} x(t) \\ \lambda(t) \end{pmatrix} = \begin{bmatrix} A & B(B^*B)^{-1}B^* \\ C^*C & -A^* \end{bmatrix} \begin{pmatrix} x(t) \\ \lambda(t) \end{pmatrix} \equiv \mathcal{H} \begin{pmatrix} x(t) \\ \lambda(t) \end{pmatrix}.$$

First, we must verify that $\varkappa(\mathcal{H}) < \infty$, i.e., the matrix of coefficients of the Hamiltonian system has no purely imaginary eigenvalues. In practice, only the inequality $\varkappa(\mathcal{H}) < \varkappa^*$ is checked with a sufficiently large $\varkappa^*$. At the same time, by using the above algorithm, it is possible to find the projection P on the invariant subspace of $\mathcal{H}$ corresponding to eigenvalues in the left half-plane and, by using the singular value decomposition, to compute the matrix

$$\begin{bmatrix} X_0 \\ \Lambda_0 \end{bmatrix}$$

whose columns form the orthonormal basis for this subspace. If the matrix pair A, B is stabilizable, then the matrix block X_0 is invertible and $\Lambda = -\Lambda_0 X_0^{-1}$ is symmetric and nonnegative definite (cf. §§6,7). Conversely, if we have computed Λ and it turns out that $\Lambda = \Lambda^* > 0$, then we can guarantee that the matrix pair A, B is stabilizable. In §7, we have shown that if the matrix pair A, B is not only detectable but is observable, then for the stabilizable matrix pair A, B the symmetric matrix Λ is positive definite ($\Lambda = \Lambda^* > 0$).

Having computed Λ, by setting $M = -(B^*B)^{-1}B^*\Lambda$ we can construct the control $u(t) = Mx(t)$ which minimizes the integral

$$\int_0^\infty [\|Cx(t)\|^2 + \|Bu(t)\|^2]dt$$

over trajectories of the system (1).

To conclude the section, we make some remarks.

REMARK 1. In control theory, the problems of minimizing general integrals

$$\int_0^\infty [(FCx, Cx) + (GBu, Bu)]dt$$

are often considered. This integral differs from the integrals studied above in that the quadratic forms (Cx, Cx), (Bu, Bu) are replaced with the quadratic forms (FCx, Cx), (GBu, Bu), where F and G are positive definite matrices. The theory of such modified variational problems is constructed in the same way as above. But we must incorporate F and G in the definition of the Hamiltonian matrix

$$\mathcal{H} = \begin{bmatrix} A & B(B^*GB)^{-1}B^* \\ C^*FC & -A^* \end{bmatrix}$$

and connect the optimal control $u(t)$ with $\lambda(t)$ by means of the feedback $u(t) = Mx(t)$, where $M = (B^*GB)^{-1}B^*\Lambda$ and Λ is a solution to the matrix Lur′e–Riccati equation of the form

$$\Lambda A + A^*\Lambda = \Lambda B(B^*GB)^{-1}B^*\Lambda - C^*FC.$$

The corresponding modifications are not of principal importance and we omit them.

REMARK 2. We discuss a rule according to which the Hamiltonian system corresponding to a minimized functional can be written out. The derivation given in §2 contains a detailed justification, but it is rather laborious. As a rule, the same equation can be obtained by formal application of a standard procedure of calculus of variations to the functional which contains additional terms with the so-called Lagrange multipliers. Thus, if we want to minimize the functional

$$\int_0^T [(FCx, Cx) + (GBu, Bu)]dt$$

over trajectories $x(t)$ of the system (1), we must replace this functional with the following:

$$\begin{aligned}\int_0^T &[(FCx, Cx) + (GBu, Bu) + 2(\lambda(t), \left[\frac{dx}{dt} - Ax - Bu\right])]dt \\ &= 2(\lambda(T), x(T)) - 2(\lambda(0), x(0)) \\ &+ \int_0^T \left[(FCx, Cx) + (GBu, Bu) - 2\left(x(t), \frac{d\lambda(t)}{dt}\right) - 2(x(t), A^*\lambda) - 2(Bu, \lambda(t))\right] dt.\end{aligned}$$

In accordance with calculus of variations, along extremals of the last functional all partial derivatives of the expression under the sign of the integral must vanish:

$$2C^*FCx - 2\frac{d\lambda}{dt} - 2A^*\lambda = 0.$$

It is obvious that this equation differs from the second equation of the Hamiltonian system

$$\frac{dx}{dt} = Ax + B(B^*GB)^{-1}B^*\lambda,$$
$$\frac{d\lambda}{dt} = C^*FCx - A^*\lambda$$

only by notation, whereas the first equation is derived from the original equation (1) after applying the relation $2B^*GBu - 2B^*\lambda = 0$. This relation also occurs in the extremum conditions for the functional; it can be obtained by equating to zero partial derivatives of the expression under the integral sign. For the formalism of calculus of variations we refer to [**5**].

EXAMPLE 1. We give an example which illustrates how the theory developed in this chapter and the corresponding computational method based on it can be used for some regularization procedures. Suppose it is possible to observe the vector $y(t) = Cx(t)$, where $x(t)$ is a trajectory of (1). We also assume that a control $u(t)$ may correct the trajectory $z(t)$ of some observed system described by the equation

$$\frac{dz}{dt} = Az(t) + f(t) + Bu(t).$$

The problem is to determine, for known $y(t)$, the control $u(t)$ such that the trajectories $z(t)$ and $x(t)$ come close as t increases, i.e., $\|z(t) - x(t)\| \to 0$. The matrix pairs A, C and A, B are assumed to be detectable and stabilizable respectively.

The solution to this problem is based on the use of the so-called dynamic compensator, i.e., on some construction which is described by an equation similar to equations for $x(t)$, $z(t)$:

$$\frac{dy}{dt} = Ay(t) + f(t) + v(t).$$

For a given observation $\varphi(t)$ (or, more exactly, the difference $Cy(t) - \varphi(t)$), the correcting control $v(t)$ is chosen so that the trajectories $z(t)$ and $x(t)$ come close. The choice of the control $u(t)$ is based on the vector $\psi(t) = z(t) - y(t)$.

We assume that $v(t)$ and $u(t)$ are defined by the equalities

$$\begin{aligned} v(t) &= M_1[Cy - \varphi] = M_1 C[y - x], \\ u(t) &= M_2[z - y], \end{aligned}$$

where M_1 and M_2 are constant matrices with corresponding numbers of rows and columns. Moreover, the behavior of the observed system and the dynamic compensator are described by the equations

$$\begin{aligned} \frac{dx}{dt} &= Ax(t) + f(t), \\ \frac{dy}{dt} &= Ay(t) + M_1 C[y(t) - x(t)] + f(t), \\ \frac{dz}{dt} &= Az(t) + BM_2[z(t) - y(t)] + f(t) \end{aligned}$$

which imply to the homogeneous linear system for the differences $y - x$ and $z - y$

$$\frac{d}{dt}\begin{pmatrix} y - x \\ z - y \end{pmatrix} = \begin{bmatrix} A + M_1 C & 0 \\ -M_1 C & A + BM_2 \end{bmatrix} \begin{pmatrix} y - x \\ z - y \end{pmatrix}.$$

Since the matrix pair A, C is detectable, then we can compute (cf. §6) a matrix M_1 such that $A + M_1 C$ is a Hurwitz matrix, i.e., its spectrum lies in the left half-plane. Similarly, the stabilizability of the matrix pair A, B implies the existence of a matrix M_2 such that $A + BM_2$ is a Hurwitz matrix (cf. §5 for the method of computation of M_2). It is clear that all eigenvalues of the composite matrix

$$\begin{bmatrix} A + M_1 C & 0 \\ -M_1 C & A + BM_2 \end{bmatrix}$$

lie in the left half-plane. Therefore, $\|y(t) - x(t)\|$, $\|x(t) - y(t)\|$, and consequently, $\|z(t) - x(t)\|$ tends to zero as $t \to \infty$. Thus, we have shown the possibility of a constructive choice of controls $v(t)$ and $u(t)$, which provides observation of the trajectory $x(t)$ on the basis of information about $x(t)$ in the form $\varphi(t) = Cx(t)$.

To conclude our exposition, we mention that numerical characteristics of the notions of controllability and stabilizability, as well as new criteria for solvability of the matrix Lur′e–Riccati equation, are contained in [**3, 10, 11**].

References

1. A. Ya. Bulgakov, *An effectively calculable parameter for the stability property of a system of linear differential equations with constant coefficients*, Sibirsk. Mat. Zh. **21** (1980), no. 3, 32–41; English transl., Siberian Math. J. **21** (1980), 339–346.
2. ———, *The basis of guaranteed accuracy in the problem of separation of invariant subspaces for non-self-adjoint matrices*, Trudy Inst. Mat. Sibirsk. Otdel. Akad. Nauk SSSR **15** (1989), 12–92; English transl., Siberian Advances in Math. **1** (1991), no. 1,2, 64–108, 1–56.
3. ———, *An algorithm for solving the Lur′e–Riccati matrix equation with guaranteed accuracy*, Trudy Inst. Mat. Sibirsk. Otdel. Rus. Akad. Nauk **24** (1994), 63–104; English transl., Siberian Advances in Math. **5** (1995), no. 4, 1–49.
4. F. R. Gantmacher, *The theory of matrices*, vol. I, II, Chelsea Publishing Company, New York, 1959.
5. I. M. Gelfand and S. V. Fomin, *Calculus of variations*, Fizmatgiz, Moscow, 1961; English transl., Prentice-Hall, Englewood Cliffs, NJ, 1963.
6. I. M. Gelfand and G. E. Shilov, *Some questions of the theory of differential equations*, Fizmatgiz, Moscow, 1958; English transl., Academic Press, New York and London, 1967.
7. S. K. Godunov, *Problem of the dichotomy of the spectrum of a matrix*, Sibirsk. Mat. Zh. **27** (1986), no. 5, 24–37; English transl., Siberian Math. J. **27** (1986), no. 5, 649–660.
8. ———, *Quadratic Lyapunov functions*, Novosibirsk State University, Novosibirsk, 1982.
9. ———, *Matrix exponential, the Green matrix, and the Lopatinskii conditions*, Novosibirsk State University, Novosibirsk, 1983.
10. ———, *An estimate for Green's matrix of a Hamiltonian system in the optimal control problem*, Sibirsk. Mat. Zh. **34** (1993), no. 4, 70–80; English transl., Siberian Math. J. **34** (1993), no. 4, 653–662.
11. ———, *Norms of solutions to the Lur′e–Riccati matrix equation as a criterion for quality of stabilizability and detectability*, Trudy Inst. Mat. Sibirsk. Otdel. Rus. Akad. Nauk **22** (1992), 3–22; English transl., Siberian Advances in Math. **2** (1992), no. 3, 135–158.
12. ———, *Current aspects of linear algebra*, "Nauchnaya kniga", Novosibirsk, 1997.
13. S. K. Godunov, A. G. Antonov, O. P. Kirilyuk, and V. I. Kostin, *Guaranteed accuracy for solutions of system of linear equations in Euclidean spaces*, "Nauka", Novosibirsk, 1988; English transl., Kluwer, Dordrecht, 1993.
14. S. K. Godunov and V. M. Gordienko, *The Green matrix of the boundary-value problem for ordinary differential equations*, Uspekhi Mat Nauk **39** (1984), no. 1, 39–76; English transl. in Russian Math. Surveys **39** (1984).
15. A. N. Kolmogorov and S. V. Fomin, *Introductory real analysis*, "Nauka", Moscow, 1981; English transl. of the 2nd edition, Dover, New York, 1975.
16. A. G. Kurosh, *Higher algebra*, "Nauka", Moscow, 1971. (Russian)
17. S. V. Kuznetsov, *Solution of boundary-value problems for ordinary differential equations*, Trudy Inst. Mat. Sibirsk. Otdel. Akad. Nauk SSSR **6** (1985), 85–110.
18. L. A. Lyusternik and V. I. Sobolev, *A short course in functional analysis*, "Vyssh. Shkola", Moscow, 1982; English transl. of the 1st edition, Halsted Press, New York, 1974.
19. A. N. Malyshev, *Guaranteed accuracy in spectral problems of linear algebra*, Trudy Inst. Mat. Sibirsk. Otdel. Rus. Akad. Nauk **17** (1991), 19–103; English transl., Siberian Advances in Math. **2** (1992), no. 1, 144–197.
20. ———, *Introduction in linear algebra*, "Nauka", Novosibirsk, 1991. (Russian)
21. G. I. Petrovskiĭ, *Lectures on the theory of ordinary differential equations*, "Nauka", Moscow, 1964; English transl., Prentice-Hall, Englewood Cliffs, NJ, 1966.
22. J. H. Wilkinson, *The algebraic eigenvalue problem*, Clarendon Press, Oxford, 1965.

Index

Selected Titles in This Series

(Continued from the front of this publication)

130 **V. V. Vershinin,** Cobordisms and spectral sequences, 1993

129 **Mitsuo Morimoto,** An introduction to Sato's hyperfunctions, 1993

128 **V. P. Orevkov,** Complexity of proofs and their transformations in axiomatic theories, 1993

127 **F. L. Zak,** Tangents and secants of algebraic varieties, 1993

126 **M. L. Agranovskiĭ,** Invariant function spaces on homogeneous manifolds of Lie groups and applications, 1993

125 **Masayoshi Nagata,** Theory of commutative fields, 1993

124 **Masahisa Adachi,** Embeddings and immersions, 1993

123 **M. A. Akivis and B. A. Rosenfeld,** Élie Cartan (1869–1951), 1993

122 **Zhang Guan-Hou,** Theory of entire and meromorphic functions: deficient and asymptotic values and singular directions, 1993

121 **I. B. Fesenko and S. V. Vostokov,** Local fields and their extensions: A constructive approach, 1993

120 **Takeyuki Hida and Masuyuki Hitsuda,** Gaussian processes, 1993

119 **M. V. Karasev and V. P. Maslov,** Nonlinear Poisson brackets. Geometry and quantization, 1993

118 **Kenkichi Iwasawa,** Algebraic functions, 1993

117 **Boris Zilber,** Uncountably categorical theories, 1993

116 **G. M. Fel′dman,** Arithmetic of probability distributions, and characterization problems on abelian groups, 1993

115 **Nikolai V. Ivanov,** Subgroups of Teichmüller modular groups, 1992

114 **Seizô Itô,** Diffusion equations, 1992

113 **Michail Zhitomirskiĭ,** Typical singularities of differential 1-forms and Pfaffian equations, 1992

112 **S. A. Lomov,** Introduction to the general theory of singular perturbations, 1992

111 **Simon Gindikin,** Tube domains and the Cauchy problem, 1992

110 **B. V. Shabat,** Introduction to complex analysis Part II. Functions of several variables, 1992

109 **Isao Miyadera,** Nonlinear semigroups, 1992

108 **Takeo Yokonuma,** Tensor spaces and exterior algebra, 1992

107 **B. M. Makarov, M. G. Goluzina, A. A. Lodkin, and A. N. Podkorytov,** Selected problems in real analysis, 1992

106 **G.-C. Wen,** Conformal mappings and boundary value problems, 1992

105 **D. R. Yafaev,** Mathematical scattering theory: General theory, 1992

104 **R. L. Dobrushin, R. Kotecký, and S. Shlosman,** Wulff construction: A global shape from local interaction, 1992

103 **A. K. Tsikh,** Multidimensional residues and their applications, 1992

102 **A. M. Il′in,** Matching of asymptotic expansions of solutions of boundary value problems, 1992

101 **Zhang Zhi-fen, Ding Tong-ren, Huang Wen-zao, and Dong Zhen-xi,** Qualitative theory of differential equations, 1992

100 **V. L. Popov,** Groups, generators, syzygies, and orbits in invariant theory, 1992

99 **Norio Shimakura,** Partial differential operators of elliptic type, 1992

98 **V. A. Vassiliev,** Complements of discriminants of smooth maps: Topology and applications, 1992 (revised edition, 1994)

97 **Itiro Tamura,** Topology of foliations: An introduction, 1992

96 **A. I. Markushevich,** Introduction to the classical theory of Abelian functions, 1992

95 **Guangchang Dong,** Nonlinear partial differential equations of second order, 1991

94 **Yu. S. Il′yashenko,** Finiteness theorems for limit cycles, 1991

(See the AMS catalog for earlier titles)